AF353118

Numerical Simulation of Convective-Radiative Heat Transfer

Numerical Simulation of Convective-Radiative Heat Transfer

Editor

Mikhail Sheremet

MDPI • Basel • Beijing • Wuhan • Barcelona • Belgrade • Manchester • Tokyo • Cluj • Tianjin

Editor
Mikhail Sheremet
Tomsk State University
Russia

Editorial Office
MDPI
St. Alban-Anlage 66
4052 Basel, Switzerland

This is a reprint of articles from the Special Issue published online in the open access journal *Energies* (ISSN 1996-1073) (available at: https://www.mdpi.com/journal/energies/special_issues/ Convective-Radiative_Heat_Transfer).

For citation purposes, cite each article independently as indicated on the article page online and as indicated below:

LastName, A.A.; LastName, B.B.; LastName, C.C. Article Title. *Journal Name* **Year**, *Article Number*, Page Range.

ISBN 978-3-03943-194-6 (Hbk)
ISBN 978-3-03943-195-3 (PDF)

Contents

About the Editor

Mikhail Sheremet is Head of the Laboratory on Convective Heat and Mass Transfer and Head of the Department of Theoretical Mechanics at National Research Tomsk State University. There, he received a Ph.D. (Russia, Candidate of Science in Physics and Mathematics degree) in (2006) and habilitation (Russia, Doctor of Science in Physics and Mathematics) (2012). Dr. Sheremet has published over 300 papers in peer-reviewed journals and conference proceedings, and contributed to several books. He received the Web of Science Award 2017 in the category of Highly Cited Researcher in Russia. He is a member of Editorial Boards of the *International Journal of Numerical Methods for Heat & Fluid Flow, Journal of Magnetism and Magnetic Materials, Journal of Applied and Computational Mechanics, Nanomaterials, Energies*, and *Coatings*. He is also a Scientific Council Member of the International Centre for Heat and Mass Transfer.

Preface to "Numerical Simulation of Convective-Radiative Heat Transfer"

Heat transfer is a major transport phenomenon that occurs in various engineering and natural systems. The development of modern engineering apparatuses and natural bio- and geosystems requires deep insight into the processes that have evolved within these systems. Convective and radiative heat-transfer mechanisms are the main processes in the systems under consideration. Therefore, an in-depth study of them is very important and useful for both the growth of industry and the preservation of natural resources.

There are three main methods for investigating heat-transfer phenomena: theoretical methods, experimental methods, and computational approaches. Theoretical methods generally involve an analytical description of thermal processes using the laws of conservation of mass, momentum, angular momentum, and energy. Experimental analysis includes an investigation of heat-transfer phenomena using experimental techniques and measurements. The development of computer engineering involves using many types of numerical simulation to obtain a description and an understanding of heat-transfer processes. Such an approach has the advantages of theoretical methods in which analysis can be performed in a wide range of governing parameters and the benefits of experimental methods where deep insight of considered phenomena is possible. Therefore, numerical simulation of convective and radiative heat transfer is a very useful and important topic for different industry sectors and various natural systems. This book therefore seeks to open various engineering and natural fields where convective–radiative heat transfer plays a vital role, and the results can be used for the development and optimization of these systems.

Mikhail Sheremet
Editor

Article

Mixed Convection Stagnation-Point Flow of a Nanofluid Past a Permeable Stretching/Shrinking Sheet in the Presence of Thermal Radiation and Heat Source/Sink

Anuar Jamaludin [1,2], Roslinda Nazar [2,*] and Ioan Pop [3]

[1]   Department of Mathematics, Universiti Pertahanan Nasional Malaysia, 57000 Kuala Lumpur, Malaysia; mohdanuar@upnm.edu.my
[2]   School of Mathematical Sciences, Faculty of Science and Technology, Universiti Kebangsaan Malaysia, 43600 UKM Bangi, Selangor, Malaysia
[3]   Department of Mathematics, Babeş-Bolyai University, R-400084 Cluj-Napoca, Romania; popm.ioan@yahoo.co.uk
*   Correspondence: rmn@ukm.edu.my; Tel.: +603-8921-3371

Received: 27 December 2018; Accepted: 12 February 2019; Published: 27 February 2019

Abstract: In this study we numerically examine the mixed convection stagnation-point flow of a nanofluid over a vertical stretching/shrinking sheet in the presence of suction, thermal radiation and a heat source/sink. Three distinct types of nanoparticles, copper (Cu), alumina (Al_2O_3) and titania (TiO_2), were investigated with water as the base fluid. The governing partial differential equations were converted into ordinary differential equations with the aid of similarity transformations and solved numerically by utilizing the bvp4c programme in MATLAB. Dual (upper and lower branch) solutions were determined within a particular range of the mixed convection parameters in both the opposing and assisting flow regions and a stability analysis was carried out to identify which solutions were stable. Accordingly, solutions were gained for the reduced skin friction coefficients, the reduced local Nusselt number, along with the velocity and temperature profiles for several values of the parameters, which consists of the mixed convection parameter, the solid volume fraction of nanoparticles, the thermal radiation parameter, the heat source/sink parameter, the suction parameter and the stretching/shrinking parameter. Furthermore, the solutions were presented in graphs and discussed in detail.

Keywords: mixed convection; nanofluids; thermal radiation; heat source/sink; dual solutions; stability analysis

---

1. Introduction

Mixed convection flows or a combination of forced and free convections exists in numerous transport operations, both naturally occurring and in engineering applications. Such applications for example, include heat exchangers, solar collectors, nuclear reactors, atmospheric boundary layer flow, nanotechnology, electronic apparatus, etc. These operations occur during the effects of buoyancy forces in forced convections or the effects of forced flow in free convections become substantial. Over the past several decades, most research in mixed convection flow analysis has emphasised the occurrence of dual solutions for a particular range of the buoyancy (mixed convection) parameter in the opposing flow region, such as in the research by Ramachandran et al. [1], Merkin and Mahmood [2], Devi et al. [3] and Lok et al. [4]. In contrast to [1–4], Ridha and Curie [5] continued the study by Merkin and Mahmood [2] by establishing the existence of dual solutions in both the opposing and assisting flow regions. Furthermore, by implementing a stability analysis of the dual solutions for mixed convection flow in a saturated porous medium, Merkin [6] demonstrated that the upper branch of the solutions is stable whereas the lower branch shows instability. Accordingly, various other researches have also stated the occurrence of dual solutions in the mixed convection flow in different configurations, namely by Roşca et al. [7], Rahman et al. [8] and recently by Abbasbandy et al. [9]. An inclusive account of the theoretical research prior to 1987 for both laminar and turbulent mixed convection boundary layer flows may be found in the books by Gebhart et al. [10], Schlichting and Gersten [11], Pop and Ingham [12] and Bejan [13], for example.

The innovative idea of nanofluids was first brought up by Choi et al. [14] in 1995, when the authors suggested a path for exceeding the performance of heat transfer fluids which were currently available. An extraordinary enhancement in the thermal properties of base fluids may be achieved just by utilizing a minimal amount of nanoparticles scattered uniformly and suspended stably in a base fluid. Nanofluids, as colloidal mixtures of nanoparticles (1–100 nm) along with a base liquid (nanoparticle fluid suspensions) are known, provide access to a new class of nanotechnology-based heat transfer media (Das et al. [15]). Numerous techniques and methodologies, such as rising either the heat transfer surface or the heat transfer coefficient between the fluid and the surface that allows high heat transfer rates in a small volume, may be utilized to promote heat transfer. Notwithstanding, cooling turns out to be one of the most critical technical challenges faced by numerous and diverse industries, including microelectronics, transportation, solid-state lighting, and manufacturing. The addition of micrometre- or millimetre-sized solid metal or metal oxide particles to base fluids produces an increase in the thermal conductivity of the resultant fluids. On the other hand, apart from being applied in the field of heat transfer, nanofluids (nanometre-sized particles in a fluid) may also be synthesised for unique magnetic, electrical, chemical, and biological applications (see Manca et al. [16]). Nanoparticles are produced from various materials such as copper (Cu), alumina (Al_2O_3), titania (TiO_2), copper oxide (CuO) as well as silver (Ag) (see Oztop and Abu-Nada [17]). References on nanofluids are mentioned in the books written by Das et al. [15], Nield and Bejan [18], Minkowycz et al. [19] and Shenoy et al. [20], and also in the review papers written by Buongiorno et al. [21], Kakaç and Pramuanjaroenkij [22], Fan and Wang [23], Mahian et al. [24], Sheikholeslami and Ganji [25], Groşan et al. [26], Myers et al. [27], etc. These review papers elaborate specifically on the production of nanofluids, the theoretical and experimental exploration of the thermal conductivity and viscosity of nanofluids, as well as the work conducted on the convective transport of nanofluids.

Interestingly, many studies investigating the boundary layer problem of mixed convection flow in a nanofluid are reported in the literature. Tamim et al. [28] examined the effects of the magnetic field, suction/injection and solid volume fraction of nanoparticles on mixed convection about the stagnation-point flow of a nanofluid. On the other hand, Subhashini et al. [29] investigated the mixed convection flow about the stagnation-point region over an exponentially stretching/shrinking sheet in a nanofluid for both suction and injection cases. Later, Mustafa et al. [30] extended the study conducted by Tamim et al. [28] in consideration of the combined effects of viscous dissipation and the magnetic field by gaining a unique solution for assisting and opposing flow cases. Recently, Ibrahim et al. [31], Mabood et al. [32] and Othman et al. [33], similarly investigated the problem of mixed convection boundary layer flow in nanofluids under different physical conditions.

The impact of thermal radiation on heat transfer becomes increasingly important in the design of advanced energy conversion systems operating at high temperature. Moreover, thermal radiation has applications in numerous technological problems such as combustion, nuclear reactor safety, solar collectors, furnace design and many others (see Ozisik [34]). Furthermore, the study of thermal radiation on flow and heat transfer characteristics in a nanofluid have attracted immense interest because nanofluids have different properties than those found in either the particles or the base fluid. Given this fact, many researchers have explored the impact of thermal radiation on flow and heat transfer in a nanofluid along with other various aspects. An important analysis by Hady et al. [35] studied the boundary layer viscous flow and heat transfer characteristics of a nanofluid over a nonlinearly stretching sheet in the presence of thermal radiation in a single-phase model. In a separate study, Ibrahim and Shankar [36] investigated the influences of thermal radiation, magnetic fields and slip boundary conditions on boundary layer flow and heat transfer past a permeable stretching sheet in a nanofluid. Notwithstanding, Haq et al. [37] discussed the combined effects of thermal radiation, magnetohydrodynamic (MHD), velocity and thermal slip on the boundary layer stagnation-point flow of nanofluid and the effects over a stretching sheet. More recently, Daniel et al. [38] investigated the effects of thermal radiation, magnetic fields, electrical fields, Ohmic dissipation, thermal and concentration stratifications on the flow and heat transfer of electrically conducting nanofluid past a permeable stretching sheet. In another recent study, Sreedevi et al. [39] analysed the effect of thermal radiation, magnetic field and the chemical reaction on flow, heat and mass transfer of nanofluid over a linear and nonlinear stretching sheet saturated by the porous medium. Accordingly, several other studies have been undertaken on mixed convection boundary layer flow in nanofluids in the presence of thermal radiation, including works by Yazdi et al. [40], Pal and Mandal [41] and Ayub et al. [42].

The heat source/sink effect in addition to the thermal radiation effect plays a vital role in governing the heat transfer in industrial operations in which the attributes of the output are dependent on the factors of heat control. Accordingly, many researchers have studied the impacts of a heat source/sink on the boundary layer flow and heat transfer of nanofluids along with different aspects. Rana and Bhargava [43] numerically investigated the impact of the various types of nanoparticles on mixed convection flow of nanofluid along the vertical plate with a heat source/sink. Furthermore, Pal et al. [44] analysed the combined impacts of internal heat generation/absorption, thermal radiation and suction/injection on mixed convection stagnation point flow of nanofluids over a stretching/shrinking sheet in a porous medium. In addition, Pal and Mandal [45] discussed the impacts of microrotation and nanoparticles on boundary layer flow in nanofluids in the occurrence of non-uniform heat source/sink, suction, thermal radiation and magnetic fields. In another paper, Mondal et al. [46] considered the influence of heat generation/absorption and thermal radiation on hydromagnetic three-dimensional mixed convection flow of nanofluid over a vertical stretching surface. Sharma and Gupta [47] further investigated the effect of heat generation/absorption, MHD, thermal radiation, viscous dissipation on flow and heat transfer of Jeffrey nanofluids.

Interestingly, previous studies did not include the combined effects of thermal radiation, heat source/sink and suction on mixed convection flow of a nanofluid. Therefore, the primary aim of this article is to examine the impact of thermal radiation, heat source/sink and suction on mixed convection stagnation point flow over a stretching/shrinking sheet in a nanofluid, by applying a mathematical nanofluid model suggested by Tiwari and Das [48]. In our opinion, the problem is relatively new, novel with no such articles reported at this stage in the literature. Suitable similarity transformations are employed to transform nonlinear partial differential equations into nonlinear ordinary differential equations. The equations are then solved numerically with the assistance of the bvp4c programme in MATLAB, and the results are graphically plotted and displayed in tables. The results from the study indicates that dual solutions exist for a particular range of parameters, namely, the mixed convection parameter, solid volume fraction of nanoparticles, thermal radiation parameter, heat source/sink parameter, suction parameter and stretching/shrinking parameter. Further, it is useful to mention, that the stability analysis of the dual solutions is conducted to investigate which solution is stable.

2. Mathematical Formulation

This study considers the two-dimensional steady mixed convection flow of a viscous and incompressible nanofluid near the stagnation-point past a permeable vertical stretching/shrinking surface with the velocity $u_w(x)$ and free stream velocity $u_e(x)$, as illustrated in Figure 1, where x and y denote the Cartesian coordinates evaluated along the surface of the stretching/shrinking sheet and normal to it, respectively. The fluid consists of a water-based nanofluid comprising three distinct types of nanoparticles which are copper (Cu), alumina (Al_2O_3) and titania (TiO_2). The thermophysical properties of water (the base fluid) and nanoparticles are shown in Table 1. These thermophysical properties will be used in the numerical computations of this study. Also, it is assumed that the flow was subjected to the combined impact of thermal radiation and a heat source/sink. Another assumption made are such that the temperature of the stretching/shrinking sheet, $T_w(x)$, and the temperature of the ambient nanofluid adopt a constant value T_∞. Furthermore, it is also assumed that the water-based fluid and the nanoparticles are in thermal equilibrium and that no slip exists among them. The mathematical nanofluid model suggested by Tiwari and Das [48] is applied in this case. It should be mentioned that this nanofluid model is a single-phase approach where the nanoparticles are assumed to have a uniform shape and size, and the interactions between nanoparticles and surrounding fluid are also neglected (Pang et al. [49], Ebrahimi et al. [50] and Sheremet et al. [51]). This assumption is practical when the base fluid is easily fluidized, so it can be considered to behave as a single fluid, hence it applies to the justification of using single phase model in this study.

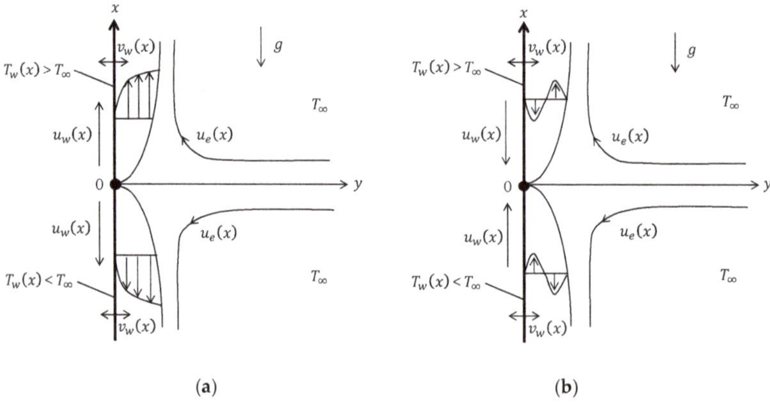

Figure 1. Physical model and coordinate system. (**a**) Stretching surface; (**b**) Shrinking surface.

Table 1. Thermophysical properties of water and nanoparticles (Oztop and Abu-Nada [17]).

Physical Properties	Water	Cu	Al$_2$O$_3$	TiO$_2$
C_p (J·kg^{-1}·K^{-1})	4179	385	765	686.2
ρ (kg·m^{-3})	997.1	8933	3970	4250
k (W·m^{-1}·K^{-1})	0.613	400	40	8.9538
$\beta \times 10^{-5}$ (K^{-1})	21	1.67	0.85	0.9

By taking into considerations of these assumptions together with the Boussinesq and the boundary layer approximations, the governing boundary layer equations of continuity, momentum and thermal energy in the existence of thermal radiation and the heat source or sink, are given as shown below:

$$\frac{\partial u}{\partial x} + \frac{\partial v}{\partial y} = 0, \tag{1}$$

$$u\frac{\partial u}{\partial x} + v\frac{\partial u}{\partial y} = u_e\frac{du_e}{dx} + \frac{\mu_{nf}}{\rho_{nf}}\frac{\partial^2 u}{\partial y^2} + \frac{(\rho\beta)_{nf}}{\rho_{nf}}(T - T_\infty)g, \tag{2}$$

$$u\frac{\partial T}{\partial x} + v\frac{\partial T}{\partial y} = \alpha_{nf}\frac{\partial^2 T}{\partial y^2} - \frac{1}{(\rho C_p)_{nf}}\frac{\partial q_r}{\partial y} + \frac{Q_0}{(\rho C_p)_{nf}}(T - T_\infty), \tag{3}$$

and the associated boundary conditions to present the flow are:

$$u = u_w(x), \; v = v_w(x), \; T = T_w(x) \text{ at } y = 0,$$
$$u \to u_e(x), \; T \to T_\infty(x) \text{ as } y \to \infty, \tag{4}$$

where u and v represent the velocity elements along the x and y directions, respectively; g stands for the acceleration caused by gravity, T denotes the temperature of the nanofluid, Q_0 denotes the heat source/sink coefficient, with $Q_0 > 0$ corresponding to the heat source and $Q_0 < 0$ corresponding to the heat sink. Further, $v_w(x)$ represents the wall mass flux, with $v_w(x) < 0$ corresponding to the suction. Moreover, μ_{nf} represents the dynamic viscosity of the nanofluid, ρ_{nf} refers to the density of the nanofluid, $(\rho\beta)_{nf}$ denotes the thermal expansion coefficient of the nanofluid as described in the Brinkman's model, $(\rho C_p)_{nf}$ denotes the heat capacitance of the nanofluid, α_{nf} denotes the thermal diffusivity of the nanofluid, q_r represents the radiation heat flux, and lastly, v_{nf} reflects the kinematic viscosity of the nanofluid. The relations of μ_{nf}, α_{nf}, ρ_{nf}, $(\rho\beta)_{nf}$, $(\rho C_p)_{nf}$ and k_{nf} are described in the following equations (see Oztop and Abu-Nada [17]):

$$\mu_{nf} = \frac{\mu_{nf}}{(1-\phi)^{2.5}}, \ \alpha_{nf} = \frac{k_{nf}}{(\rho C_p)_{nf}}, \ \rho_{nf} = (1-\phi)\rho_f + \phi\rho_s,$$

$$(\rho\beta)_{nf} = (1-\phi)(\rho\beta)_f + \phi(\rho\beta)_s, \ (\rho C_p)_{nf} = (1-\phi)(\rho C_p)_f + \phi(\rho C_p)_s, \tag{5}$$

$$\frac{k_{nf}}{k_f} = \frac{k_s + 2k_f - 2\phi(k_f - k_s)}{k_s + 2k_f + \phi(k_f - k_s)},$$

where μ_f refers to the dynamic viscosity of the base fluid, ϕ denotes the solid volume fraction of the nanoparticles, ρ_f and ρ_s represent the density of the base fluid and the density of the solid nanoparticle, respectively, k_{nf} represents the thermal conductivity of the nanofluid, as approximated by the Maxwell-Garnett's model, the subscript 'f' represents the base fluid, and lastly, 's' reflects the solid nanoparticle.

Meanwhile, upon employing the Rosseland's approximation, the radiation heat flux, q_r is given by Zheng [52], which adopts the following form:

$$q_r = -\frac{4\sigma^*}{3k^*}\frac{\partial T^4}{\partial y}, \tag{6}$$

where σ^* represents the Stefan-Boltzmann constant and k^* is the Rosseland mean spectral absorption coefficient. Furthermore, it is assumed that the temperature difference between the flow is such that T^4 can be expanded using Taylor's series as a linear combination of the temperature. Next, after the expansion of T^4 into the Taylor's series for T_∞, the approximation was obtained by omitting the higher order terms, obtaining $T^4 = 4T_\infty^3 T - 3T_\infty^4$. Therefore, upon substituting Equations (5) and (6) into Equation (3), the following equation is obtained:

$$u\frac{\partial T}{\partial x} + v\frac{\partial T}{\partial y} = \alpha_{nf}\frac{\partial^2 T}{\partial y^2} + \frac{16\sigma^* T_\infty^3}{3k^*(\rho C_p)_{nf}}\frac{\partial^2 T}{\partial y^2} + \frac{Q_0}{(\rho C_p)_{nf}}(T - T_\infty). \tag{7}$$

To determine the similar forms of the Equations (1), (2) and (7), with boundary conditions (4), the terms are defined; $u_w(x)$, $v_w(x)$, $T_w(x)$ and $u_e(x)$ in the following form:

$$u_w(x) = bx, \ v_w(x) = -\sqrt{av_f}s, \ T_w(x) = T_\infty + T_0 x, \ u_e(x) = ax. \tag{8}$$

Here, a and b are constants, s denotes the suction parameter and T_0 represents the constant characteristic temperature, with $T_0 < 0$ indicating the cooled surface (opposing flow) while $T_0 > 0$ signifies the heated surface (assisting flow).

Furthermore, the governing Equations (1), (2) and (7) together with the boundary conditions (4) have been transformed into ordinary differential equations by the dimensionless functions u, v and θ, in relation to the suitable similarity variable η as follows:

$$u = axf'(\eta), \ v = -\sqrt{av_f}f(\eta), \ \theta(\eta) = \frac{T - T_\infty}{T_w - T_\infty}, \ \eta = \sqrt{\frac{a}{v_f}}y. \tag{9}$$

Note that $f(\eta)$ denotes the dimensionless stream function, $f'(\eta)$ be the dimensionless velocity profile, $\theta(\eta)$ represents the dimensionless temperature profile and the prime indicates the differentiation with respect to η.

Equation (1) is therefore satisfied identically with the given similarity transformation (9). After substituting similarity transformation (9) into Equations (2) and (7), we obtain the following coupled nonlinear ordinary differential equations:

$$\frac{1}{(1-\phi)^{2.5}}f''' + \left(1-\phi+\phi\frac{\rho_s}{\rho_f}\right)\left(ff''+1-(f')^2\right) + \left(1-\phi+\phi\frac{(\rho\beta)_s}{(\rho\beta)_f}\right)\lambda\theta = 0, \tag{10}$$

$$\frac{1}{Pr}\left(\frac{k_{nf}}{k_f}+\frac{4}{3}Nr\right)\theta'' + \left(1-\phi+\phi\frac{(\rho C_p)_s}{(\rho C_p)_f}\right)(f\theta'-f'\theta+K\theta) = 0, \tag{11}$$

while the boundary conditions (4) adopt the new form:

$$f(\eta) = s,\ f'(\eta) = c, \theta(\eta) = 1 \text{ at } \eta = 0,$$
$$f'(\eta) = 1,\ \theta(\eta) = 0 \text{ as } \eta \to \infty, \tag{12}$$

where Pr denotes the Prandtl number, λ represents the mixed convection parameter with the case of $\lambda < 0$ corresponds to the opposing flow, whereas $\lambda > 0$ corresponds to the assisting flow. Moreover, Nr denotes the thermal radiation parameter, K represents the heat source/sink parameter with the case $K > 0$ refers to the heat source and $K < 0$ refers to the heat sink. Further c denotes the stretching/shrinking parameter, with $c > 0$ for a stretching sheet and $c < 0$ for a shrinking sheet, and s is the constant mass flux parameter, with $s > 0$ for suction and $s < 0$ for injection or withdrawal of the fluid, The parameters Pr, λ, Nr, K, s and c can be expressed in the following equations as:

$$Pr = \frac{\nu_f}{\alpha_f}, \lambda = \frac{Gr_x}{Re_x^2}, Nr = \frac{4T_\infty^3\sigma^*}{k_f k^*}, K = \frac{Q_0}{a(\rho C_p)_{nf}}, s = -\frac{v_w(x)}{\sqrt{a\nu_f}}, c = \frac{b}{a}. \tag{13}$$

Here, the local Grashof number Gr_x and the local Reynolds number Re_x are given by:

$$Gr_x = \frac{g\beta_f(T_w - T_\infty)x^3}{\nu_f^2}, Re_x = \frac{u_e x}{\nu_f}. \tag{14}$$

The interested physical quantities are the skin friction coefficient C_f and the local Nusselt number Nu_x which are expressed by:

$$C_f = \frac{\tau_w}{\rho_f u_e^2}, Nu_x = \frac{x q_w}{k_f(T_w - T_\infty)}, \tag{15}$$

where the shear stress at wall τ_w and the constant surface heat flux q_w are expressed as:

$$\tau_w = \mu_{nf}\left(\frac{\partial u}{\partial y}\right)_{y=0}, q_w = -k_{nf}\left(\frac{\partial T}{\partial y}\right)_{y=0} + (q_r)_{y=0}. \tag{16}$$

Substituting (9) into (16) and using (15), the following is obtained:

$$Re_x^{1/2}C_f = \frac{1}{(1-\phi)^{2.5}}f''(0), Re_x^{-1/2}Nu_x = -\left(\frac{k_{nf}}{k_f}+\frac{4}{3}Nr\right)\theta'(0). \tag{17}$$

3. Stability Analysis

The numerical results of the nonlinear ordinary differential equations given in Equations (10) and (11) together with the boundary conditions in Equation (12) indicates that for a particular range of the mixed convection parameter λ, there exist dual solutions (upper and lower branch solutions) for the various values of the selected governing parameters. Therefore, to validate which solution is in the stable flow, the stability of the dual solutions is tested by accommodating the stability analysis shown in Merkin [53]. To perform this, an unsteady form of the problem was considered. Equation (1) was retained, while Equations (2) and (7) were substituted by the following:

$$\frac{\partial u}{\partial t} + u\frac{\partial u}{\partial x} + v\frac{\partial u}{\partial y} = u_e\frac{du_e}{dx} + \frac{\mu_{nf}}{\rho_{nf}}\frac{\partial^2 u}{\partial y^2} + \frac{(\rho\beta)_{nf}}{\rho_{nf}}(T - T_\infty)g, \tag{18}$$

$$\frac{\partial u}{\partial t} + u\frac{\partial T}{\partial x} + v\frac{\partial T}{\partial y} = \alpha_{nf}\frac{\partial^2 T}{\partial y^2} + \frac{16\sigma^* T_\infty^3}{3(\rho C_p)_{nf}k^*}\frac{\partial^2 T}{\partial y^2} + \frac{Q_0}{(\rho C_p)_{nf}}(T - T_\infty), \tag{19}$$

where t represents the time. Analogous to the similarity transformation (9), the following new dimensionless functions u, v and θ have been introduced in conjunction to the similarity variable η which is the same as defined in (9), and the new similarity variable τ as follows:

$$u = ax\frac{\partial f}{\partial \eta}, \ v = -\sqrt{av_f}f(\eta,\tau), \ \theta(\eta,\tau) = \frac{T - T_\infty}{T_w - T_\infty}, \ \eta = \sqrt{\frac{a}{v_f}}y, \ \tau = at. \tag{20}$$

Of note, with variables u and v given in the above, the equation of continuity (1) is identically satisfied. Next, after substituting the new similarity transformation (20) into Equations (18) and (19), we obtained the following equations:

$$\frac{1}{(1-\phi)^{2.5}}\frac{\partial^3 f}{\partial \eta^3} + \left(1 - \phi + \phi\frac{\rho_s}{\rho_f}\right)\left(f\frac{\partial^2 f}{\partial \eta^2} + 1 - \left(\frac{\partial f}{\partial \eta}\right)^2 - \frac{\partial^2 f}{\partial \eta \partial \tau}\right) + \left(1 - \phi + \phi\frac{(\rho\beta)_s}{(\rho\beta)_f}\right)\lambda\theta = 0, \tag{21}$$

$$\frac{1}{Pr}\left(\frac{k_{nf}}{k_f} + \frac{4}{3}Nr\right)\frac{\partial^2 \theta}{\partial \eta^2} + \left(1 - \phi + \phi\frac{(\rho C_p)_s}{(\rho C_p)_f}\right)\left(f\frac{\partial \theta}{\partial \eta} - \frac{\partial f}{\partial \eta}\theta + K\theta - \frac{\partial \theta}{\partial \tau}\right) = 0, \tag{22}$$

and were subjected to the boundary conditions:

$$f(\eta,\tau) = s, \ \frac{\partial f}{\partial \eta} = c, \theta(\eta,\tau) = 1 \text{ at } \eta = 0, \frac{\partial f}{\partial \eta} = 1, \ \theta(\eta,\tau) = 0 \text{ as } \eta \to \infty. \tag{23}$$

Next, to study the stability of the dual solutions, small disturbances of the growth (or decay) rate γ or better known as the unknown eigenvalue parameter, are taken in the form (see Weidman et al. [54]):

$$f(\eta,\tau) = f_0(\eta) + e^{-\gamma\tau}F(\eta,\tau), \ \theta(\eta,\tau) = \theta_0(\eta) + e^{-\gamma\tau}G(\eta,\tau), \tag{24}$$

where $f_0(\eta)$ and $\theta_0(\eta)$ satisfied the problem (10)–(12). Besides, $F(\eta,\tau)$, $G(\eta,\tau)$ and all of the respective derivatives were assumed to be smaller when compared to $f_0(\eta)$, $\theta_0(\eta)$ and its derivatives. By means of using (24), hence Equations (21) and (22) can be given as:

$$\frac{1}{(1-\phi)^{2.5}}\frac{\partial^3 F}{\partial \eta^3} + \left(1 - \phi + \phi\frac{\rho_s}{\rho_f}\right)\left(f_0\frac{\partial^2 F}{\partial \eta^2} + f_0''F - 2f_0'\frac{\partial F}{\partial \eta} + \gamma\frac{\partial F}{\partial \eta} - \frac{\partial^2 F}{\partial \eta \partial \tau}\right) + \left(1 - \phi + \phi\frac{(\rho\beta)_s}{(\rho\beta)_f}\right)\lambda\theta = 0, \tag{25}$$

$$\frac{1}{Pr}\left(\frac{k_{nf}}{k_f} + \frac{4}{3}Nr\right)\frac{\partial^2 G}{\partial \eta^2} + \left(1 - \phi + \phi\frac{(\rho C_p)_s}{(\rho C_p)_f}\right)\left(f_0\frac{\partial G}{\partial \eta} + F\theta_0' - f_0'G - \theta_0\frac{\partial F}{\partial \eta} + KG + \gamma G - \frac{\partial G}{\partial \tau}\right) = 0, \tag{26}$$

together with the following boundary conditions:

$$F(\eta,\tau) = 0, \ \frac{\partial F}{\partial \eta} = 0, G(\eta,\tau) = 0 \text{ at } \eta = 0,$$
$$\frac{\partial F}{\partial \eta} = 0, G(\eta,\tau) = 0 \text{ as } \eta \to \infty. \tag{27}$$

As proposed by Weidman et al. [54], the initial growth or decay of the solutions (24) is identified, by setting $\tau = 0$, thus, giving $F = F_0(\eta)$ and $G = G_0(\eta)$. In this respect, the following linear eigenvalue problem was solved:

$$\frac{1}{(1-\phi)^{2.5}}F_0''' + \left(1 - \phi + \phi\frac{\rho_s}{\rho_f}\right)\left(f_0 F_0'' + f_0'' F_0 - 2f_0' F_0' + \gamma F_0'\right) + \left(1 - \phi + \phi\frac{(\rho\beta)_s}{(\rho\beta)_f}\right)\lambda G_0 = 0, \quad (28)$$

$$\frac{1}{\text{Pr}}\left(\frac{k_{nf}}{k_f} + \frac{4}{3}Nr\right)G_0'' + \left(1 - \phi + \phi\frac{(\rho C_p)_s}{(\rho C_p)_f}\right)\left(f_0 G_0' + F_0\theta_0' - f_0' G_0 - \theta_0 F_0' + KG_0 + \gamma G_0\right) = 0, \quad (29)$$

with the boundary conditions given by:

$$F_0(\eta) = 0,\ F_0'(\eta) = 0, G_0(\eta) = 0 \text{ at } \eta = 0,$$
$$F_0'(\eta) = 0, G_0(\eta) = 0 \text{ as } \eta \to \infty. \quad (30)$$

Indeed, it should be stated at this point, that the solutions $f_0(\eta)$ and $\theta_0(\eta)$ were determined from the problem depicted in Equations (10)–(12). Upon obtaining the results, $f_0(\eta)$ and $\theta_0(\eta)$ were again applied to Equations (28) and (29), and the linear eigenvalue problem (28)–(30) were solved. Harris et al. [55] proposed to relax a suitable boundary condition on $F_0'(\infty) = 0$ or $G_0(\infty) = 0$ to determine a better range of γ. In the current study, the condition $F_0'(\infty) = 0$ is relaxed and for a fixed value of γ, the linear eigenvalue problem (28)–(30) are solved, together with the new boundary condition; $F_0''(0) = 1$. Notably, it is worth mentioning that the solutions of the linear eigenvalue problem (28)–(30) provides an infinite set of eigenvalues $\gamma_1 < \gamma_2 < \gamma_3 < \dots$, where γ_1 refers to the smallest eigenvalue. Furthermore, a positive γ_1 reflects to an initial decay of disturbances and a stable flow. In contrast, a negative γ_1 indicates an initial growth of disturbances and unstable flow.

4. Results and Discussion

The derived nonlinear ordinary differential equations given in Equations (10) and (11) along with the boundary conditions given in (12) were solved numerically and were obtained using the bvp4c programme in MATLAB (Matlab R2015a, MathWorks, Natick, MA, USA) for the selected values of the mixed convection parameter λ, solid volume fraction of nanoparticles ϕ, thermal radiation parameter Nr, heat source/sink parameter K, suction parameter s and stretching/shrinking parameter c. The range of ϕ values was taken as $0 \le \phi \le 0.2$, where $\phi = 0$ indicates a regular base fluid, while the value of the Prandtl number was considered as $\text{Pr} = 6.2$ (water), except for comparisons with the prior case. The correlative output of the results obtained for $f''(0)$, with the ones obtained in Bachok et al. [56] for some values of c and ϕ with $\lambda = Nr = K = s = 0$ for Cu-water nanofluid, are presented in Table 2. Also, it was achieved that the present results were in very good alliance, which confirms that the numerical approach applied in this study is perfect, and therefore, the obtained results were believed to be accurate and correct.

The variations of the reduced skin friction coefficient $f''(0)$ and the reduced local Nusselt number $-\theta'(0)$ against λ are shown in Figures 2–13 for several values of ϕ, Nr, K, s and c. It was observed that dual solutions (upper and lower branch solutions) occurred for Equations (10) and (11) subject to the boundary conditions (12) in the range of $\lambda > \lambda_c$, where λ_c denotes the critical value of λ. Note that no solutions exist for $\lambda < \lambda_c$ while a unique solution exists when $\lambda = \lambda_c$. Also, it is obvious from these figures that the values of $|\lambda_c|$ increase as the parameters ϕ, s and c increase, therefore suggesting that these parameters widen the range of occurrence of dual solutions. Accordingly, it is further confirmed that the presence of nanoparticles, heat sink, suction and stretching sheet could decelerate the separation of the boundary layer, while the presence of thermal radiation, heat source and shrinking sheet could accelerate the separation of the boundary layer. Also, the values of $-\theta'(0)$ are always positive for the upper branch solution which is due to the heat being transferred from the hot surface of the stretching/shrinking sheet to the cold fluid. The reverse trend is observed in the case of the lower branch solution, i.e., $-\theta'(0)$ becomes unbounded as $\lambda \to 0^+$ and $\lambda \to 0^-$.

Table 2. Values of $f''(0)$ for various values of the stretching/shrinking parameter c and the solid volume fraction of nanoparticles ϕ for Cu-water nanofluid.

c	Bachok et al. [56]			Present Results		
	$\phi = 0$	$\phi = 0.1$	$\phi = 0.2$	$\phi = 0$	$\phi = 0.1$	$\phi = 0.2$
2	−1.887307	−2.217106	−2.298822	−1.887307	−2.217106	−2.298822
1	0	0	0	0	0	0
0.5	0.713295	0.837940	0.868824	0.713295	0.837940	0.868824
0	1.232588	1.447977	1.501346	1.232588	1.447977	1.501346
−0.5	1.495670	1.757032	1.821791	1.495670	1.757032	1.821791
−1.15	1.082231	1.271347	1.318205	1.082231	1.271347	1.318205
	[0.116702]	[0.137095]	[0.142148]	[0.116702]	[0.137095]	[0.142148]
−1.2	0.932473	1.095419	1.135794	0.932473	1.095419	1.135793
	[0.233650]	[0.274479]	[0.284596]	[0.233650]	[0.274479]	[0.284596]

"[ ]" refers to the lower branch solution.

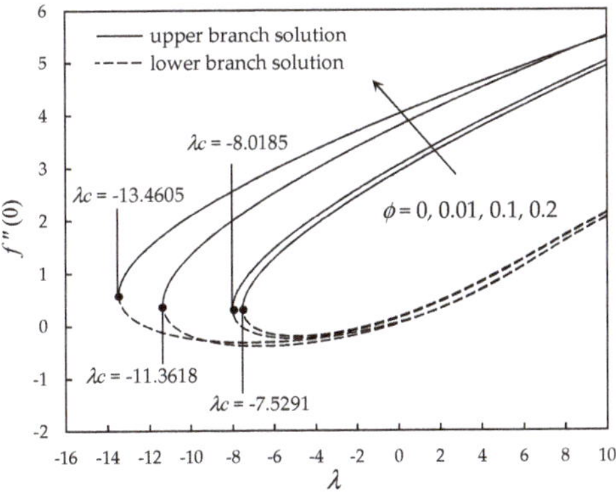

Figure 2. Variation of the reduced skin friction coefficient $f''(0)$ with the mixed convection parameter λ for several values of the volume fraction of nanoparticles ϕ when $Nr = 0.1$, $K = -0.1$, $c = -1$ and $s = 1$ for Cu-water nanofluid.

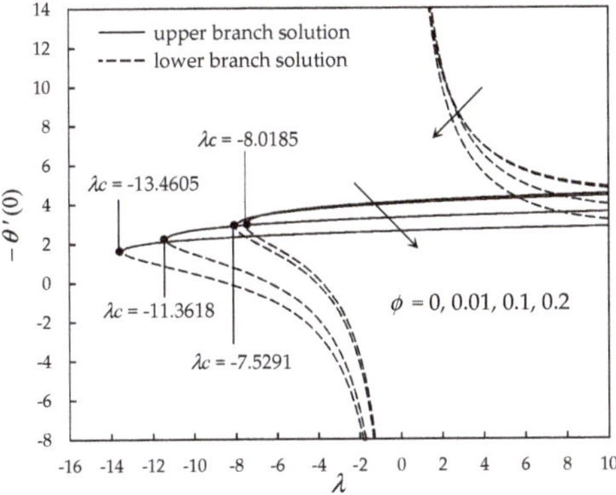

Figure 3. Variation of the reduced local Nusselt number $-\theta'(0)$ with the mixed convection parameter λ for several values of the volume fraction of nanoparticles ϕ when $Nr = 0.1$, $K = -0.1$, $c = -1$ and $s = 1$ for Cu-water nanofluid.

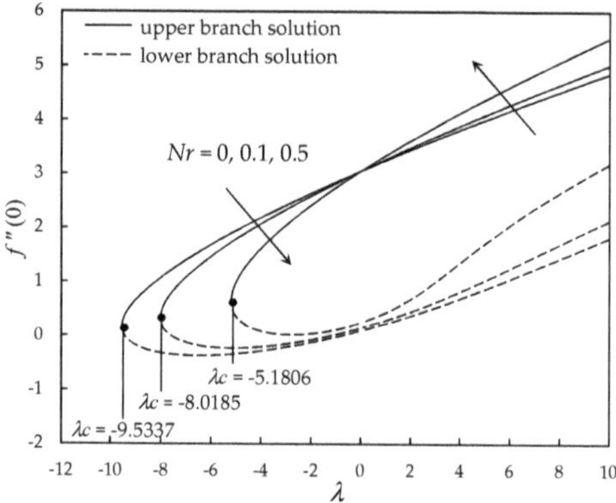

Figure 4. Variation of the reduced skin friction coefficient $f''(0)$ with the mixed convection parameter λ for several values of the radiation parameter Nr when $\phi = 0.01$, $K = -0.1$, $s = 1$ and $c = -1$ for Cu-water nanofluid.

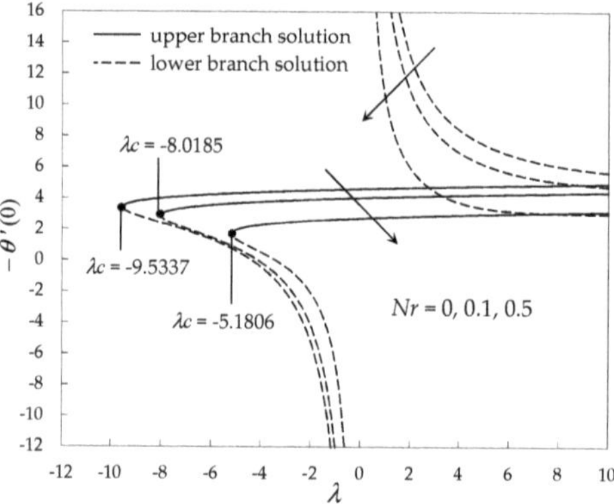

Figure 5. Variation of the reduced local Nusselt number $-\theta'(0)$ with the mixed convection parameter λ for several values of the radiation parameter Nr when $\phi = 0.01$, $K = -0.1$, $s = 1$ and $c = -1$ for Cu-water nanofluid.

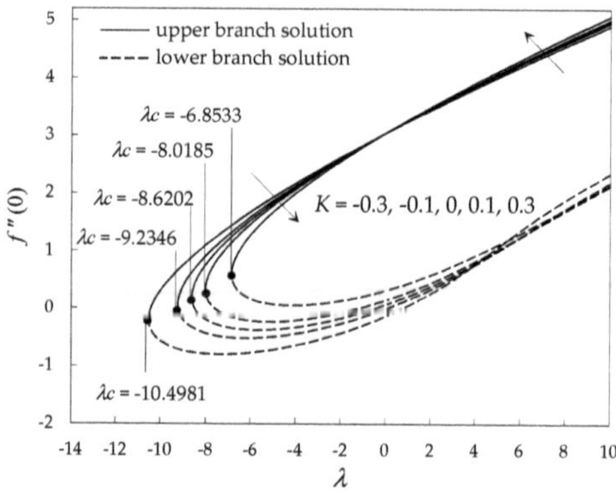

Figure 6. Variation of the reduced skin friction coefficient $f''(0)$ with the mixed convection parameter λ for various values of the heat source/sink parameter K when $\phi = 0.01$, $Nr = 0.1$, $s = 1$ and $c = -1$ for Cu-water nanofluid.

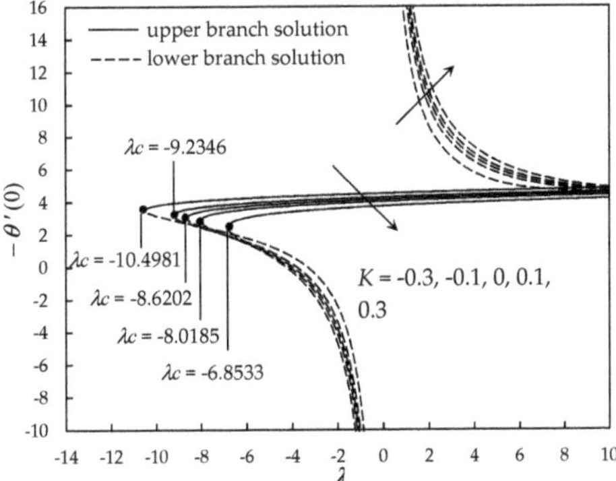

Figure 7. Variation of the reduced local Nusselt number $-\theta'(0)$ with the mixed convection parameter λ for various values of the heat source/sink parameter K when $\phi = 0.01$, $Nr = 0.1$, $s = 1$ and $c = -1$ for Cu-water nanofluid.

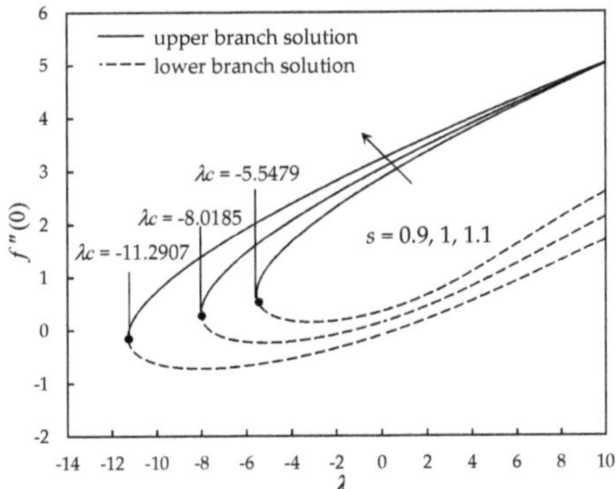

Figure 8. Variation of the reduced skin friction coefficient of $f''(0)$ with mixed convection parameter λ for various values of the suction parameter s when $\phi = 0.01$, $Nr = 0.1$, $K = 0.1$ and $c = -1$ for Cu-water nanofluid.

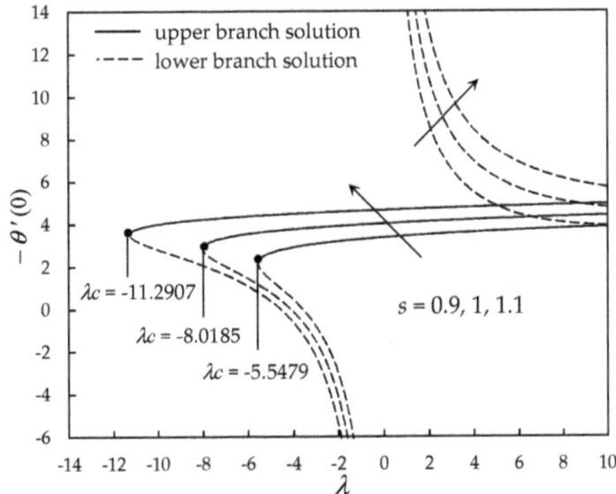

Figure 9. Variation of the reduced local Nusselt number $-\theta'(0)$ with the mixed convection parameter λ for various values of the suction parameter s when $\phi = 0.01$, $Nr = 0.1$, $K = 0.1$ and $c = -1$ for Cu-water nanofluid.

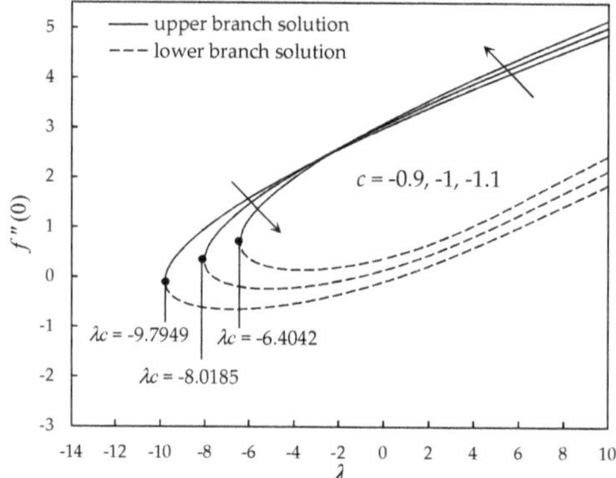

Figure 10. Variation of the reduced skin friction coefficient of $f''(0)$ with the mixed convection parameter λ for several values of c (shrinking sheet) when $\phi = 0.01$, $Nr = 0.1$, $K = 0.1$ and $s = 1$ for Cu-water nanofluid.

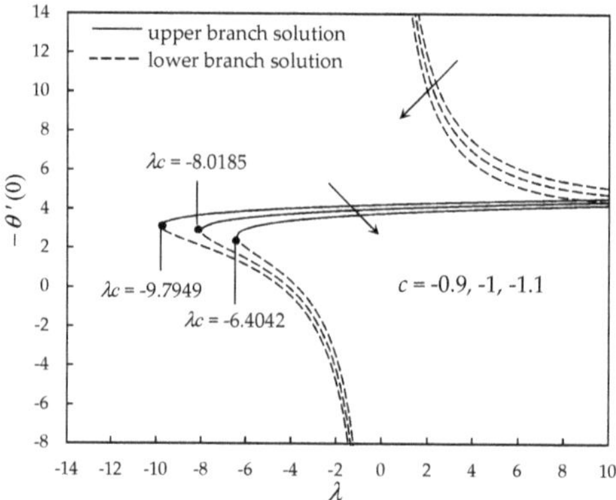

Figure 11. Variation of the reduced local Nusselt number $-\theta'(0)$ with the mixed convection parameter λ for several values of c (shrinking sheet) when $\phi = 0.01$, $Nr = 0.1$, $K = 0.1$ and $s = 1$ for Cu-water nanofluid.

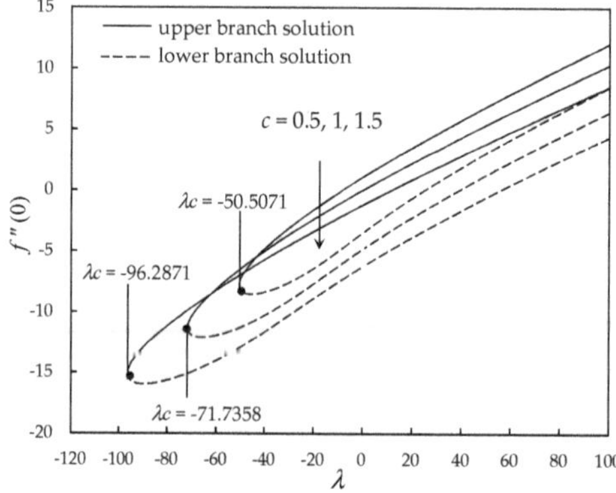

Figure 12. Variation of the reduced skin friction coefficient of $f''(0)$ with the mixed convection parameter λ for several values of c (stretching sheet) when $\phi = 0.01$, $Nr = 0.1$, $K = 0.1$ and $s = 1$ for Cu-water nanofluid.

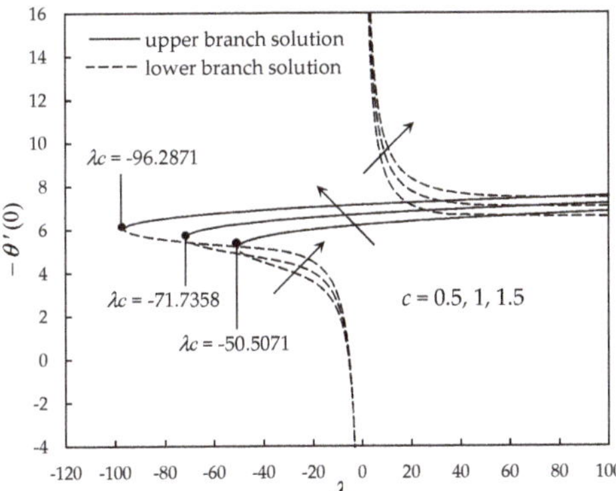

Figure 13. Variation of the reduced local Nusselt number $-\theta'(0)$ with the mixed convection parameter λ for various values of c (stretching sheet) when $\phi = 0.01$, $Nr = 0.1$, $K = 0.1$ and $s = 1$ for Cu-water nanofluid.

The influences of the solid volume fraction of nanoparticles ϕ on $f''(0)$ and $-\theta'(0)$ are illustrated in Figures 2 and 3. Figure 2 illustrates that $f''(0)$ increases smoothly with an increasing value of ϕ. Brinkman [57] explained that an increment in the nanoparticle volume fraction increases the fluid's viscosity. Hence, this situation contributed towards increasing the skin friction along the surface. Figure 3 depicts that as the volume fraction of nanoparticles increases which consequently decreases the rate of heat transfer at the surface of $-\theta'(0)$. This occurs because the nanoparticles increase the viscosity, density and conductivity. However, they may decrease the heat capacitance of the nanofluid $(\rho C_p)_{nf}$ (see MacDevette et al. [58]). Therefore, there is a trade-off between the enhanced properties, increased viscosity, decreased $(\rho C_p)_{nf}$ and possible decrease in the heat transfer coefficient.

Figures 4 and 5 display the impacts of the thermal radiation parameter Nr on $f''(0)$ and $-\theta'(0)$, respectively. Figure 4 also illustrates that the reduced skin friction coefficient increases as Nr is reduced in the case of the opposing flow. The reverse trend is noted in the event of the assisting flow and an increment in thermal radiation enhances the transmission of energy between the nanoparticles on a heated surface (assisting flow). Eventually, the thickness of the momentum boundary layer becomes thinner and increases the wall shear stress which raise the values of $f''(0)$. The decrement trend in the rate of heat transfer at the surface leads to an increment in the thermal radiation. Notably, this is in accordance with the result shown in Figure 5 where the reduced local Nusselt number $-\theta'(0)$ decreases with increasing Nr. Notwithstanding, this fact can be explained as follows. As the influence of the thermal radiation becomes stronger, the thermal boundary layer thickness increases and further decreases the values of the rate of heat transfer. Therefore, the usage of nanofluids having thermal radiation cannot improve the cooling of the heated sheet.

Figure 6 displays the variations of $f''(0)$ for various values of the heat source/sink parameter K. It is noticeable that the decrement of $f''(0)$ as the heat source/sink parameter increases from the negative (heat sink) to the positive values (heat source) in the event of the opposing flow. A contrary trend is noticeable in the event of the assisting flow. The influence of the heat source/sink parameter K on the reduced local Nusselt number $-\theta'(0)$ is illustrated in Figure 7. The results presented in this figure indicate that the rate of heat transfer at the surface decreases as the heat source/sink parameter increases from negative (heat sink) to positive values (heat source). Indeed, this is because the higher heat source effect can increase the thermal boundary layer thickness that reduces the rate of heat transfer.

Figure 8 illustrates the effect of the suction on the reduced skin friction coefficient. Further, it is observed that the values of $f''(0)$ increase with the suction. This is because the influence of suction at the boundary that slows down the nanofluid motion and increases the velocity gradient at the surface. By observing Figure 9, it is evident that the values of $-\theta'(0)$ that represents the heat transfer rate at the surface, increases with the suction. Precisely, the increase in the magnitude of the suction parameter consequently increased the rate of heat transfer. This is because the increasing suction decreases the thermal boundary layer thickness and in return increases the temperature gradient at the surface.

Figures 10–13 are shown to present the impacts of stretching/shrinking parameter c on $f''(0)$ and $-\theta'(0)$. Notably, it is evident from these figures that $f''(0)$ and $-\theta'(0)$ were higher in the event of the stretching sheet, in comparison with the case of the shrinking sheet. Hence, the stretching parameter provided the most significant effects upon the skin friction along the surface and the heat flux at the surface. Indeed, this suggests that the stretching sheet enhances the rate of heat transfer, whereas the shrinking sheet inhibits the effect of the heat transfer rate. Also, the increment in the value of the stretching parameter further increases the effect of free convection.

The variations of $f''(0)$ and $-\theta'(0)$ for three distinct types of nanofluids with nanoparticles containing Cu, Al_2O_3 and TiO_2 are shown in Figures 14 and 15 From the figures, it is evident that there is little difference in the value of the reduced skin friction coefficient and the reduced local Nusselt number for TiO_2- and Al_2O_3-water nanofluids. Further, it is also discovered that the values of $f''(0)$ and $-\theta'(0)$ are highest for Cu-water nanofluid, followed by TiO_2- and Al_2O_3-water nanofluids which is due to Cu having the highest value of thermal conductivity in comparison to the other nanoparticles.

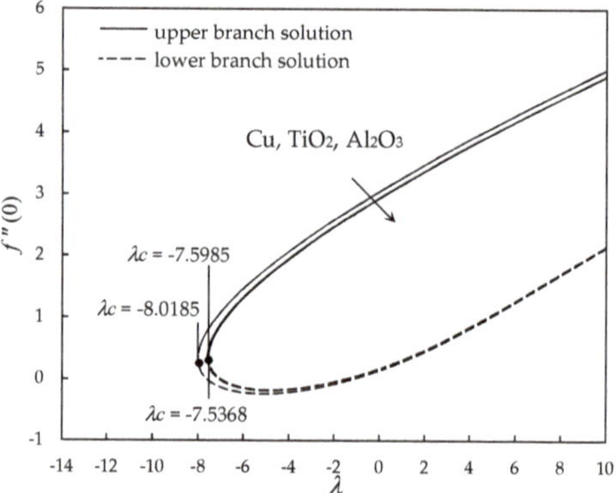

Figure 14. Variation of the reduced skin friction coefficient of $f''(0)$ with the mixed convection parameter λ when $\phi = 0.01$, $Nr = 0.1$, $K = 0.1$, $s = 1$ and $c = -1$ for Cu-water, Al_2O_3-water and TiO_2-water nanofluids.

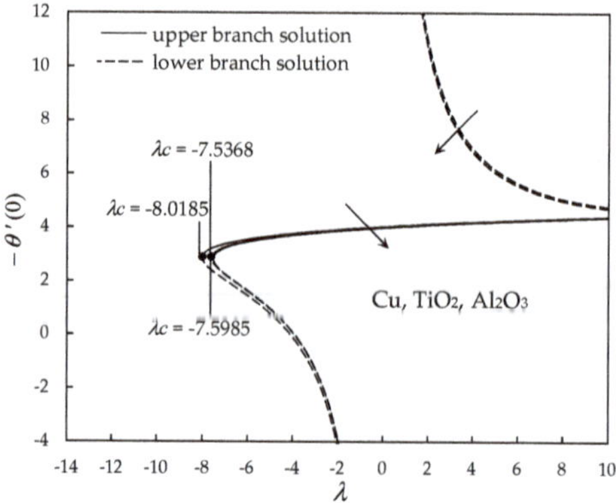

Figure 15. Variation of the reduced local Nusselt number $-\theta'(0)$ with the mixed convection parameter λ when $\phi = 0.01$, $Nr = 0.1$, $K = 0.1$, $s = 1$ and $c = -1$ for Cu-water, Al_2O_3-water and TiO_2-water nanofluids.

The velocity and temperature profiles for Cu-water nanofluid with different values of the volume fraction of the nanoparticles ϕ are displayed in Figures 16 and 17. It was discovered that an increment in the value of ϕ, increased the velocity of the fluid. Since an increment in ϕ is believed to increase the fluid's viscosity as suggested by Brinkman [57], hence it increases the skin friction along the permeable vertical shrinking flat plate. This statement can be proved by the velocity profiles as shown in Figure 16, as an increment in ϕ reduces the momentum boundary layer thickness. Furthermore, from Figure 17, we can see that the temperature of the fluid increases with an increase in ϕ, therefore, suggesting that the temperature of the nanofluids can be managed by increasing or decreasing the volume fraction of the nanoparticles in the base fluid.

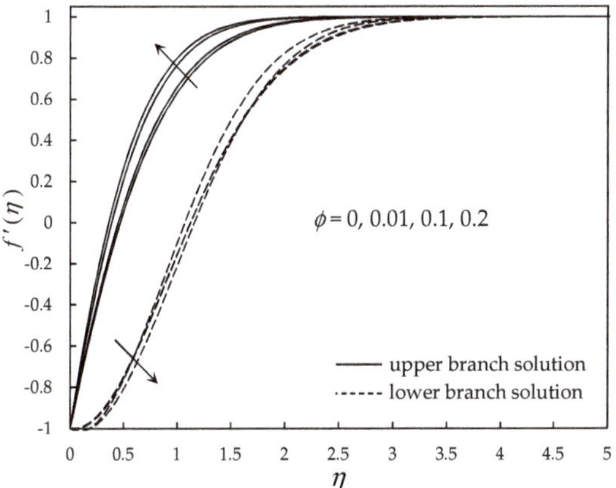

Figure 16. Velocity profiles $f'(\eta)$ for various values of the volume fraction of the nanoparticles ϕ when $Nr = 0.1, K = 0.1, s = 1, c = -1$ and $\lambda = -2$ (opposing flow) for Cu-water nanofluid.

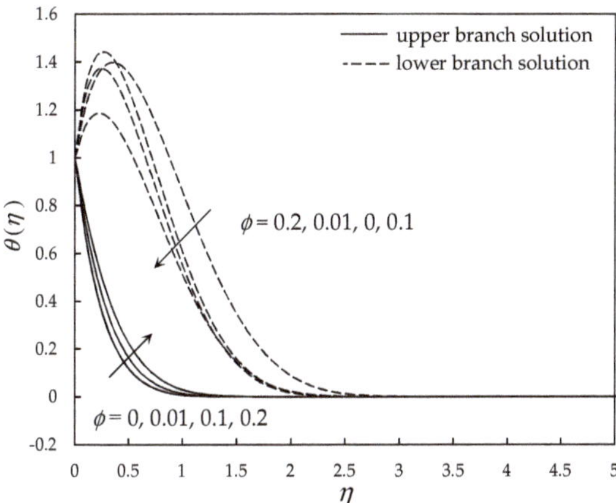

Figure 17. Temperature profiles $\theta(\eta)$ for various values of the volume fraction of the nanoparticles ϕ when $Nr = 0.1, K = 0.1, s = 1, c = -1$ and $\lambda = -2$ (opposing flow) for Cu-water nanofluid.

The velocity and temperature profiles for Cu-, Al_2O_3- and TiO_2-water nanofluids are shown in Figures 18 and 19, which indicate that by utilizing distinct types of nanofluids, the values of velocity and temperature change. Furthermore, we discovered that Cu-water nanofluid has higher velocity distribution and lower temperature distribution in comparison to the other two nanofluids for the upper branch solution. Also, it is discovered that the velocity and temperature profiles for Al_2O_3-water and TiO_2-water nanofluids nearly coincide with each other. Besides, the Cu-water nanofluid (compared to Al$_2$O$_3$-water and TiO$_2$-water nanofluids) has thinner momentum and thermal boundary layer thickness which is due to the fact Cu nanoparticles have the highest thermal conductivity value in comparison to the other two kinds of nanoparticles. Accordingly, the reduced value of thermal diffusivity causes higher temperature gradients and therefore, the higher enhancement in heat transfers. Notwithstanding, the Cu nanoparticle possess high values of thermal diffusivity, and therefore, lowers the temperature gradients that affects the performance of Cu nanoparticle.

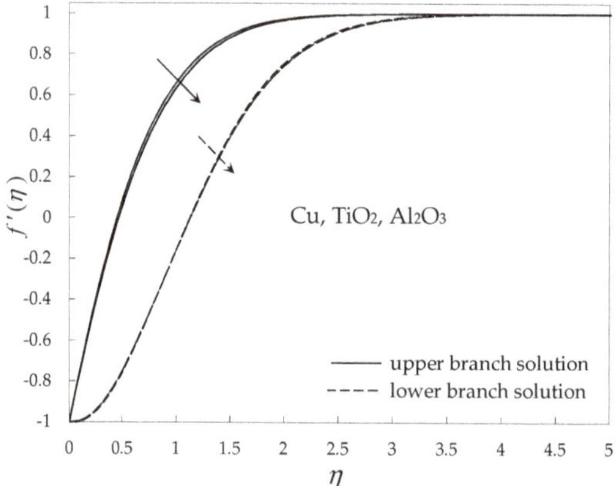

Figure 18. Velocity profiles $f'(\eta)$ for Cu-water, Al$_2$O$_3$-water and TiO$_2$-water nanofluids when ϕ = 0.01, Nr = 0.1, K = 0.1, s = 1, c = -1 and λ = -2 (opposing flow).

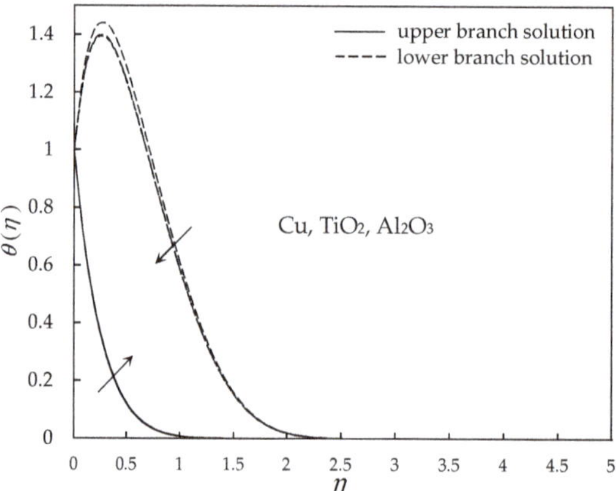

Figure 19. Temperature profiles $\theta(\eta)$ for Cu-water, Al$_2$O$_3$-water and TiO$_2$-water nanofluids when ϕ = 0.01, Nr = 0.1, K = 0.1, s = 1, c = -1 and λ = -2 (opposing flow).

From observing Figures 16–19, is apparent that the velocity and temperature profiles for both the upper and lower branch solutions satisfied the far field boundary conditions (12) asymptotically. Thus, aids in validating and supporting the numerical results retrieved for the boundary value problem (10)–(12) and proved the occurrence of the dual nature of the solutions as shown in Figures 2–15. Also, the lower branch solution for the corresponding profiles displayed a larger boundary layer thickness when compared to the upper branch solution.

A stability analysis was undertaken to test the stability of the dual solutions. Hence, the smallest eigenvalue γ_1 was determined by solving the linear eigenvalue problem (28)–(30) using the bvp4c programme in MATLAB. Table 3 depicts the smallest eigenvalues for Cu-water nanofluid for some values of parameters K, Nr and λ when $s = 1$ and $c = -1$. The results in Table 3 further indicate that γ_1 is negative for the lower branch solution, while γ_1 is positive for the upper branch solution due to their correspondence to the initial decay of disturbances, thus signifying a stable flow. Furthermore, for both the upper and lower branch solutions, the values of $|\gamma_1| \to 0$ as λ approaches λ_c, is consistent with Merkin [53], Weidman et al. [54] and Harris et al. [55].

Table 3. Smallest eigenvalues γ_1 for various values of K, Nr and λ.

K	Nr	λ	γ_1 (Upper Branch)	γ_1 (Lower Branch)
		-9	0.3230	-0.3084
-0.1	0.1	-9.1	0.2430	-0.2347
		-9.23	0.0412	-0.0410
		-8	0.5352	-0.4950
0	0.1	-8.5	0.2299	-0.2222
		-8.62	0.0016	-0.0003
		-9	0.4609	-0.4299
0.1	0	-9.4	0.2265	-0.2188
		-9.53	0.0343	-0.0341
		-5	0.3497	-0.3323
0.1	0.5	-5.1	0.2311	-0.2234
		-5.17	0.0805	-0.0796

5. Conclusions

In this paper, the mixed convection flow near a stagnation-point over a vertical stretching/shrinking sheet with suction, thermal radiation and heat source/sink in Cu-, Al_2O_3- and TiO_2-water nanofluids has been investigated numerically. The main limitation of this study is the mathematical nanofluid model used which considers boundary layer approximations, thus, we are unable to get the dual solutions beyond the separation point of a laminar boundary layer by utilizing the boundary layer approximations. To achieve the dual solutions beyond this point, the full Navier-Stokes equations are solved. Unfortunately, this case is way beyond the scope of the present research. Another limitation is that we utilized the mathematical nanofluid model suggested by Tiwari and Das [49], which does not consider the impacts of the Brownian motion and thermophoresis on nanofluids. It is worth acknowledging that a mathematical nanofluid model that refers to the Brownian motion and thermophoresis has been suggested by Buongiorno [59]. The results from this study may be summarised as follows:

- Dual solutions (upper and lower branch solutions) occur within the specific range of mixed convection parameters, whereby boundary layer separation occurs in the opposing flow region.
- The effects of thermal radiation, heat source and shrinking sheet could accelerate the separation of the boundary layer, while the presence of nanoparticles and the impacts of the heat sink, suction and stretching sheet could decelerate the separation of the boundary layer.
- The increment of the nanoparticle volume fraction in the nanofluid increased the velocity distribution, the temperature distribution and the reduced skin friction coefficient. But, decreases the rate of heat transfer at the surface.
- Cu-water nanofluid has the highest values of the reduced skin friction coefficient, reduced the local Nusselt number and velocity distribution, but a lower temperature distribution as compared with the others.
- A stability analysis has been performed to confirm that the upper branch solution is stable, while the lower branch solution is unstable.

It is observed from Table 2 that the values of $f''(0)$ for $\phi = 0$ (classical viscous fluid) are lower than those for a nanofluid ($\phi \neq 0$). $f''(0)$ increases almost monotonically with increasing the nanoparticle volume fraction ϕ.

Author Contributions: Numerical analysis were performed by A.J. and R.N. These authors also explained the results and wrote the manuscript. The literature review was written by I.P. and he co-wrote the manuscript. All authors originated and developed the problem and reviewed the manuscript.

Funding: This research was funded by the research university grant from the Universiti Kebangsaan Malaysia (UKM), grant number DIP-2017-009. The work of Ioan Pop was supported from Unitatea Executivă pentru Finanțarea Învățământului Superior, a Cercetării, Dezvoltării și Inovării (UEFISCDI), Romania, grant number PN-III-P4-ID-PCE-2016-0036.

Acknowledgments: The authors would like to thank the reviewers for their constructive comments which beautify the quality of the manuscript.

Conflicts of Interest: The authors declare no conflict of interest.

References

1. Ramachandran, N.; Chen, T.S.; Armaly, B.F. Mixed convection in stagnation flows adjacent to vertical surfaces. *J. Heat Transf.* **1988**, *110*, 373–377. [CrossRef]
2. Merkin, J.H.; Mahmood, T. Mixed convection boundary layer similarity solutions: Prescribed wall heat flux. *Z. Angew. Math. Phys.* **1989**, *40*, 51–68. [CrossRef]
3. Devi, C.D.S.; Takhar, H.S.; Nath, G. Unsteady mixed convection flow in stagnation region adjacent to a vertical surface. *Heat Mass Transf.* **1991**, *26*, 71–79. [CrossRef]
4. Lok, Y.Y.; Amin, N.; Campean, D.; Pop, I. Steady mixed convection flow of a micropolar fluid near the stagnation point on a vertical surface. *Int. J. Numer. Methods Heat Fluid Flow* **2005**, *15*, 654–670. [CrossRef]
5. Ridha, A.; Curie, M. Aiding flows non-unique similarity solutions of mixed-convection boundary-layer equations. *Z. Angew. Math. Phys.* **1996**, *47*, 341–352. [CrossRef]
6. Merkin, J.H. On dual solutions occurring in mixed convection in a porous medium. *J. Eng. Math.* **1986**, *20*, 171–179. [CrossRef]
7. Roşca, A.V.; Roşca, N.C.; Pop, I. Note on dual solutions for the mixed convection boundary layer flow close to the lower stagnation point of a horizontal circular cylinder: Case of constant surface heat flux. *Sains Malays.* **2014**, *43*, 1239–1247.
8. Rahman, M.M.; Merkin, J.H.; Pop, I. Mixed convection boundary-layer flow past a vertical flat plate with a convective boundary condition. *Acta Mech.* **2014**, *226*, 2441–2460. [CrossRef]
9. Abbasbandy, S.; Shivanian, E.; Vajravelu, K.; Kumar, S. A new approximate analytical technique for dual solutions of nonlinear differential equations arising in mixed convection heat transfer in a porous medium. *Int. J. Numer. Methods Heat Fluid Flow* **2017**, *27*, 486–503. [CrossRef]
10. Gebhart, B.; Jaluria, Y.; Mahajan, R.L.; Sammakia, B. *Buoyancy-Induced Flows and Transport*; Hemisphere: New York, NY, USA, 1988.
11. Schlichting, H.; Gersten, K. *Boundary Layer Theory*; Springer: New York, NY, USA, 2000.
12. Pop, I.; Ingham, D.B. *Convective Heat Transfer: Mathematical and Computational Viscous Fluids and Porous Media*; Pergamon: Oxford, UK, 2001.
13. Bejan, A. *Convective Heat Transfer*, 3rd ed.; Wiley: New York, NY, USA, 2013.
14. Choi, S.U.S. Enhancing thermal conductivity of fluids with nanoparticles. In Proceedings of the 1995 ASME International Mechanical Engineering Congress and Exposition, San Francisco, CA, USA, 12–17 November 1995; FED 231/MD. Volume 66, pp. 99–105.
15. Das, S.K.; Choi, S.U.S.; Yu, W.; Pradeep, T. *Nanofluids: Science and Technology*; Wiley: Hoboken, NJ, USA, 2007.
16. Manca, O.; Jaluria, Y.; Poulikakos, D. Heat transfer in nanofluids. *Adv. Mech. Eng.* **2010**, *2010*, 380826. [CrossRef]
17. Oztop, H.F.; Abu-Nada, E. Numerical study of natural convection in partially heated rectangular enclosures filled with nanofluids. *Int. J. Heat Fluid Flow* **2008**, *29*, 1326–1336. [CrossRef]
18. Nield, D.A.; Bejan, A. *Convection in Porous Media*, 4th ed.; Springer: New York, NY, USA, 2013.
19. Minkowycz, W.J.; Sparrow, E.M.; Abraham, J.P. (Eds.) *Nanoparticle Heat Transfer and Fluid Flow*; CRC Press: New York, NY, USA, 2012.
20. Shenoy, A.; Sheremet, M.; Pop, I. *Convective Flow and Heat Transfer from Wavy Surfaces: Viscous Fluids, Porous Media and Nanofluids*; CRC Press: New York, NY, USA, 2016.

21. Buongiorno, J.; Venerus, D.C.; Prabhat, N.; McKrell, T.; Townsend, J.; Christianson, R.; Tolmachev, Y.V.; Keblinski, P.; Hu, L.W.; Alvarado, J.L.; et al. A benchmark study on the thermal conductivity of nanofluids. *J. Appl. Phys.* **2009**, *106*, 094312. [CrossRef]

22. Kakaç, S.; Pramuanjaroenkij, A. Review of convective heat transfer enhancement with nanofluids. *Int. J. Heat Mass Transf.* **2009**, *52*, 3187–3196. [CrossRef]

23. Fan, J.; Wang, L. Review of heat conduction in nanofluids. *ASME J. Heat Transf.* **2011**, *133*, 040801. [CrossRef]

24. Mahian, O.; Kianifar, A.; Kalogirou, S.A.; Pop, I.; Wongwises, S. A review of the applications of nanofluids in solar energy. *Int. J. Heat Mass Transf.* **2013**, *57*, 582–594. [CrossRef]

25. Sheikholeslami, M.; Ganji, D.D. Nanofluid convective heat transfer using semi analytical and numerical approaches: A review. *J. Taiwan Inst. Chem. Eng.* **2016**, *65*, 43–77. [CrossRef]

26. Groşan, T.; Sheremet, M.A.; Pop, I. Heat transfer enhancement in cavities filled with nanofluids. In *Advances in Heat Transfer Fluids: From Numerical to Experimental Techniques*; Minea, A.A., Ed.; CRC Press: New York, NY, USA, 2017; pp. 267–284.

27. Myers, T.G.; Ribera, H.; Cregan, V. Does mathematics contribute to the nanofluid debate? *Int. J. Heat Mass Transf.* **2017**, *111*, 279–288. [CrossRef]

28. Tamim, H.; Dinarvand, S.; Hosseini, R.; Pop, I. MHD mixed convection stagnation-point flow of a nanofluid over a vertical permeable surface: A comprehensive report of dual solutions. *Heat Mass Transf.* **2014**, *50*, 639–650. [CrossRef]

29. Subhashini, S.V.; Sumathi, R.; Momoniat, E. Dual solutions of a mixed convection flow near the stagnation point region over an exponentially stretching/shrinking sheet in nanofluids. *Meccanica* **2014**, *49*, 2467–2478. [CrossRef]

30. Mustafa, I.; Javed, T.; Majeed, A. Magnetohydrodynamic (MHD) mixed convection stagnation point flow of a nanofluid over a vertical plate with viscous dissipation. *Can. J. Phys.* **2015**, *93*, 1365–1374. [CrossRef]

31. Ibrahim, S.M.; Lorenzini, G.; Kumar, P.V.; Raju, C.S.K. Influence of chemical reaction and heat source on dissipative MHD mixed convection flow of a Casson nanofluid over a nonlinear permeable stretching sheet. *Int. J. Heat Mass Transf.* **2017**, *111*, 346–355. [CrossRef]

32. Mabood, F.; Ibrahim, S.M.; Kumar, P.V.; Khan, W.A. Viscous dissipation effects on unsteady mixed convective stagnation point flow using Tiwari-Das nanofluid model. *Results Phys.* **2017**, *7*, 280–287. [CrossRef]

33. Othman, N.A.; Yacob, N.A.; Bachok, N.; Ishak, A.; Pop, I. Mixed convection boundary-layer stagnation point flow past a vertical stretching/shrinking surface in a nanofluid. *Appl. Therm. Eng.* **2017**, *115*, 1412–1417. [CrossRef]

34. Ozisik, M.N. Interaction of Radiation with Convection. In *Handbook of Single-Phase Convective Heat Transfer*; Wiley: New York, NY, USA, 1987; pp. 19.1–19.34.

35. Hady, F.M.; Ibrahim, F.S.; Abdel-Gaied, S.M.; Eid, M.R. Radiation effect on viscous flow of a nanofluid and heat transfer over a nonlinearly stretching sheet. *Nanoscale Res. Lett.* **2012**, *7*, 229. [CrossRef] [PubMed]

36. Ibrahim, W.; Shankar, B. MHD boundary layer flow and heat transfer of a nanofluid past a permeable stretching sheet with velocity, thermal and solutal slip boundary conditions. *Comput. Fluids* **2013**, *75*, 1–10. [CrossRef]

37. Haq, R.U.; Nadeem, S.; Khan, Z.H.; Akbar, N.S. Thermal radiation and slip effects on MHD stagnation point flow of nanofluid over a stretching sheet. *Physica E* **2015**, *65*, 17–23. [CrossRef]

38. Daniel, Y.S.; Aziz, Z.A.; Ismail, Z.; Salah, F. Effects of thermal radiation, viscous and Joule heating on electrical MHD nanofluid with double stratification. *Chin. J. Phys.* **2017**, *55*, 630–651. [CrossRef]

39. Sreedevi, P.; Reddy, P.S.; Chamkha, A.J. Heat and mass transfer analysis of nanofluid over linear and non-linear stretching surfaces with thermal radiation and chemical reaction. *Powder Technol.* **2017**, *315*, 194–204. [CrossRef]

40. Yazdi, M.; Moradi, A.; Dinarvand, S. MHD mixed convection stagnation-point flow over a stretching vertical plate in porous medium filled with a nanofluid in the presence of thermal radiation. *Arab. J. Sci. Eng.* **2014**, *39*, 2251–2261. [CrossRef]

41. Pal, D.; Mandal, G. Influence of thermal radiation on mixed convection heat and mass transfer stagnation-point flow in nanofluids over stretching/shrinking sheet in a porous medium with chemical reaction. *Nucl. Eng. Des.* **2014**, *273*, 644–652. [CrossRef]

42. Ayub, S.; Hayat, T.; Asghar, S.; Ahmad, B. Thermal radiation impact in mixed convective peristaltic flow of third grade nanofluid. *Results Phys.* **2017**, *7*, 3687–3695. [CrossRef]

43. Rana, P.; Bhargava, R. Numerical study of heat transfer enhancement in mixed convection flow along a vertical plate with heat source/sink utilizing nanofluids. *Commun. Nonlinear Sci. Numer. Simul.* **2011**, *16*, 4318–4334. [CrossRef]
44. Pal, D.; Mandal, G.; Vajravalu, K. Mixed convection stagnation-point flow of nanofluids over a stretching/shrinking sheet in a porous medium with internal heat generation/absorption. *Commun. Numer. Anal.* **2015**, *2015*, 30–50. [CrossRef]
45. Pal, D.; Mandal, G. Thermal radiation and MHD effects on boundary layer flow of micropolar nanofluid past a stretching sheet with non-uniform heat source/sink. *Int. J. Mech. Sci.* **2017**, *126*, 308–318. [CrossRef]
46. Mondal, H.; De, P.; Chatterjee, S.; Sibanda, P.; Roy, P.K. MHD three-dimensional nanofluid flow on a vertical stretching surface with heat generation/absorption and thermal radiation. *J. Nanofluids* **2017**, *6*, 189–195. [CrossRef]
47. Sharma, K.; Gupta, S. Viscous dissipation and thermal radiation effects in MHD flow of Jeffrey nanofluid through impermeable surface with heat generation/absorption. *Nonlinear Eng.* **2017**, *6*, 153–166. [CrossRef]
48. Tiwari, R.K.; Das, M.K. Heat transfer augmentation in a two-sided lid-driven differentially heated square cavity utilizing nanofluids. *Int. J. Heat Mass Transf.* **2007**, *50*, 2002–2018. [CrossRef]
49. Pang, C.; Jung, J.Y.; Kang, Y.T. Aggregation based model for heat conduction mechanism in nanofluids. *Int. J. Heat Mass Transf.* **2014**, *72*, 392–399. [CrossRef]
50. Ebrahimi, A.; Rikhtegar, F.; Sabaghan, A.; Roohi, E. Heat transfer and entropy generation in a microchannel with longitudinal vortex generators using nanofluids. *Energy* **2016**, *101*, 190–201. [CrossRef]
51. Sheremet, M.A.; Pop, I.; Bachok, N. Effect of thermal dispersion on transient natural convection in a wavy-walled porous cavity filled with a nanofluid: Tiwari and Das' nanofluid model. *Int. J. Heat Mass Transf.* **2016**, *92*, 1053–1060. [CrossRef]
52. Zheng, L.; Zhang, C.; Zhang, X.; Zhang, J. Flow and radiation heat transfer of a nanofluid over a stretching sheet with velocity slip and temperature jump in porous medium. *J. Frankl. Inst.* **2013**, *350*, 990–1007. [CrossRef]
53. Merkin, J.H. Mixed convection boundary layer flow on a vertical surface in a saturated porous medium. *J. Eng. Math.* **1980**, *14*, 301–313. [CrossRef]
54. Weidman, P.D.; Kubitschek, D.G.; Davis, A.M. The effect of transpiration on self-similar boundary layer flow over moving surfaces. *Int. J. Eng. Sci.* **2006**, *44*, 730–737. [CrossRef]
55. Harris, S.D.; Ingham, D.B.; Pop, I. Mixed convection boundary-layer flow near the stagnation point on a vertical surface in a porous medium: Brinkman model with slip. *Trans. Porous Media* **2009**, *77*, 267–285. [CrossRef]
56. Bachok, N.; Ishak, A.; Pop, I. Stagnation-point flow over a stretching/shrinking sheet in a nanofluid. *Nanoscale Res. Lett.* **2011**, *6*, 623. [CrossRef] [PubMed]
57. Brinkman, H.C. The viscosity of concentrated suspensions and solutions. *J. Chem. Phys.* **1952**, *20*, 571. [CrossRef]
58. MacDevette, M.M.; Myers, T.G.; Wetton, B. Boundary layer analysis and heat transfer of a nanofluid. *Microfluid Nanofluid* **2014**, *17*, 401–412. [CrossRef]
59. Buongiorno, J. Convective transport in nanofluids. *J. Heat Transf.* **2006**, *28*, 240–250. [CrossRef]

Article

The Influence of Surface Radiation on the Passive Cooling of a Heat-Generating Element

Igor V. Miroshnichenko [1,*], Mikhail A. Sheremet [2] and Abdulmajeed A. Mohamad [3]

[1] Regional Scientific and Educational Mathematical Centre, Tomsk State University, 634050 Tomsk, Russia
[2] Laboratory on Convective Heat and Mass Transfer, Tomsk State University, 634050 Tomsk, Russia; sheremet@math.tsu.ru
[3] Department of Mechanical and Manufacturing Engineering, Schulich School of Engineering, CEERE, The University of Calgary, Calgary, AB T2N 1N4, Canada; mohamad@ucalgary.ca
*   Correspondence: miroshnichenko@mail.tsu.ru; Tel.: +7-3822-529740

Received: 21 February 2019; Accepted: 9 March 2019; Published: 13 March 2019

Abstract: Low-power electronic devices are suitably cooled by thermogravitational convection and radiation. The use of modern methods of computational mechanics makes it possible to develop efficient passive cooling systems. The present work deals with the numerical study of radiative-convective heat transfer in enclosure with a heat-generating source such as an electronic chip. The governing unsteady Reynolds-averaged Navier–Stokes (URANS) equations were solved using the finite difference method. Numerical results for the stream function–vorticity formulation are shown in the form of isotherm and streamline plots and average Nusselt numbers. The influence of the relevant parameters such as the Ostrogradsky number, surface emissivity, and the Rayleigh number on fluid flow characteristics and thermal transmission are investigated in detail. The comparative assessment clearly emphasizes the effect of surface radiation on the overall energy balance and leads to change the mean temperature inside the heat generating element. The results of the present study can be applied to the design of passive cooling systems.

Keywords: convection; local heat-generating element; surface radiation; Ostrogradsky number; finite difference method

---

1. Introduction

Numerical and experimental studies on turbulent thermogravitational convection with surface radiation represent a highly topical issue for investigators due to its vast applicability in various technological applications that include electronic cooling, heat exchangers, thermal insulation systems, etc. The cooling of electronic components which are located in closed electronic cabinets remains a large problem for researchers due to the constant miniaturization of these components and the increasing level of operating temperatures. Thermogravitational convection cooling is very suitable, because it does not assume any fans which may break down. Moreover, passive convective cooling is also reliable, inexpensive, and quiet. The efficient optimization and thorough design of electronic devices require modern experimental and numerical approaches.

In the last decades, a significant number of papers have been published in specialized literature concerning the problem under consideration [1–4]. An excellent review on thermogravitational convection in enclosures for engineering applications was presented by Baïri et al. [5]. They studied a wide variety of configurations of cavities with different inclinations and shapes, heat source distributions, initial conditions, thermal boundary conditions, radiative properties, and nature of the fluid. In many practical situations, if either the characteristic dimension of the enclosure or the temperature difference is large enough, the fluid motion becomes turbulent in nature. Miroshnichenko and Sheremet [6] presented a comprehensive review of main results in the field of turbulent thermogravitational convection with and without radiation in rectangular cavities. That study also indicated that the effects of radiation on thermal transmission and liquid circulation required investigation. A detailed review of the literature on the entropy generation analysis for heat transfer processes involving various practical applications was performed by Biswal and Basak [7].

Baudoin et al. [8] investigated the optimized distribution of a significant number of power electronic devices cooled by turbulent thermogravitational convection. The authors tried to evaluate the distribution of up to 36 flush-mounted rectangular heat sources to give the best possible cooling capacity. They studied the effect of turbulence on the optimal (i.e., minimal temperature rise for a given area) distribution of heaters. Tou et al. [9] performed a numerical study of thermogravitational convection cooling on a 3-by-3 array of discrete heat sources on one vertical wall of a rectangular enclosure and cooled by the opposite wall. It was found that heat transfer from discrete heaters was non-uniform and should be accounted for by applying the averaging techniques. They also found that the effects of the Prandtl number were negligible in the range from 5 to 130. The paper presented by Bondarenko et al. [10] was devoted to the numerical simulation of free convection of the nanofluid cooling of heat-generating and heat-conducting sources using the heatline visualization technique. They concluded that the addition of nanoparticles could enhance the cooling process for the electronic devices for various distances between the heat source and cold wall. A numerical investigation of laminar unsteady thermogravitational convection in a closed enclosure having a local heat source of different geometric shapes was performed by Gibanov and Sheremet [11]. They considered heat source shapes which illustrated a smooth transition from rectangular cross-section to a triangular one through trapezoidal forms. Thermogravitational convection combined with thermal radiation in an air-filled rectangular enclosure with a discrete heater was studied by Saravanan and Sivaraj [12]. It was found that the global heat transfer rate was enhanced with an increase in the Rayleigh number and of the emissivity of the surface for both the heat generating and the isothermal heat sources. Sheremet et al. [13] investigated natural convection thermal transmission in a partially open alumina-water nanoliquid area under the effect of vertical solid mural and local heat-generating source. They concluded that the influence of nano-sized alumina particles was more significant in the case of low intensive liquid circulation and when the heat source was located near the cooling wall.

To simplify modeling of heat and mass transfer processes, radiation is generally ignored owing to the expected low temperatures. Nevertheless, over the past decade, refined computations as well as some reduced scale experiments have shown that thermal radiation can have a big impact on fluid flow characteristics and thermal transmission even at relatively insignificant temperature levels. A series of numerical experiments considering thermogravitational convection with radiative thermal transmission at large temperature differences in a rectangular cavity with a heated cylinder at its center were performed by Parmananda et al. [14]. They showed that square geometry was the optimum design of a heated cylinder as it had minimum entropy generation and maximum heat transfer. A similar numerical study of combined radiative-convective heat transfer in a three-dimensional differentially heated cavity was carried out by Parmananda et al. [15]. Dehbi et al. [16] investigated the effect of thermal radiation on thermogravitational convection inside enclosures. Numerical experiments were conducted with and without consideration of the gas radiation heat transfer. It should be mentioned that including radiation significantly improves the prediction of the flow field. Turbulent modes of convection and radiation in an air-filled differentially heated enclosure at $Ra = 1.5 \times 10^9$ were studied by Ibrahim et al. [17]. They conducted a comparative analysis for four cases (only gas radiation, only wall radiation, combined gas with wall radiation, and without radiation) and compared them in terms of temperature and velocity fields as well as turbulent quantities (turbulence intensity and kinetic energy). Their findings indicated that gas radiation had a little influence on the flow structure, at least when considered alone (without wall radiation). The work of Sharma et al. [18] was devoted to the turbulent thermogravitational convection in a rectangular enclosure with symmetrical cooling from the vertical side walls and localized heating from below. It should be noted that correlations have been developed to evaluate the Nusselt number in terms of the Rayleigh number and the heated width.

The effect of rotating a square cavity with a heater was investigated by Mikhailenko et al. [19]. They tried to find optimal conditions for the heated devices in order to decrease the working temperature of these elements. One main result of that work was that an intensive rotation could essentially reduce the average temperature inside a local heat-generating source.

The effect of enclosure shapes on thermal transmission and fluid flow has received considerable attention in the recent past [20]. Das et al. [21] presented a review of the results of studies on free convection in enclosures of various shapes. That work summarized the research on convection heat transfer in trapezoidal, triangular, and parallelogrammic cavities and cavities with wavy and curved walls. The mentioned results demonstrated possible strategies for improving the efficiency of convective heat transfer. Kang et al. [22] experimentally investigated vertical tubes with inverted triangular fins under thermogravitational convection. The influence of different fin heights, fin numbers, and heat inputs was studied. They proposed the Nusselt number correlation, which has the potential of being used in the tube design of various cooling devices. Free convection inside a square system containing two vertical thin heat-generating baffles was analyzed by Saravanan and Vidhya Kumar [23]. They focused mainly on investigating the effect of various positions of the baffles and different boundary conditions on the vertical walls.

To the best of the authors' knowledge, the problem of turbulent convective-radiative heat transfer inside cavities with heat-generating sources has not been well understood. The cavity configuration under consideration is quite important for the electronics industry where similar basic designs are used. Numerical simulations were performed for various surface emissivities of walls, Rayleigh numbers, and Ostrogradsky numbers. The aim of the analyses of thermal transmission was also to investigate the Nusselt number distribution on the heat source surface.

2. Governing Equations and Numerical Method

The present work considered a turbulent flow of Newtonian fluid (air) under the condition of a transparent medium. Solid walls and a heater were assumed to be opaque, gray, and diffuse emitters. The Boussinesq approximation was used to describe the density changes. It was assumed in the analysis that the fluid was viscous, heat conducting, and Newtonian. Using air as the working fluid inside the enclosure and the real wall material allowed us to consider that the thermophysical properties of the fluid were independent of temperature. The physical system is defined in Figure 1. In the present model, a heat-generating source is located at the bottom wall of the enclosure with length of $0.2L$ and has an internal constant volumetric heat flux Q. The external surfaces of the top wall $(y = L + 2l)$, right wall $(x = L + 2l)$, and left wall $(x = 0)$ are maintained under conditions of convective heat exchange with an environment. Remaining external surface $(y = 0)$ is adiabatic. No-slip boundary conditions are supposed for the internal walls of the cavity.

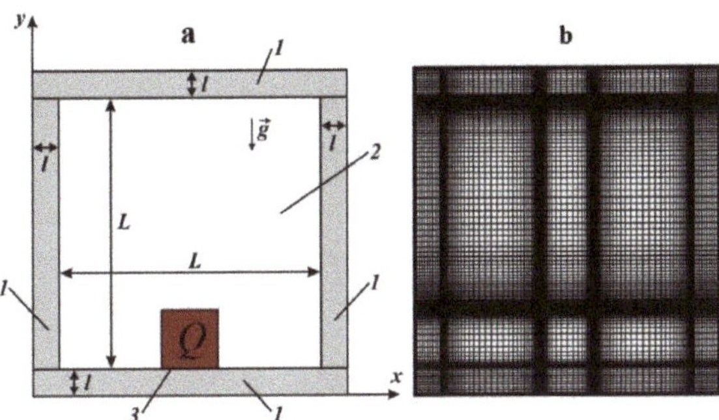

Figure 1. (a) A schematic of the system: *1*—solid walls, *2*—air, *3*—heat-generating element; (b) computational domain with a non-uniform grid.

From time $t > 0$, there is a temperature difference between the fluid and various surfaces. The thermodynamic equilibrium is disturbed by the appearance of buoyancy forces. Bearing in mind the above assumptions, the Reynolds-averaged Navier–Stokes equations can be written in the following form:

$$\frac{\partial G}{\partial t} + \frac{\partial S_1}{\partial x_1} + \frac{\partial S_2}{\partial x_2} = R, \tag{1}$$

$$G = \begin{pmatrix} 0 \\ u_1 \\ u_2 \\ T_f \\ k \\ \epsilon \\ T_w \\ T_{hs} \end{pmatrix}, \tag{2}$$

$$S_i = \begin{pmatrix} u_i \\ p\delta_{i1} - (\nu + \nu_t)\sigma_{i1} + u_i u_1 \\ p\delta_{i2} - (\nu + \nu_t)\sigma_{i2} + u_i u_2 \\ -(\alpha + \alpha_t)\partial T_f / \partial x_i + u_i T_f \\ -(\nu + \nu_t/\sigma_k)\partial k / \partial x_i + u_i k \\ -(\nu + \nu_t/\sigma_\epsilon)\partial \epsilon / \partial x_i + u_i \epsilon \\ \alpha_w \partial T_w / \partial x_i \\ \alpha_{hs} \partial T_{hs} / \partial x_i \end{pmatrix}, \forall\, i = 1, 2, \tag{3}$$

$$R = \begin{pmatrix} 0 \\ 0 \\ g\beta\Delta T \\ 0 \\ P_k + G_k - \epsilon \\ (c_{1\epsilon}(P_k + c_{3\epsilon}G_k) - c_{2\epsilon}\epsilon)\frac{\epsilon}{k} \\ 0 \\ q_v \end{pmatrix}, \tag{4}$$

$$\sigma_{ij} = \frac{\partial u_i}{\partial x_j} + \frac{\partial u_j}{\partial x_i}, \tag{5}$$

where P_k describes the production of k, G_k defines the generation or dissipation of turbulent kinetic energy due to buoyancy, and these terms are expressed as

$$P_k = \nu_t \left[2\left(\frac{\partial u_1}{\partial x_1}\right)^2 + 2\left(\frac{\partial u_2}{\partial x_2}\right)^2 + \left(\frac{\partial u_1}{\partial x_2} + \frac{\partial u_2}{\partial x_1}\right)^2 \right], \; G_k = -\frac{g\beta\nu_t}{Pr_t}\frac{\partial T}{\partial x_2}. \tag{6}$$

In the present study, the k-ϵ model was used to describe the Reynolds stress terms [24,25]. For completeness of information, the following expressions adopted to define the closure coefficients of the turbulence model (i.e., c_μ, $c_{1\epsilon}$, $c_{2\epsilon}$, $c_{3\epsilon}$, σ_k, σ_ϵ, and Pr_t) are summarized in Table 1. To make this problem dimensionless, the reference parameters $V_0 = \sqrt{g\beta\Delta TL}$, $t_0 = \sqrt{g\beta\Delta T/L}$, and L which, respectively, represent the velocity, time, and length are used. The following dimensionless variables are given as follows:

$$X = x_1/L, Y = x_2/L, \tau = t\sqrt{g\beta\Delta T/L}, U = u_1/\sqrt{g\beta\Delta TL}, V = u_2/\sqrt{g\beta\Delta TL}, \Theta = (T - T^e)/(T_0 - T^e),$$
$$\Psi = \psi/\sqrt{g\beta\Delta TL^3}, \Omega = \omega\sqrt{L/g\beta\Delta T}, K = k/(g\beta\Delta TL), E = \epsilon/\sqrt{g^3\beta^3(\Delta T)^3 L}$$

Table 1. Empirical constants for the turbulence model.

Parameters	c_μ	$c_{1\epsilon}$	$c_{2\epsilon}$	$c_{3\epsilon}$	σ_k	σ_ϵ	Pr_t
Values	0.09	1.44	1.92	0.8	1.0	1.3	1.0

In order to investigate the fluid flow in terms of streamlines, the dimensionless vorticity and stream function can be calculated in the usual way as follows:

$$\Omega = \frac{\partial V}{\partial X} - \frac{\partial U}{\partial Y}, \; U = \frac{\partial \Psi}{\partial Y}, \; V = -\frac{\partial \Psi}{\partial X}. \tag{7}$$

The thermal transmission and liquid circulation are determined by the following governing parameters: the Ostrogradsky number (Os), the Prandtl number (Pr), and the Rayleigh number (Ra). They are defined, respectively, in the following form:

$$Os = \frac{q_v L^2}{\lambda_{hs} \Delta T}, \; Pr = \frac{\nu}{\alpha_f}, \; Ra = \frac{g \beta \Delta T L^3}{\nu \alpha_f}. \tag{8}$$

The initial conditions for the non-dimensional governing equations are considered in the following form:

$$\Psi(X,Y,0) = \Omega(X,Y,0) = K(X,Y,0) = E(X,Y,0) = 0, \; \Theta(X,Y,0) = 1.0 \text{ at } \tau = 0.$$

The boundary conditions are expressed as follows:

For the energy equations: $\frac{\partial \Theta}{\partial Y} = 0$ at $Y = 0$; $\frac{\partial \Theta}{\partial n} = Bi \cdot \Theta$ at $X = 0$, $X = 1 + 2l/L$, $Y = 1 + 2l/L$; $\Theta_1 = \Theta_2$, $\frac{\partial \Theta_w}{\partial n} = \frac{\lambda_f}{\lambda_w} \frac{\partial \Theta_f}{\partial n} - N_{rad} Q_{rad}$ at internal solid–fluid interfaces; and $\frac{\partial \Theta_{hs}}{\partial n} = \frac{\lambda_f}{\lambda_{hs}} \frac{\partial \Theta_f}{\partial n} - N_{rad} Q_{rad}$ at the heat-generating source surface.

For the momentum equations: $\Psi = \frac{\partial \Psi}{\partial n} = 0$, $\Omega = -\frac{\partial^2 \Psi}{\partial n^2}$ at internal solid–fluid interfaces.

The boundary conditions for the turbulent parameters have been described in detail previously in [24]. To change a non-uniform grid in physical domain to a uniform grid in computational domain, a special algebraic coordinate transformation has been applied [24,25].

In this paper, the surfaces were assumed to be gray. The process of radiative heat exchange between such surfaces (due to absorption and reflection) is complicated compared to the same process for absolutely black bodies. Radiation analysis has been carried out using the balance method [24–26]. To receive the dimensionless net radiative heat flux Q_{rad}, it is necessary to solve the following equations:

$$Q_{rad,k} = R_k - \sum_{i=1}^{N} F_{k-i} R_{i,,} \tag{9}$$

$$R_k = (1 - \tilde{\epsilon}_k) \sum_{i=1}^{N} F_{k-i} R_i + \tilde{\epsilon}_k (1 - \zeta)^4 \left(\Theta_k + 0.5 \frac{1+\zeta}{1-\zeta} \right)^4. \tag{10}$$

The contribution of radiation is a significant topic for heat transfer in enclosures. Radiation heat transfer depends on various parameters such as the wall temperature, surface emissivity, cavity geometry, and thermophysical properties of the internal medium. In studies focused on thermogravitational convection, the radiative mode of heat transfer is sometimes neglected due to the suppressing number of computational resources it demands. At the same time, as practice shows, radiation has a significant impact on the system under consideration and cannot be neglected. Thus, real materials will emit (and consequently absorb) less thermal radiation than predicted for a black body. In the present work, the internal surfaces of the heat source and all walls were considered to be both gray diffusive emitters and reflectors of radiation. The air flow inside the enclosure was considered radiatively non-participating. The parameter $\widetilde{\epsilon}$(which varies in this work) was also reasonable from a physical point of view.

The finite difference method was used to solve the set of governing equations. The diffusion terms were discretized using the second-order accurate central difference scheme, whereas the accurate upwind difference scheme was used for the convective terms. The Samarskii locally one-dimensional scheme was adopted to solve the parabolic equations. The first-order upwind scheme was used for the discretization of the transient term. The resulting systems of linear equations were solved using the Thomas algorithm. The elliptical equation was discretized using the second-order accurate central differencing scheme. The successive over-relaxation method was used to solve the obtained difference equation. The time step used here was chosen to be $\Delta t = 10^{-4}$. A numerical code was written in C++ programming language. The case of surface radiation and thermogravitational convection in a differentially heated air-filled square cavity was investigated to validate the numerical procedure. The obtained results were compared with the numerical data of Wang et al. [27]. Figures 2 and 3 show a good agreement between the data under consideration (maximum deviation in parameters is 10%).

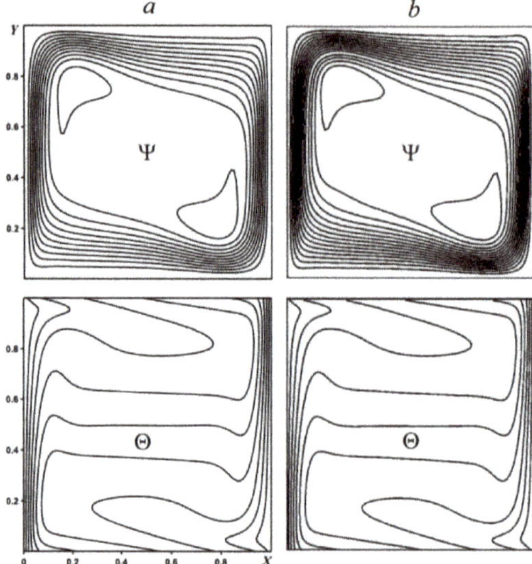

Figure 2. (**a**) Streamlines Ψ and isotherms Θ at $Ra = 10^6$, $\widetilde{\epsilon} = 0.8$ (**b**) numerical data of Wang et al. [27].

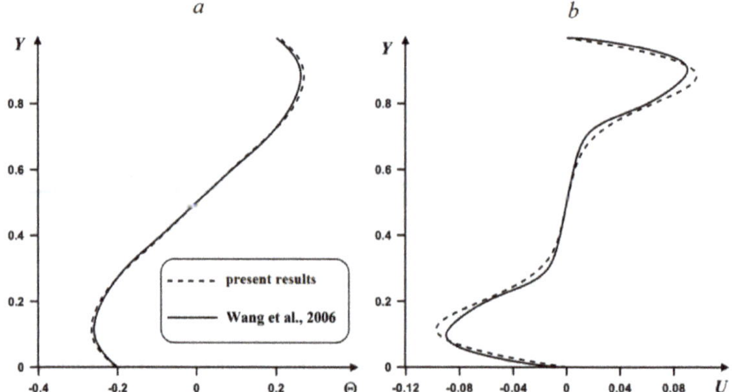

Figure 3. Profiles of (**a**) temperature and (**b**) horizontal velocity at $X = 0.5$ at $Ra = 10^6$, $\widetilde{\epsilon} = 0.2$ in comparison with numerical data of Wang et al. [27].

In the case of thermogravitational convection in the enclosure, the developed computational code was validated successfully using the numerical results of Dixit and Babu [28], Zhuo and Zhong [29], and Le Quéré [30] (see Table 2). The numerical simulation was carried out on finer and coarser meshes to check the grid independence. These computational tests with various grids allowed choosing the suitable grid size selection without compromising CPU time and accuracy. The average convective Nusselt number on the heat source surface, fluid flow rate, and average heater temperature are shown in Table 3. Therefore, the grid of 120×120 nodes was chosen for further investigation.

Table 2. Variation of the average Nusselt numbers.

Ra	Dixit et al. [28]	Zhuo et al. [29]	Le Quéré [30]	Present Data
10^7	16.79	16.523	–	17.13
10^8	30.506	30.225	28.78	33.06
10^9	57.35	–	62.0	60.54

Table 3. Grid independence study for $Ra = 10^9, \widetilde{\epsilon} = 0, \tau = 1000$.

Grid Size	δ	Nu_{conv}	$|\psi|_{max}$	Θ_{hs}
60×60	5.7×10^{-3}	48.78	0.031	0.764
120×120	2.5×10^{-3}	61.54	0.028	0.783
156×156	1.9×10^{-3}	64.974	0.027	0.786

It should be noted that the optimum over-relaxation parameter for the elliptical equation was found to be 1.9. Simple, but at the same time thorough criteria for each of the variables were used to obtain converged solutions at each time step. The convergence condition $\left| \beta_{ij}^{k+1} - \beta_{ij}^{k} \right| < 10^{-6}$ must be satisfied by each variable β_{ij}^{k} at any grid point (i, j); here, k is a given iteration parameter.

3. Results

The main aim of this work was to analyze the effect of surface radiation as well as the Ostrogradsky and Rayleigh numbers variation on fluid flow characteristics and thermal transmission. In this section, the numerical results are reported for $0.1 < Os < 2$, $Ra = 10^8$, $Ra = 10^9$, $Ra = 10^{10}$. The various values of surface emissivity (0, 0.3, 0.6, and 0.9) of the heat-generating source and the solid wall surfaces were used to examine the influence of radiation on convective heat exchange. Air ($Pr = 0.71$) was the working fluid. It should be noted that the thermal conductivity of the heater material λ_{hs} corresponded to the value of silicon. The thermal conductivity of the solid walls λ_w was specially selected with a small value of 0.7. Numerical results for the various Nusselt numbers (Nu_{con} and Nu_{rad}) for different conditions are represented and discussed. The geometry was selected in order to simulate the cooling of an electronic device located on a bottom horizontal surface. The results were obtained using one Intel Core i7 processor of 3.30 GHz with 16GB of RAM memory.

Figure 4 shows two-dimensional temperature fields at $Ra = 10^8$, $Os = 1$, $\widetilde{\epsilon} = 0.6$ for various values of the dimensionless time. At the initial time ($\tau = 3$), the distributions of temperature reflect the formation of two small thermal plumes over the heat-generating source. It is worth noting that heat conduction is the governing mechanism of heat transfer near the heater due to low convective velocity at initial time. At $\tau = 10$, the thermal plumes' height is about a third of the enclosure height, and further on ($\tau = 15$) a single thermal plume is formed. The ascending hot air reaches an internal surface of the top wall, while close to the internal surfaces of two vertical walls the descending flow is formed. Subsequently, the dimensionless time increasing leads to more intensive low temperature penetration from the environment.

The effect of surface-to-surface radiation (non-participating gas medium) was investigated. Isotherms Θ and streamlines Ψ are shown in Figure 5 for the various surface emissivities of the heater and solid walls. It can be seen that two convective cells (with counterclockwise and clockwise rotating) are observed for all the values of $\widetilde{\epsilon}$. The fluid due to heat transport from the heat-generating source rises up and collides at the top adiabatic wall; further on, it bifurcates and flows to the solid vertical walls. Thus, the convective cells have nearly equal strength. The strength of these vortices increases (see Figure 5) with a decreasing value of surface emissivity. Therefore, an insignificant intensification of the fluid flow in the enclosure $|\Psi|_{max}^{\widetilde{\epsilon}=0.9} = 0.021$, $|\Psi|_{max}^{\widetilde{\epsilon}=0.3} = 0.022$, $|\Psi|_{max}^{\widetilde{\epsilon}=0.0} = 0.023$ is observed with a reduction of $\widetilde{\epsilon}$. At the same time, the temperature fields are changed more significantly. Inside the enclosure, an increase in Θ is observed. In particular, it can be seen in isotherm ($\Theta = 0.6$) which is located above the heater ($\widetilde{\epsilon} = 0$) that it is further ascended with a thermal plume ($\widetilde{\epsilon} = 0.9$). These numerical results are confirmed by the observations made by Martyushev and Sheremet [31] in the laminar case.

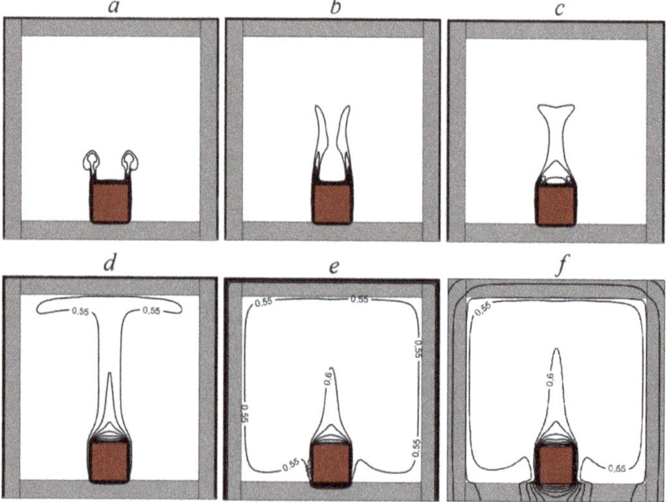

Figure 4. Isotherms Θ at $Ra = 10^9$, $Os = 1$, $\widetilde{\epsilon} = 0.3$: (**a**) $\tau = 3$, (**b**) $\tau = 10$, (**c**) $\tau = 15$, (**d**) $\tau = 50$, (**e**) $\tau = 200$, and (**f**) $\tau = 2000$.

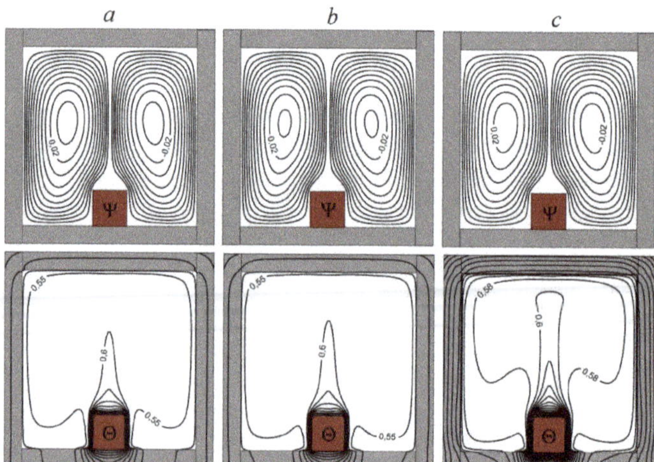

Figure 5. Isotherms Θ and streamlines Ψ at $Ra = 10^9$, $Os = 1$, $\tau = 2000$: (**a**), (**b**) $\widetilde{\epsilon} = 0.3$, and (**c**) $\widetilde{\epsilon} = 0.9$, $\widetilde{\epsilon} = 0$.

In order to study the effect of radiation on thermal transmission, the total heat transfer from the heater was evaluated using the mean radiative and convective Nusselt numbers. Figure 6 depicts Nu_{conv} and Nu_{rad} at the fluid–solid interfaces for various values of surface emissivity. The presented results indicate that there is no time point which characterizes the stationary distribution of heat transfer coefficients. It is caused by permanent heat generation at the local heater. Thus, surface emissivity enhances the mean radiative Nusselt number by 2.84 times upon the change of $\tilde{\epsilon}$ from 0.3 to 0.9. Due to a decrease in the temperature gradient, a growth of $\tilde{\epsilon}$ leads to the reduction of the mean convective Nusselt number. However, this change is minor.

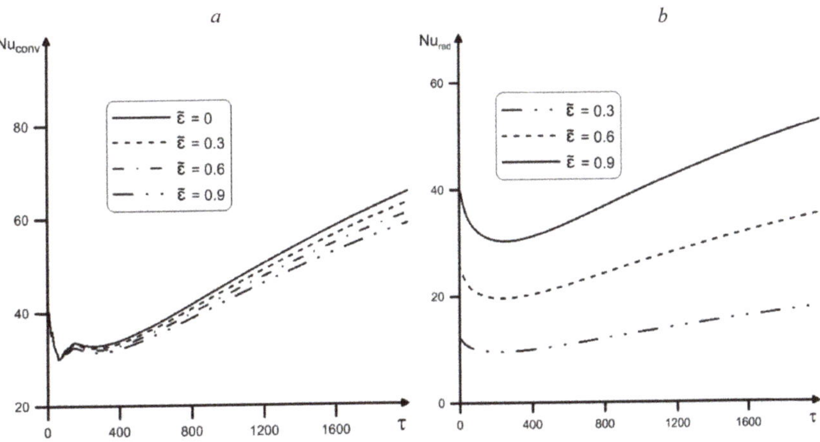

Figure 6. Dependences of the mean (**a**) convective and (**b**) radiative Nusselt numbers at the heat-generating source surface vs. surface emissivity and dimensionless time at $Ra = 10^9$, $Os = 1$.

A more detailed impact of the surface emissivity on the temperature profiles at the middle cross-section $X = 0.6$ is depicted in Figure 7. The main focus was on investigating the effect of radiation on the temperature field in a heat-generating source. An in-depth analysis of this problem allowed us to better understand the principles of passive cooling of electronic elements. It should be noted that the growth of surface emissivity was manifested in a noticeable decrease in temperature inside the local heater. This fact was also expressed in a more intensive cooling of the top heat-conducting wall and, consequently, an increase in the temperature gradient on the surface of this wall. It is clear that the consideration of surface radiation modifies the temperature fields substantially.

Table 4 demonstrates variations of various considered parameters (average temperature inside the air cavity, maximum absolute value of the stream function, average temperature inside the heat-generating source) for $Ra = 10^8$, $Os = 1$. An increment of surface emissivity value reduces $|\psi|_{max}$ due to a decrease in the convective flow in the enclosure. A growth of $\tilde{\epsilon}$ allows the reduction of the average temperature inside the heater and, at the same time, leads to a slight increase in the mean temperature inside the enclosure.

Table 4. Variations of different considered parameters for $Ra = 10^8$, $\tau = 2000$.

Surface Emissivity Value	Θ_{cavity}	$|\psi|_{max}$	Θ_{hs}
0	0.426	0.0349	0.9425
0.3	0.428	0.0344	0.9295
0.6	0.431	0.0339	0.9172
0.9	0.433	0.0334	0.9050

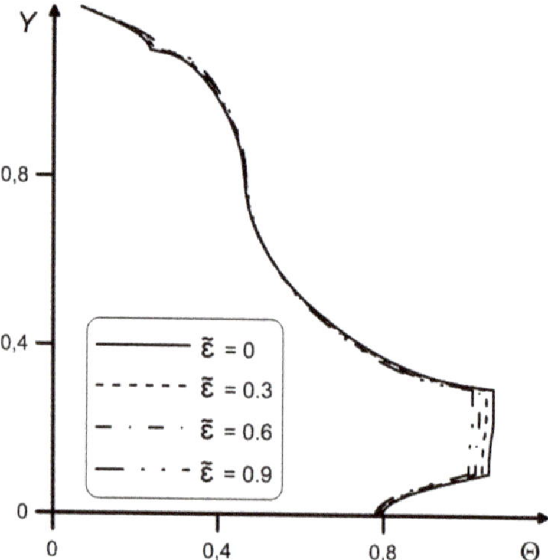

Figure 7. Temperature profiles at $X = 0.6$, $Ra = 10^9$, $Os = 1$ and different values of surface emissivity.

When designing electronic devices, the primary interest is to always control and predict the temperature inside the enclosure. The development of new types of electronics tends to miniaturize individual heat-generating elements, which, as a rule, leads to an increase in operating temperatures and, accordingly, affects the reliability of the equipment. The Ostrogradsky number characterizes a volumetric heat source in the system. An increase in Os leads to a growth of the density of heat flux. Figure 8 shows the development of the isotherms Θ for various values of the Ostrogradsky number. Just to clarify again, the heat transfer processes are caused by both the impact of the heat-generating source and the cooling of the given area owing to the convective heat exchange with an environment. It should be noted that the ambient temperature is less than the initial temperature of the area under consideration. A thermal plume is formed in the central part of the cavity, which characterizes the location of the ascending flows of hot air. Due to heat dissipation from the heater, the bottom solid wall is a zone least affected by the environment. It is clear that an increase in the Ostrogradsky number leads to an essential temperature growth. In particular, the isotherm of 0.5 at $Os = 0.1$ is located above the heat-generating source in the air zone, whereas at $Os = 2$ the considered isotherm is located close to the solid walls (due to more substantial heating).

Values of the mean radiative and convective Nusselt numbers at the heat source surface, mean temperature inside the heat source, and air flow rate inside the enclosure for different values of the Ostrogradsky number are presented in Table 5. At $\tau = 2000$, the mean radiative Nusselt number increases up to 2.01 times upon the change of Os from 0.5 to 2. It is worth noting that an increase in the Ostrogradsky number enhances the average temperature inside the heater up to 26%. In this regard, if the power of the energy source is high, its cooling becomes a non-trivial task. Nu_{conv} and $|\psi|_{max}$ are also increased with the density of heat flux, which is obvious. Finally, a parameter such as the Ostrogradsky number affects the reliability of the device, which is the main aspect for practical applications.

Table 5. Variations of different considered parameters for $Ra = 10^9$, $\tilde{\varepsilon} = 0.6$, $\tau = 2000$.

Ostrogradsky Number	Nu_{conv}	Nu_{rad}	$|\psi|_{max}$	Θ_{hs}
0.5	41.67	33.84	0.0196	0.7097
2	87.59	68.30	0.0250	0.8966

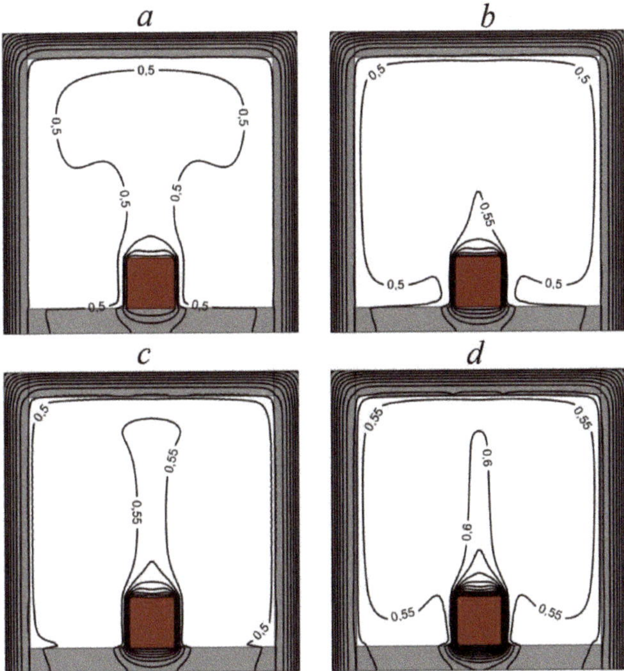

Figure 8. Isotherms Θ at $Ra = 10^9$, $\widetilde{e} = 0.6$: (**a**) $Os = 0.1$, (**b**) $Os = 0.5$, (**c**) $Os = 1$, and (**d**) $Os = 2$.

4. Conclusions

The present work was devoted to the study of the interaction of turbulent natural convection with surface radiation in an air-filled enclosure with a heat-generating source. From an engineering point of view, the considered problem is quite important for the electronics industry where similar basic designs are also used. Governing equations of energy, mass, and momentum were solved by the finite difference method. This research covered a broad range of the dimensionless parameters that were previously presented in the study. The effect of the Ostrogradsky number and surface radiation on fluid flow characteristics and thermal transmission was analyzed in detail for various combinations of the dimensionless parameters. A change in the value of Os led to either a decrease or an increase in the average heat-source temperature. It was found that the average radiative Nusselt number increased up to 2.84 times upon the change of the surface emissivity value from 0.3 to 0.9. Therefore, the surface radiation has a substantial impact on the total heat exchange and may reach 50% of the total heat flux, particularly if the heater and wall surfaces have high emissivity. In accordance with the results, it is possible to conclude that in a sealed electronic case with one heat-generating element under conditions of convective heat exchange with an environment, the heat removal from the heater can be enhanced even by a slight increase in $\widetilde{e}$.

Author Contributions: I.V.M., M.A.S., and A.A.M. conceived the main concept. I.V.M., M.A.S., and A.A.M. contributed to the investigation and data analysis. I.V.M., M.A.S., and A.A.M. wrote the manuscript. All authors contributed to the writing of the final manuscript.

Funding: This work was supported by the Ministry of Science and Higher Education of Russia (state assignment No. 1.13557.2019/13.1).

Acknowledgments: The authors wish to express their thanks to the very competent reviewers for the valuable comments and suggestions.

Abbreviations

$Bi = hL/\lambda_w$	Biot number
E	dimensionless dissipation rate of turbulent kinetic energy
F_{k-i}	view factor from k-th element to the i-th element of an enclosure
g	acceleration of gravity (m/s^2)
G_k	dimensionless generation/destruction of buoyancy turbulent kinetic energy
h	heat-transfer coefficient (W/m^2 K)
k	dimensional turbulence kinetic energy (m^2/s^2)
K	dimensionless turbulent kinetic energy
l	thickness of walls (m)
L	air cavity size (m)
$N_{rad} = \sigma T_{hs}^4 L/\left[\lambda_f(T_{hs} - T^e)\right]$	radiation number (or Stark number)
Nu_{con}	average convective Nusselt number
Nu_{rad}	average radiative Nusselt number
$Os = q_v L^2/\lambda_{hs}\Delta T$	Ostrogradsky number
p	pressure (N/m^2)
P_k	dimensionless shearing production
$Pr = \nu/\alpha_f$	Prandtl number
$Pr_t = \nu_t/\alpha_t$	turbulent Prandtl number
q_v	volume density of heat flux (W/m^3)
Q_{rad}	dimensionless net radiative heat flux
R_k	dimensionless radiosity of the k-th element of an enclosure
$Ra = g\beta(T_{hs} - T^e)L^3/\nu\alpha_f$	Rayleigh number
t	dimensional time (s)
T	dimensional temperature (K)
T^e	environmental temperature (K)
T_f	dimensional fluid temperature (K)
T_{hs}	dimensional heater temperature (K)
T_w	dimensional wall temperature (K)
u_1, u_2	dimensional velocity components along x and y axes (m/s)
U, V	dimensionless velocity components along X and Y axes
$x_1, x_2\ (x, y)$	dimensional Cartesian coordinates (m)
X, Y	dimensionless Cartesian coordinates
Greek symbols	
α_w	thermal diffusivity of the wall material (m^2/s)
α_f	air thermal diffusivity (m^2/s)
α_{hs}	thermal diffusivity of the heater material (m^2/s)
β	coefficient of volumetric thermal expansion (1/K)
δ	the smallest size of the mesh element
ϵ	dimensional dissipation rate of turbulent kinetic energy (m^2/s^3)
$\widetilde{\epsilon}$	surface emissivity
$\zeta = T^e/T_0$	temperature parameter
Θ	dimensionless temperature
Θ_{hs}	dimensionless heater temperature
Θ_f	dimensionless temperature of fluid
Θ_w	dimensionless temperature of wall
λ_w	thermal conductivity of the wall material (W/m K)
λ_f	air thermal conductivity (W/m K)
λ_{hs}	thermal conductivity of the heater (W/m K)
ν	kinematic viscosity (m^2/s)
ν_t	turbulent viscosity (m^2/s)
σ	Stefan–Boltzmann constant (W/m^2 K^4)

τ	dimensionless time
ψ	dimensional stream function (m^2/s)
Ψ	dimensionless stream function
ω	dimensional vorticity (s^{-1})
Ω	dimensionless vorticity

References

1. Xu, G.; Hu, X.; Liao, Z.; Xu, C.; Yang, C.; Deng, Z. Experimental and Numerical Study of an Electrical Thermal Storage Device for Space Heating. *Energies* **2018**, *11*, 2180. [CrossRef]
2. Durgam, S.; Venkateshan, S.P.; Sundararajan, T. Experimental and Numerical Investigations on Optimal Distribution of Heat Source Array under Natural and Forced Convection in a Horizontal Channel. *Int. J. Therm. Sci.* **2017**, *115*, 125–138. [CrossRef]
3. Kondrashov, A.; Sboev, I.; Dunaev, P. Evolution of Convective Plumes Adjacent to Localized Heat Sources of Various Shapes. *Int. J. Heat Mass Transf.* **2016**, *103*, 298–304. [CrossRef]
4. Ridouane, E.H.; Hasnaoui, M.; Amahmid, A.; Raji, A. Interaction between Natural Convection and Radiation in a Square Cavity Heated from Below. *Numer. Heat Transf. Part A Appl.* **2004**, *45*, 289–311. [CrossRef]
5. Baïri, A.; Zarco-Pernia, E.; Garcia De Maria, J.M. A Review on Natural Convection in Enclosures for Engineering Applications. the Particular Case of the Parallelogrammic Diode Cavity. *Appl. Therm. Eng.* **2014**, *63*, 304–322. [CrossRef]
6. Miroshnichenko, I.V.; Sheremet, M.A. Turbulent Natural Convection Heat Transfer in Rectangular Enclosures Using Experimental and Numerical Approaches: A Review. *Renew. Sustain. Energy Rev.* **2018**, *82*, 40–59. [CrossRef]
7. Biswal, P.; Basak, T. Entropy Generation vs Energy Efficiency for Natural Convection Based Energy Flow in Enclosures and Various Applications: A Review. *Renew. Sustain. Energy Rev.* **2017**, *80*, 1412–1457. [CrossRef]
8. Baudoin, A.; Saury, D.; Boström, C. Optimized Distribution of a Large Number of Power Electronics Components Cooled by Conjugate Turbulent Natural Convection. *Appl. Therm. Eng.* **2017**, *124*, 975–985. [CrossRef]
9. Tou, S.K.W.; Tso, C.P.; Zhang, X. 3-D Numerical Analysis of Natural Convective Liquid Cooling of a 3 $\times$ 3 Heater Array in Rectangular Enclosures. *Int. J. Heat Mass Transf.* **1999**, *42*, 3231–3244. [CrossRef]
10. Bondarenko, D.S.; Sheremet, M.A.; Oztop, H.F.; Ali, M.E. Natural Convection of Al$_2$O$_3$/H$_2$O Nanofluid in a Cavity with a Heat-Generating Element. Heatline Visualization. *Int. J. Heat Mass Transf.* **2019**, *130*, 564–574. [CrossRef]
11. Gibanov, N.S.; Sheremet, M.A. Natural Convection in a Cubical Cavity with Different Heat Source Configurations. *Therm. Sci. Eng. Prog.* **2018**, *7*, 138–145. [CrossRef]
12. Saravanan, S.; Sivaraj, C. Combined Thermal Radiation and Natural Convection in a Cavity Containing a Discrete Heater: Effects of Nature of Heating and Heater Aspect Ratio. *Int. J. Heat Fluid Flow* **2017**, *66*, 70–82. [CrossRef]
13. Sheremet, M.; Oztop, H.; Gvozdyakov, D.; Ali, M. Impacts of Heat-Conducting Solid Wall and Heat-Generating Element on Free Convection of Al$_2$O$_3$/H$_2$O Nanofluid in a Cavity with Open Border. *Energies* **2018**, *11*, 3434. [CrossRef]
14. Parmananda, M.; Khan, S.; Dalal, A.; Natarajan, G. Critical Assessment of Numerical Algorithms for Convective-Radiative Heat Transfer in Enclosures with Different Geometries. *Int. J. Heat Mass Transf.* **2017**, *108*, 627–644. [CrossRef]
15. Parmananda, M.; Dalal, A.; Natarajan, G. Unified Framework for Buoyancy Induced Radiative-Convective Flow and Heat Transfer on Hybrid Unstructured Meshes. *Int. J. Heat Mass Transf.* **2018**, *126*, 908–925. [CrossRef]
16. Dehbi, A.; Kelm, S.; Kalilainen, J.; Mueller, H. The Influence of Thermal Radiation on the Free Convection inside Enclosures. *Nucl. Eng. Des.* **2019**, *341*, 176–185. [CrossRef]
17. Ibrahim, A.; Saury, D.; Lemonnier, D. Coupling of Turbulent Natural Convection with Radiation in an Air-Filled Differentially-Heated Cavity at Ra = 1.5 $\times$ 109. *Comput. Fluids* **2013**, *88*, 115–125. [CrossRef]
18. Sharma, A.K.; Velusamy, K.; Balaji, C. Turbulent Natural Convection in an Enclosure with Localized Heating from Below. *Int. J. Therm. Sci.* **2007**, *46*, 1232–1241. [CrossRef]

19. Mikhailenko, S.A.; Sheremet, M.A.; Mohamad, A.A. Convective-Radiative Heat Transfer in a Rotating Square Cavity with a Local Heat-Generating Source. *Int. J. Mech. Sci.* **2018**, *142–143*, 530–540. [CrossRef]

20. Shenoy, A.; Sheremet, M.; Pop, I. Convective Flow and Heat Transfer from Wavy Surfaces. In *Viscous Fluids, Porous Media, and Nanofluids*; CRC Press: Boca Raton, FL, USA, 2017.

21. Das, D.; Roy, M.; Basak, T. Studies on Natural Convection within Enclosures of Various (Non-Square) Shapes—A Review. *Int. J. Heat Mass Transf.* **2017**, *106*, 356–406. [CrossRef]

22. Kang, B.D.; Kim, H.J.; Kim, D.K. Nusselt Number Correlation for Vertical Tubes with Inverted Triangular Fins under Natural Convection. *Energies* **2017**, *10*, 1183. [CrossRef]

23. Saravanan, S.; Vidhya Kumar, A.R. Natural Convection in Square Cavity with Heat Generating Baffles. *Appl. Math. Comput.* **2014**, *244*, 1–9. [CrossRef]

24. Miroshnichenko, I.V.; Sheremet, M.A. Turbulent Natural Convection Combined with Thermal Surface Radiation inside an Inclined Cavity Having Local Heater. *Int. J. Therm. Sci.* **2018**, *124*, 122–130. [CrossRef]

25. Miroshnichenko, I.V.; Sheremet, M.A. Turbulent Natural Convection and Surface Radiation in a Closed Air Cavity with a Local Energy Source. *J. Eng. Phys. Thermophys.* **2017**, *90*, 557–563. [CrossRef]

26. Howell, J.R.; Menguc, M.P.; Siegel, R. Thermal Radiation Heat Transfer, 6th Edition. *J. Heat Transf.* **1970**. [CrossRef]

27. Wang, H.; Xin, S.; Le Quere, P. Numerical study of natural convection-surface radiation coupling in air-filled square cavities. *C. R. Mecanique* **2006**, *334*, 48–57. [CrossRef]

28. Dixit, H.N.; Babu, V. Simulation of High Rayleigh Number Natural Convection in a Square Cavity Using the Lattice Boltzmann Method. *Int. J. Heat Mass Transf.* **2006**, *49*, 727–739. [CrossRef]

29. Zhuo, C.; Zhong, C. LES-Based Filter-Matrix Lattice Boltzmann Model for Simulating Turbulent Natural Convection in a Square Cavity. *Int. J. Heat Fluid Flow* **2013**, *42*, 10–22. [CrossRef]

30. Le Quéré, P. Accurate Solutions to the Square Thermally Driven Cavity at High Rayleigh Number. *Comput. Fluids* **1991**, *20*, 29–41. [CrossRef]

31. Martyushev, S.G.; Sheremet, M.A. Conjugate Natural Convection Combined with Surface Thermal Radiation in an Air Filled Cavity with Internal Heat Source. *Int. J. Therm. Sci.* **2014**, *76*, 51–67. [CrossRef]

Article

A Stability Analysis for Magnetohydrodynamics Stagnation Point Flow with Zero Nanoparticles Flux Condition and Anisotropic Slip

Najiyah Safwa Khashi'ie [1,2,*], Norihan Md Arifin [1,3], Roslinda Nazar [4], Ezad Hafidz Hafidzuddin [5], Nadihah Wahi [3] and Ioan Pop [6]

[1] Institute for Mathematical Research, Universiti Putra Malaysia, Serdang 43400 UPM, Selangor, Malaysia; norihana@upm.edu.my

[2] Fakulti Teknologi Kejuruteraan Mekanikal dan Pembuatan, Universiti Teknikal Malaysia Melaka, Hang Tuah Jaya, Durian Tunggal 76100, Melaka, Malaysia

[3] Department of Mathematics, Faculty of Science, Universiti Putra Malaysia, Serdang 43400 UPM, Selangor, Malaysia; nadihah@upm.edu.my

[4] School of Mathematical Sciences, Faculty of Science and Technology, Universiti Kebangsaan Malaysia, Bangi 43600 UKM, Selangor, Malaysia; rmn@ukm.edu.my

[5] Centre of Foundation Studies for Agricultural Science, Universiti Putra Malaysia, Serdang 43400 UPM, Selangor, Malaysia; ezadhafidz@upm.edu.my

[6] Department of Mathematics, Babeş-Bolyai University, R-400084 Cluj-Napoca, Romania; popm.ioan@yahoo.co.uk

* Correspondence: najiyah@utem.edu.my

Received: 12 January 2019; Accepted: 19 February 2019; Published: 2 April 2019

Abstract: The numerical study of nanofluid stagnation point flow coupled with heat and mass transfer on a moving sheet with bi-directional slip velocities is emphasized. A magnetic field is considered normal to the moving sheet. Buongiorno's model is utilized to assimilate the mixed effects of thermophoresis and Brownian motion due to the nanoparticles. Zero nanoparticles' flux condition at the surface is employed, which indicates that the nanoparticles' fraction are passively controlled. This condition makes the model more practical for certain engineering applications. The continuity, momentum, energy and concentration equations are transformed into a set of nonlinear ordinary (similarity) differential equations. Using bvp4c code in MATLAB software, the similarity solutions are graphically demonstrated for considerable parameters such as thermophoresis, Brownian motion and slips on the velocity, nanoparticles volume fraction and temperature profiles. The rate of heat transfer is reduced with the intensification of the anisotropic slip (difference of two-directional slip velocities) and the thermophoresis parameter, while the opposite result is obtained for the mass transfer rate. The study also revealed the existence of non-unique solutions on all the profiles, but, surprisingly, dual solutions exist boundlessly for any positive value of the control parameters. A stability analysis is implemented to assert the reliability and acceptability of the first solution as the physical solution.

Keywords: nanofluid; stagnation sheet; three-dimensional flow; slip condition; stability analysis

---

1. Introduction

Nanofluids are a special class of fluids that have been the subject of developing research in the recent years. The dispersion of single nanoparticles like metals, oxides, carbon nanotubes or carbides in a fluid can create a modern class of fluids known as nanofluid. Water, oil and ethylene glycol are the common base fluid used in the formation of nanofluid. The invention of nanofluids that have good thermophysical properties can improve heat transfer performance for enormous futuristic applications such as in nuclear cooling systems, solar water heating, biomedical applications, lubrication, thermal storage, refrigeration, coolant in automobile radiator, and many others [1–13]. Choi et al. [14] initiated an experiment on nanotube-in-oil suspensions and measured that the thermal conductivity is inevitably greater than the theoretical predictions. An analytical model by Buongiorno [15] highlighted the importance of thermophoresis and Brownian motion that can induce a relative velocity between the nanoparticles and base fluid. Nield and Kuznetsov [16] implemented Buongiorno's model on the Cheng–Minkowycz problem for flow in a porous medium filled with nanofluid. A brief study of nanofluid and its thermal conductivity has also been examined by Buongiorno et al. [17]. Kuznetsov and Nield [18] revised their model [16] by introducing a new boundary condition that could manipulate the nanoparticles' volume fraction at the surface. Based on the report, it is assumed that the nanoparticles' volume fraction is passively controlled at the boundary which could make the new model more realistic and physically applicable as compared to the existing model. Muhammad et al. [19] also imposed coupled effects of convective heat and zero nanoparticles flux on conditions for Darcy–Forchheimer flow of Maxwell nanofluid. In addition, research works on the zero nanoparticles flux condition were also considered by Rehman et al. [20], Rahman et al. [21], Uddin et al. [22], ur Rahman et al. [23] and Jusoh et al. [24]. Furthermore, studies on the boundary layer problem utilizing Buongiorno's model of nanofluid were also conducted by these researchers [25–31].

Magnetohydrodynamics (MHD), also acknowledged as hydromagnetics and magneto-fluid dynamics, is the study on the behaviour of electrically conducting fluids including liquid metals, plasmas, electrolytes and salt water. The magnetic fields can generate currents or Lorentz force in a moving fluid, which give resistance to the fluid flow and, simultaneously, changes the magnetic field. Magnetohydrodynamics are effectively applied in many devices such as generators, power pumps, heat exchangers and electrostatic filters. The imposition of the magnetic field is also practical in maintaining the boundary layer flow. Rashidi et al. [32] introduced a new analytical method (DTM–Padé) to solve the boundary-layer equations of an MHD micropolar fluid near an isothermal-stretching sheet and concluded that the DTM–Padé is applicable for solving magnetohydrodynamic (MHD) boundary-layer equations. Rashidi et al. [33] concluded that the magnetic nanofluid flow over a porous rotating disk was beneficial in rotating MHD energy generators for new space systems. Sheikholeslami et al. [34] investigated the problem of an eccentric semi-annulus saturated with nanofluid under the influence of a magnetic field. Hayat et al. [35] studied the combined effects of magnetic field, velocity slip and nonlinear thermal radiation on the three-dimensional nanofluid flow. Kandasamy et al. [36] considered the convective condition for an MHD mixed convection flow in a nanofluid while Bhatti et al. [37] analyzed the MHD Wlliamson nanofluid over a porous shrinking sheet. An external magnetic field was imposed, while the induced magnetic field was neglected due to the negligible magnetic Reynolds number. Bhatti et al. [38] examined the coupled effects of MHD and partial slip on the blood flow using a Ree–Eyring fluid model. Makulati et al. [39] utilized the Tiwari and Das model of water-alumina nanofluid to study the impact of a magnetic field in inclined C-shaped enclosures. Hussain et al. [40] found that an upsurge of magnetic field, Hall current, rotation and chemical reaction had an impact on the fluid flow over an accelerated moving plate.

Stagnation point flows are fluid flows that approach the surface of a solid object and then separate into different streams. Stagnation region which meets the maximum pressure, heat transfer and mass deposition are essential in the industrial and technological field. Early classical works on stagnation flow that excluded the slip velocity effect were examined by Hiemenz [41] for the two dimensional problem, Homann [42] for the axisymmetric case and Howarth [43] for the flow near the stagnation region. According to Wang [44], anisotropy is reflected in the difference of bi-directional (direction) slip velocities. Previously, Wang had considered the problem of two-dimensional stagnation flow with symmetric axes on dissimilar surfaces; stationary plate with isotropic slip [45] and moving plate [46], respectively. The results indicated that the enhancement of slip parameter will change the surface resistance and velocity profile. Hussain et al. [47] studied the slip flow and heat transfer of nanofluids embedded in a Darcy-type porous medium. From the analysis, the increment in the permeability of the porous medium and the velocity slip parameter increased the heat transfer rate, whereas it decreased the momentum and thermal boundary layer thicknesses. Khan et al. [48] also pointed up the significance of wall velocity slip for a reliable design and operation of microfluidic devicesmade of hydrophobic devices. The emerging numbers of industrial and technological applications make the study of anisotropic slip that relies upon the flow direction, which is influential for these types of surfaces: hydrophobic [49–54] and porous [55,56]. Rashad [57] investigated the coupled effects of anisotropic slip and convective condition in an unsteady nanofluid saturated with porous medium. Hafidzuddin et al. [58] discussed the anisotropic slip on stagnation flow due to a permeable moving surface, which resulted in dual solutions attained with the presence of a suction parameter. Numerous studies involving three-dimensional stagnation nanofluid flow towards a moving anisotropic slip surface are conducted by Raees et al. [59], Uddin et al. [60] and Balushi et al. [61]. Very recently, Sadiq [62] concluded that the intensity of magnetic field and velocity slip cause an increase in the nanofluid velocity while the boundary layer thickness decreases near the stagnation point region.

Motivated by the literature mentioned earlier, the current work accentuates the anisotropic slip effect on the numerical solution of stagnation point flow with nanoparticles due to a moving surface. Buongiorno's model of nanofluid with an assumption of zero normal flux condition at the wall is implemented. A set of transformations is applied on the governing model to simplify them into nonlinear ordinary differential equations (ODEs). The bvp4c function in MATLAB software (R2017b, MathWorks, Natick, MA, USA) is utilized to perform the numerical computations. The numerical results are demonstrated in the graph forms of velocity, nanoparticles concentration and temperature including the skin friction coefficient, heat and mass transfer rate within the specific range of related parameters. The authors also concern about the emergence of non-unique solutions and the way of stability analysis is conducted to prove the physical solution. The pioneer works on formulation of stability analysis were conducted by Merkin [63], Weidman et al. [64], Harris et al. [65] and Roşca and Pop [66]. A brief of explanation on stability analysis was also discussed by the following literature [67–74]. To the best of the authors' knowledge, the results are new and have not been published.

2. Mathematical Formulation

Consider a steady, three-dimensional stagnation point flow of a nanofluid towards a moving sheet with the presence of anisotropic slip as illustrated in Figure 1. U_x and V_y are the moving plate velocities along the x- and y-directions while the z-axis is the axis of stagnation flow. A magnetic field of consistent strength B_0 is applied normal to the plate. The surface is kept with fixed temperature, T_w while variable nanoparticles fraction, $D_B \dfrac{\partial C}{\partial z} + \dfrac{D_T}{T_\infty} \dfrac{\partial T}{\partial z}$ is adapted at $z = 0$. This condition is applied to achieve practically applicable results [18–24]. The ambient nanofluid concentration and temperature are denoted as C_∞ and T_∞.

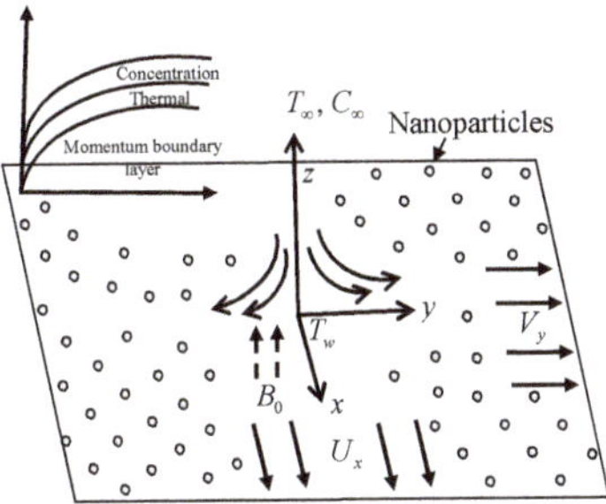

Figure 1. The coordinate system of the physical model.

The following assumptions for the physical model are also contemplated in the present work:

- The nanoparticles and the base fluid is maintained in a thermal equilibrium state.
- The Buongiorno's model of nanofluid is used to mix the combined effects of Brownian motion and thermophoresis.
- The insignificant value of the magnetic Reynolds number is assumed so that the induced magnetic field is zero.
- The Hall current effect is also omitted due to the absence of any externally applied electric field.

Under all these assumptions, the flow equations are:

$$\frac{\partial u}{\partial x} + \frac{\partial v}{\partial y} + \frac{\partial w}{\partial z} = 0, \tag{1}$$

$$u\frac{\partial u}{\partial x} + v\frac{\partial u}{\partial y} + w\frac{\partial u}{\partial z} = U_e(x)\frac{dU_e(x)}{dx} + \nu\nabla^2 u - \frac{\sigma_M B_0^2}{\rho}(u - U_e(x)), \tag{2}$$

$$u\frac{\partial v}{\partial x} + v\frac{\partial v}{\partial y} + w\frac{\partial v}{\partial z} = V_e(y)\frac{dV_e(y)}{dy} + \nu\nabla^2 v - \frac{\sigma_M B_0^2}{\rho}(v - V_e(y)), \tag{3}$$

$$u\frac{\partial w}{\partial x} + v\frac{\partial w}{\partial y} + w\frac{\partial w}{\partial z} = \nu\nabla^2 w, \tag{4}$$

$$u\frac{\partial T}{\partial x} + v\frac{\partial T}{\partial y} + w\frac{\partial T}{\partial z} = \alpha\nabla^2 T + \tau_1 D_B\left[\frac{\partial T}{\partial x}\frac{\partial C}{\partial x} + \frac{\partial T}{\partial y}\frac{\partial C}{\partial y} + \frac{\partial T}{\partial z}\frac{\partial C}{\partial z}\right] + \tau_1 \frac{D_T}{T_\infty}\left[\left(\frac{\partial T}{\partial x}\right)^2 + \left(\frac{\partial T}{\partial y}\right)^2 + \left(\frac{\partial T}{\partial z}\right)^2\right], \tag{5}$$

$$u\frac{\partial C}{\partial x} + v\frac{\partial C}{\partial y} + w\frac{\partial C}{\partial z} = D_B\nabla^2 C + \frac{D_T}{T_\infty}\nabla^2 T, \tag{6}$$

subject to the initial and boundary conditions

$$\left.\begin{aligned} u(x,y,0) = \lambda U_x + \mu S_1 \frac{\partial u}{\partial z}\Big|_{(x,y,0)}, \quad v(x,y,0) = \lambda V_y + \mu S_2 \frac{\partial v}{\partial z}\Big|_{(x,y,0)}, \quad w(x,y,0) = 0, \\ T(x,y,0) = T_w, \quad \left[D_B\frac{\partial C}{\partial z} + \frac{D_T}{T_\infty}\frac{\partial T}{\partial z}\right]_{(x,y,0)} = 0, \end{aligned}\right\}, \tag{7}$$

$$\left.\begin{aligned} u(x,y,z) \to U_e(x) = bx, \quad v(x,y,z) \to V_e(y) = by, \quad T(x,y,z) \to T_\infty, \\ C(x,y,z) \to C_\infty \quad \text{as} \quad z \to \infty, \end{aligned}\right\}, \tag{8}$$

where (u,v,w) are the respective velocities in (x,y,z) directions, T is the fluid temperature, C is the nanoparticles volume fraction, ν is the kinematic viscosity, μ is the dynamic viscosity, σ_M is the electrical conductivity of the fluid, ρ is the fluid density, τ_1 is the ratio of heat capacity of the nanoparticles to the base fluid, α is the thermal diffusivity, D_B is the Brownian diffusion coefficient, D_T is the thermophoretic diffusion coefficient, S_1 is the slip coefficient in x-direction, S_2 is the slip coefficient in y-direction, b is the strength of the stagnation flow, and λ is the moving parameter such that $\lambda > 0$ and $\lambda < 0$ refer to the moving plate, which is out and towards the origin, respectively [58,75].

The following similarity transformations which satisfy Equation (1) are employed to convert the PDEs in Equations (2)–(6) aligned with the conditions (see Equations (7) and (8)) into a set of ODEs:

$$\left.\begin{aligned} u = bxf'(\eta) + U_x h(\eta), \quad v = byg'(\eta) + V_y k(\eta), \quad w = -\sqrt{b\nu}[f(\eta) + g(\eta)], \\ T = (T_w - T_\infty)\theta(\eta) + T_\infty, \quad C = (C_w - C_\infty)\phi(\eta) + C_\infty, \quad \eta = \sqrt{\frac{b}{\nu}}z, \end{aligned}\right\} \tag{9}$$

where η is the similarity variable. Hence, the transformed nonlinear ODEs in conjunction with the conditions are:

$$f''' = (f')^2 - (f+g)f'' + M(f'-1) - 1, \tag{10}$$

$$g''' = (g')^2 - (f+g)g'' + M(g'-1) - 1, \tag{11}$$

$$h'' = hf' - (f+g)h' + Mh, \tag{12}$$

$$k'' = kg' - (f+g)k' + Mk, \tag{13}$$

$$\theta'' = -\Pr\left[(f+g)\theta' + Nb\theta'\phi' + Nt(\theta')^2\right], \tag{14}$$

$$\phi'' = -Le\Pr(f+g)\phi' - \frac{Nt}{Nb}\theta'', \tag{15}$$

$$\left.\begin{array}{l} f(0) = g(0) = 0,\ \gamma_1 f''(0) = f'(0),\ \gamma_2 g''(0) = g'(0),\ \lambda + \gamma_1 h'(0) = h(0),\ \lambda + \gamma_2 k'(0) = k(0), \\ \theta(0) = 1,\quad Nb\phi'(0) + Nt\theta'(0) = 0, \end{array}\right\} \tag{16}$$

$$f'(\eta) \to 1,\quad g'(\eta) \to 1,\quad h(\eta) \to 0,\quad k(\eta) \to 0,\quad \theta(\eta) \to 0,\quad \phi(\eta) \to 0 \quad \text{as}\quad \eta \to \infty, \} \tag{17}$$

where primes denote differentiation with respect to similarity variable η, $M = \dfrac{\sigma_M B_0{}^2}{\rho b}$ is the magnetic field parameter, $Nt = \dfrac{\tau_1 D_T(T_w - T_\infty)}{\nu T_\infty}$ is the thermophoresis parameter, $Nb = \dfrac{\tau_1 D_B(C_w - C_\infty)}{\nu}$ is the Brownian motion parameter, $\Pr = \dfrac{\nu}{\alpha}$ is the Prandtl number, $Le = \dfrac{\alpha}{D_B}$ is the Lewis number, $\gamma_1 = \mu S_1\sqrt{\dfrac{b}{\nu}}$ and $\gamma_2 = \mu S_2\sqrt{\dfrac{b}{\nu}}$ are the slip parameters in the bi-directional x- and y-axis, proportionately. The physical interests in the study are the dimensionless skin friction coefficient, local Nusselt number (heat transfer rate) and local Sherwood number (mass transfer rate) which is denoted by

$$C_f Re_x^{1/2} = f''(0),\quad \frac{Nu_x}{Re_x^{1/2}} = -\theta'(0),\quad \frac{Sh_x}{Re_x^{1/2}} = -\phi'(0), \tag{18}$$

accordingly.

3. Stability Analysis

The implementation of stability analysis is essential to affirm mathematically the stability and reliability of the dual solutions. The first non-unique solution which is asymptotically satisfying the boundary conditions will be denoted as the first or upper branch solution. For stability purposes, a time-dependent problem needs to be considered based on study in previous literature [67–73]. The unsteady form of Equations (2)–(6) is

$$\frac{\partial u}{\partial t} + u\frac{\partial u}{\partial x} + v\frac{\partial u}{\partial y} + w\frac{\partial u}{\partial z} = U_e(x)\frac{dU_e(x)}{dx} + \nu\nabla^2 u - \frac{\sigma_M B_0{}^2}{\rho}(u - U_e(x)), \tag{19}$$

$$\frac{\partial v}{\partial t} + u\frac{\partial v}{\partial x} + v\frac{\partial v}{\partial y} + w\frac{\partial v}{\partial z} = V_e(y)\frac{dV_e(y)}{dy} + \nu\nabla^2 v - \frac{\sigma_M B_0{}^2}{\rho}(v - V_e(y)), \tag{20}$$

$$\frac{\partial w}{\partial t} + u\frac{\partial w}{\partial x} + v\frac{\partial w}{\partial y} + w\frac{\partial w}{\partial z} = \nu\nabla^2 w, \tag{21}$$

$$\frac{\partial T}{\partial t} + u\frac{\partial T}{\partial x} + v\frac{\partial T}{\partial y} + w\frac{\partial T}{\partial z} = \alpha\nabla^2 T + \tau_1 D_B\left[\frac{\partial T}{\partial x}\frac{\partial C}{\partial x} + \frac{\partial T}{\partial y}\frac{\partial C}{\partial y} + \frac{\partial T}{\partial z}\frac{\partial C}{\partial z}\right] \\ + \tau_1\frac{D_T}{T_\infty}\left[\left(\frac{\partial T}{\partial x}\right)^2 + \left(\frac{\partial T}{\partial y}\right)^2 + \left(\frac{\partial T}{\partial z}\right)^2\right], \tag{22}$$

$$\frac{\partial C}{\partial t} + u\frac{\partial C}{\partial x} + v\frac{\partial C}{\partial y} + w\frac{\partial C}{\partial z} = D_B\nabla^2 C + \frac{D_T}{T_\infty}\nabla^2 T. \tag{23}$$

New transformations are applied to the unsteady problem (see Equations (19)–(23)) where τ is the non-dimensional time variable:

$$
\left.
\begin{aligned}
&u = bx\frac{\partial f(\eta,\tau)}{\partial \eta} + U_x h(\eta,\tau), \quad v = by\frac{\partial g(\eta,\tau)}{\partial \eta} + V_y k(\eta,\tau), \quad w = -\sqrt{b\nu}\left[f(\eta,\tau) + g(\eta,\tau)\right], \\
&T = (T_w - T_\infty)\,\theta(\eta,\tau) + T_\infty, \quad C = (C_w - C_\infty)\,\phi(\eta,\tau) + C_\infty, \quad \eta = \sqrt{\frac{b}{\nu}}z, \quad \tau = bt.
\end{aligned}
\right\}
\tag{24}
$$

Using Equation (24), the following equations can be attained:

$$
\frac{\partial^3 f}{\partial \eta^3} + (f+g)\frac{\partial^2 f}{\partial \eta^2} - \left(\frac{\partial f}{\partial \eta}\right)^2 - M\left(\frac{\partial f}{\partial \eta} - 1\right) + 1 - \frac{\partial^2 f}{\partial \eta \partial \tau} = 0,
\tag{25}
$$

$$
\frac{\partial^3 g}{\partial \eta^3} + (f+g)\frac{\partial^2 g}{\partial \eta^2} - \left(\frac{\partial g}{\partial \eta}\right)^2 - M\left(\frac{\partial g}{\partial \eta} - 1\right) + 1 - \frac{\partial^2 g}{\partial \eta \partial \tau} = 0,
\tag{26}
$$

$$
\frac{\partial^2 h}{\partial \eta^2} + (f+g)\frac{\partial h}{\partial \eta} - h\frac{\partial f}{\partial \eta} - Mh - \frac{\partial h}{\partial \tau} = 0,
\tag{27}
$$

$$
\frac{\partial^2 k}{\partial \eta^2} + (f+g)\frac{\partial k}{\partial \eta} - k\frac{\partial g}{\partial \eta} - Mk - \frac{\partial k}{\partial \tau} = 0,
\tag{28}
$$

$$
\frac{\partial^2 \theta}{\partial \eta^2} + \Pr\left[(f+g)\frac{\partial \theta}{\partial \eta} + Nb\frac{\partial \theta}{\partial \eta}\frac{\partial \phi}{\partial \eta} + Nt\left(\frac{\partial \theta}{\partial \eta}\right)^2 - \frac{\partial \theta}{\partial \tau}\right] = 0,
\tag{29}
$$

$$
\frac{\partial^2 \phi}{\partial \eta^2} + Le\,\Pr\left[(f+g)\frac{\partial \phi}{\partial \eta} - \frac{\partial \phi}{\partial \tau}\right] + \frac{Nt}{Nb}\frac{\partial^2 \theta}{\partial \eta^2} = 0,
\tag{30}
$$

with the transformed conditions

$$
\left.
\begin{aligned}
&f(0,\tau) = g(0,\tau) = 0, \quad \left[\gamma_1\frac{\partial^2 f}{\partial \eta^2} - \frac{\partial f}{\partial \eta}\right]_{(0,\tau)} = 0, \quad \left[\gamma_2\frac{\partial^2 g}{\partial \eta^2} - \frac{\partial g}{\partial \eta}\right]_{(0,\tau)} = 0, \\
&\lambda + \gamma_1\left.\frac{\partial h}{\partial \eta}\right|_{(0,\tau)} = h(0,\tau), \quad \lambda + \gamma_2\left.\frac{\partial k}{\partial \eta}\right|_{(0,\tau)} = k(0,\tau), \quad \theta(0,\tau) = 1, \quad \left[Nb\frac{\partial \phi}{\partial \eta} + Nt\frac{\partial \theta}{\partial \eta}\right]_{(0,\tau)} = 0,
\end{aligned}
\right\}
\tag{31}
$$

$$
\left.\frac{\partial f}{\partial \eta}\right|_{(\eta,\tau)} \to 1, \quad \left.\frac{\partial g}{\partial \eta}\right|_{(\eta,\tau)} \to 1, \quad h(\eta,\tau) \to 0, \quad k(\eta,\tau) \to 0, \quad \theta(\eta,\tau) \to 0, \quad \phi(\eta,\tau) \to 0 \text{ as } \eta \to \infty.
\tag{32}
$$

For the stability process, steady flow solutions $f(\eta) = f_0(\eta)$, $g(\eta) = g_0(\eta)$, $h(\eta) = h_0(\eta)$, $k(\eta) = k_0(\eta)$, $\theta(\eta) = \theta_0(\eta)$ and $\phi(\eta) = \phi_0(\eta)$ which have satisfied Equations (10)–(17) are examined by the following expressions:

$$
\left.
\begin{aligned}
f(\eta,\tau) &= f_0(\eta) + e^{-\sigma\tau}F(\eta) \\
g(\eta,\tau) &= g_0(\eta) + e^{-\sigma\tau}G(\eta) \\
h(\eta,\tau) &= h_0(\eta) + e^{-\sigma\tau}H(\eta) \\
k(\eta,\tau) &= k_0(\eta) + e^{-\sigma\tau}K(\eta) \\
\theta(\eta,\tau) &= \theta_0(\eta) + e^{-\sigma\tau}P(\eta) \\
\phi(\eta,\tau) &= \phi_0(\eta) + e^{-\sigma\tau}R(\eta)
\end{aligned}
\right\},
\tag{33}
$$

where σ is an eigenvalue, $F(\eta)$, $G(\eta)$, $H(\eta)$, $K(\eta)$, $P(\eta)$ and $R(\eta)$ are small relative to $f_0(\eta)$, $g_0(\eta)$, $h_0(\eta)$, $k_0(\eta)$, $\theta_0(\eta)$ and $\phi_0(\eta)$, correspondingly [66]. The linearized eigenvalue problem will be attained by replacing Equation (33) into Equations (25)–(32):

$$
F''' + (f_0 + g_0)F'' + (F+G)f_0'' - (2f_0' - \sigma + M)F' = 0,
\tag{34}
$$

$$
G''' + (f_0 + g_0)G'' + (F+G)g_0'' - (2g_0' - \sigma + M)G' = 0,
\tag{35}
$$

$$H'' + (f_0 + g_0)\, H' + (F + G)\, h'_0 - h_0 F' - (f'_0 - \sigma + M)H = 0, \tag{36}$$

$$K'' + (f_0 + g_0)\, K' + (F + G)\, k'_0 - k_0 G' - (g'_0 - \sigma + M)K = 0, \tag{37}$$

$$P'' + \Pr\left[(f_0 + g_0)\, P' + (F + G)\, \theta'_0 + Nb\,(\phi'_0 P' + \theta'_0 R') + Nt\,(2\theta'_0 P') + \sigma P\right] = 0, \tag{38}$$

$$R'' + Le\,\Pr\left[(f_0 + g_0)\, R' + (F + G)\, \phi'_0 + \sigma R\right] + \frac{Nt}{Nb} P'' = 0, \tag{39}$$

along with the conditions

$$\left.\begin{array}{l} F(0) = G(0) = 0, \qquad \gamma_1 F''(0) = F'(0), \qquad \gamma_2 G''(0) = G'(0), \\[6pt] \gamma_1 H'(0) = H(0), \qquad \gamma_2 K'(0) = K(0), \qquad P(0) = 0, \qquad Nb R'(0) + Nt P'(0) = 0 \end{array}\right\}, \tag{40}$$

$$F'(\eta) \to 0,\ G'(\eta) \to 0,\ H(\eta) \to 0,\ K(\eta) \to 0,\ P(\eta) \to 0,\ R(\eta) \to 0 \quad \text{as} \quad \eta \to \infty. \} \tag{41}$$

The stability of the solutions $f_0(\eta)$, $g_0(\eta)$, $h_0(\eta)$, $k_0(\eta)$, $\theta_0(\eta)$ and $\phi_0(\eta)$ depends on the smallest eigenvalue, σ_1 by solving the governing linearized eigenvalue model in Equations (34)–(41). Relaxation of a boundary condition is necessary to attain a possible range of eigenvalues [65]. Hence, in the present paper, the boundary condition $F'(\eta) \to 0$ as $\eta \to \infty$ (see Equation (41)) is relaxed and replaced with the normalizing boundary condition $F''(0) = 1$.

4. Results and Discussion

The similarity solutions for the transformed ODEs in Equations (10)–(15) aligned with the conditions (see Equations (16) and (17)) are found with the aid of bvp4c function in MATLAB software. The bvp4c function implements a finite difference scheme known as 3-stage Lobatto IIIa [74,76–78]. There are four separate codes in bvp4c function; code a for solving steady flow equations, code b for continuation of code a, code c and d for stability analysis of dual solutions. In the bvp4c code a, $\eta_\infty = 10$ is used for the numerical calculations, but it is found that $\eta_\infty = 5$ is sufficient enough to fulfill the boundary conditions based on all the profiles demonstrated in the present study. For the method validation, few values in the case of Newtonian fluid with the absence of magnetic parameter and heat transfer have been compared to the results by Wang [44] and Balushi et al. [61]. The present numerical values are in positive agreement with others as tabulated in Table 1.

Table 1. A comparison data with previous published results for $\lambda = 1$, $\gamma_1 = 5$ and various γ_2.

	Present		Wang [44]		Balushi et al. [61]	
	$\gamma_2 = 5$	$\gamma_2 = 2.5$	$\gamma_2 = 5$	$\gamma_2 = 2.5$	$\gamma_2 = 5$	$\gamma_2 = 2.5$
$f'(0)$	0.896418	0.895885	0.8964	0.8959	0.896418	0.895885
$g'(0)$	0.896418	0.809985	0.8964	0.8100	0.896418	0.809985
$h(0)$	0.122554	0.123428	0.123	0.123	0.122554	0.123428
$k(0)$	0.122554	0.221884	0.123	0.222	0.122554	0.221884

Figures 2–17 exhibit that dual solutions are possible in the present study. The first and second solutions are depicted by the straight and dashed lines, respectively. Generally, the solution which converges first is assumed as the first or physical solution, and in this case, the stability analysis as discussed in the previous section will affirm which solution is physically realizable. Figures 2–7 display the nanofluid velocities on x- and y-directions, temperature and concentration profiles in limiting numbers of γ_2 when $Le = \Pr = 1$, $Nb = Nt = 0.2$, $M = 0.5$, $\lambda = -1, 1$ and $\gamma_1 = 5$. Both of the velocities as revealed in Figures 2 and 3 increase because the shear stress decrease with the enlargement of the slip parameter and this is supported by the result in Figure 11. Figures 4 and 5 demonstrate the slip velocity profiles $h(\eta)$ and $k(\eta)$ in both directions, respectively. There are opposite and symmetrically

effect for both profiles at $\lambda = \pm 1$. The changes in λ only influence the slip velocity profiles since the moving parameter only remains in the initial condition at $h(0)$ and $k(0)$ (see Equation (16)). Boundary layer thickness for both solutions diminishes with an increment of γ_2 as manifested in Figures 4 and 5.

The nanofluid temperature as illustrated in Figure 6 declines with the expanding values of γ_2 while the opposite result is obtained for the nanoparticles' volume fraction profile in Figure 7. Meanwhile, Figures 8–10 present the influence of the Nb and Nt on both nanoparticles volume fraction (concentration) and temperature profiles. The nanofluid concentration diminishes as Nb increases while expands as Nt increases and these outcomes are in conjunction with the previous results by Balushi et al. [61]. An inflation of Nt seems to boost the thermal boundary layer thickness and temperature, whereas the presence of Nb give zero impact on the nanofluid temperature. The appearance of the nanoparticles in the base fluid will generate thermophoresis parameter. As Nt increases, the higher thermal conductivity in nanofluid will enhance both temperature and concentration.

Variations of $g''(0)$, $k'(0)$, $-\theta'(0)$ and $-\phi'(0)$ against the slip parameter, γ_1 for selected values of γ_2, Nb and Nt are visualized in Figures 11–17, respectively. Unlike the other studies which conduct stability analysis for non-unique solutions, no critical value or turning point is found in this analysis. Critical value is defined as a value that separates the first (physical) and second solutions. Both upper and lower branches existing boundlessly for each value of γ_1 and for the graph visualization, $0 \leq \gamma_1 \leq 30$ have been selected. There is no significant effect of γ_2 on $f''(0)$ and $h'(0)$, hence the graph is not highlighted. This is supported by the velocity and slip velocity profile that have been shown previously in Figures 2 and 4. Variations of $-\theta'(0)$ and $-\phi'(0)$ as can be seen in Figures 13 and 14 only show minimal changes as γ_2 increase but since the value of Nb and Nt is same, $-\theta'(0) = -(-\phi'(0))$ for all values of γ_1 and γ_2. An increasing values of the Brownian motion parameter Nb escalate the mass transfer rate whereas the thermophoresis parameter Nt reduces both heat and mass transfer rate as displayed in Figures 15–17.

Since dual solutions are obtained in the study, a stability analysis has been conducted by solving Equations (34)–(39) with the conditions (see Equations (40) and (41)) using Matlab bvp4c code c and d. Generally, negative σ_1 indicates an initial upsurge of disturbances, which signifies that the flow is unstable while there are opposite flow characteristics for positive σ_1. The smallest eigenvalue, σ_1 for some values of γ_1, is tabulated in Table 2. It shows that the first and second solutions have positive and negative eigenvalues, respectively, which validate the stability and reliability of the first solution.

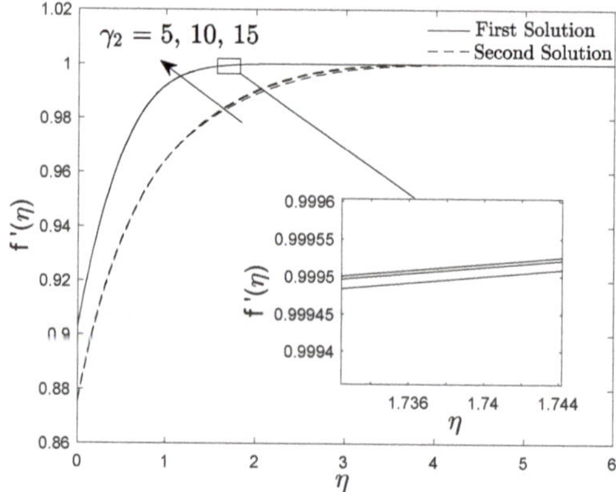

Figure 2. Nanofluid velocity along the x-direction for diverse values of γ_2 with $\lambda = -1, 1$.

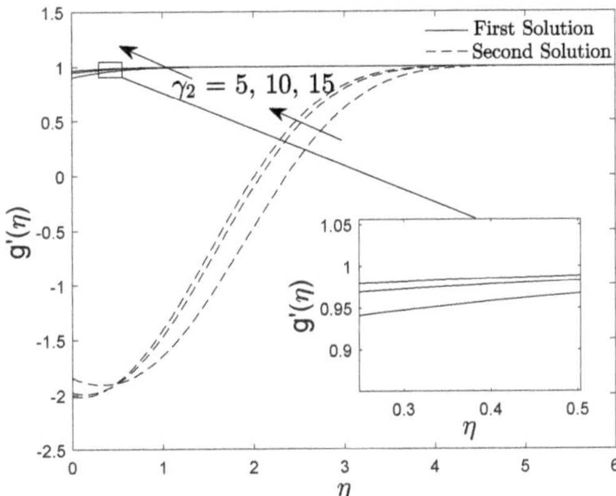

Figure 3. Nanofluid velocity along the y-direction for diverse values of γ_2 with $\lambda = -1, 1$.

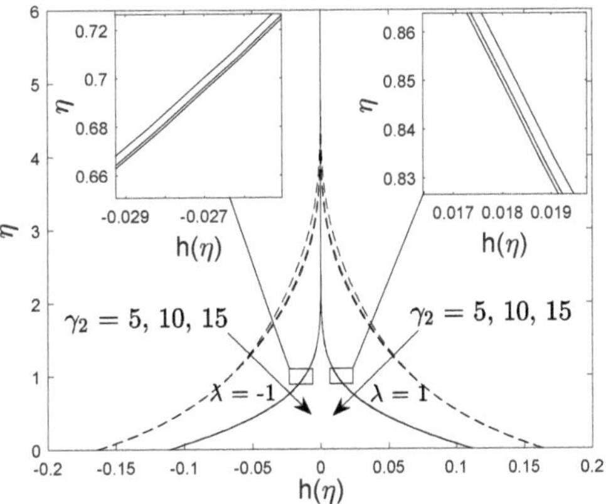

Figure 4. Slip velocity along the x-direction for diverse values of γ_2 with $\lambda = -1, 1$.

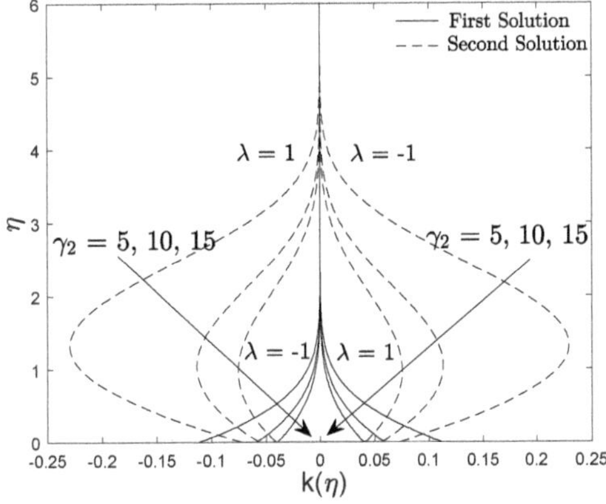

Figure 5. Slip velocity along the y-direction for diverse values of γ_2 with $\lambda = -1, 1$.

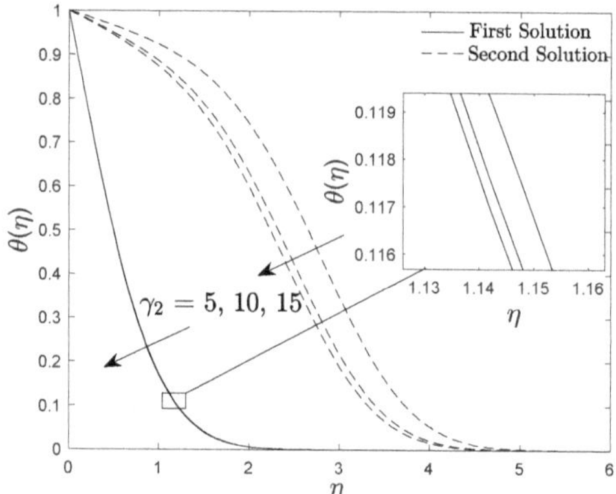

Figure 6. Nanofluid temperature in a limiting range of γ_2 with $\lambda = -1, 1$.

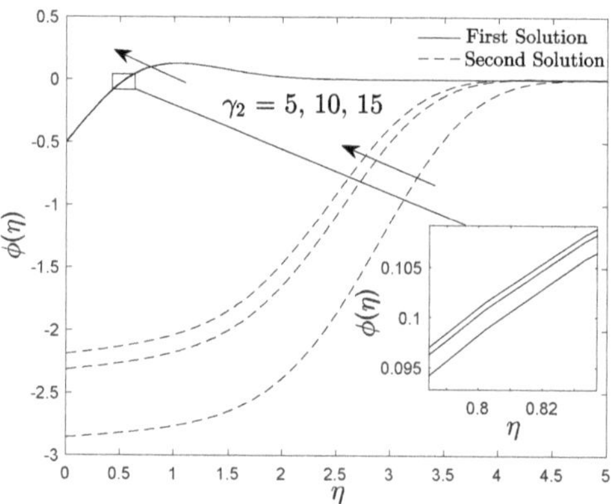

Figure 7. Nanoparticles volume fraction profile in a limiting range of γ_2 with $\lambda = -1, 1$.

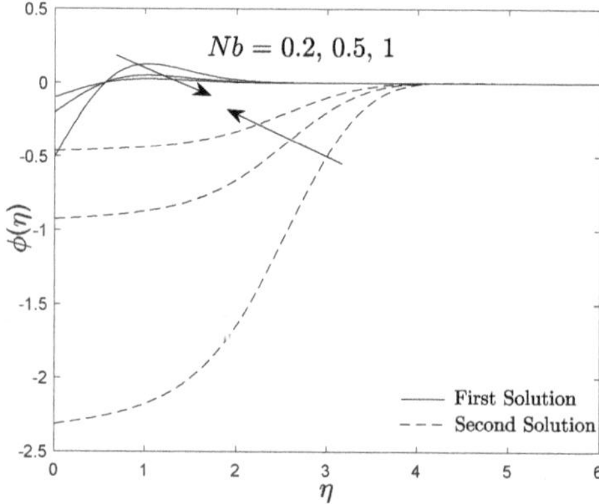

Figure 8. Nanoparticles' volume fraction profile in limiting range of Nb when $Le = \mathrm{Pr} = 1$, $Nt = 0.2$, $\gamma_1 = 5$, $\gamma_2 = 10$, $M = 0.5$ and $\lambda = -1, 1$.

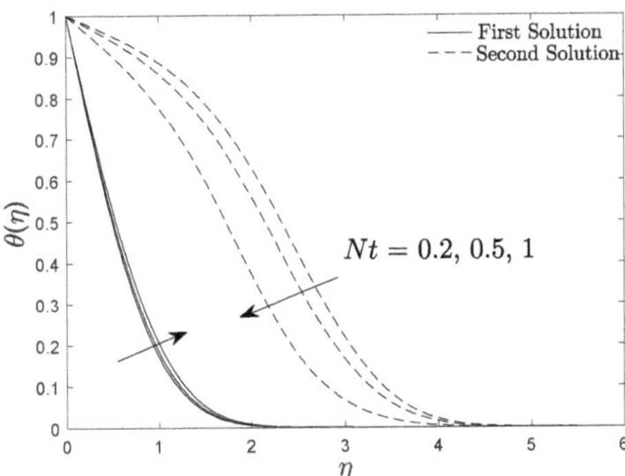

Figure 9. Nanofluid temperature in limiting range of Nt when Le = Pr = 1, $Nb = 0.2$, $\gamma_1 = 5$, $\gamma_2 = 10$, $M = 0.5$ and $\lambda = -1, 1$.

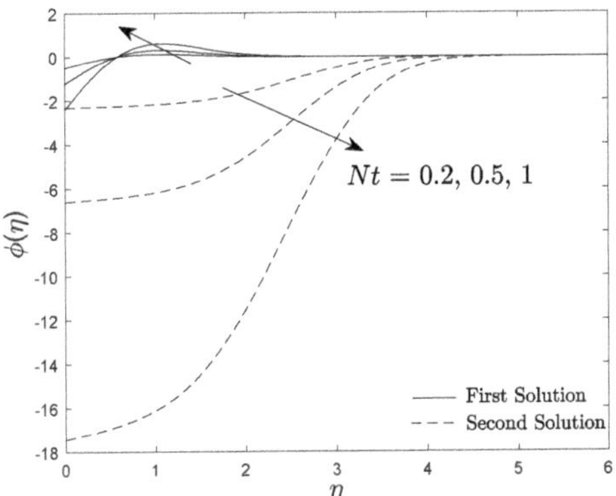

Figure 10. Nanoparticles' volume fraction profile in a limiting range of Nt when Le = Pr = 1, $Nb = 0.2$, $\gamma_1 = 5$, $\gamma_2 = 10$, $M = 0.5$ and $\lambda = -1, 1$.

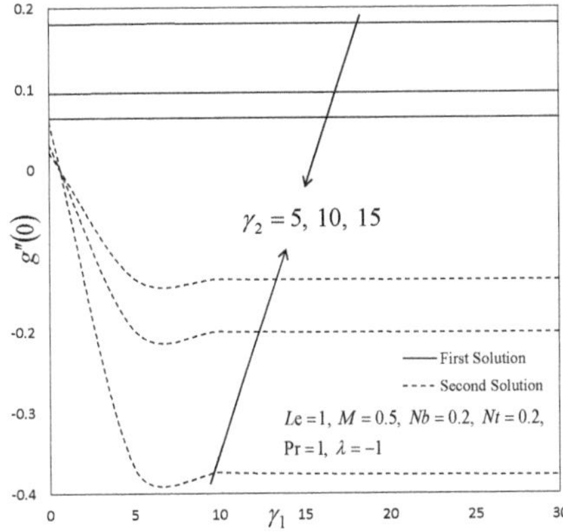

Figure 11. Variations of $g''(0)$ towards γ_1 for selected values of γ_2.

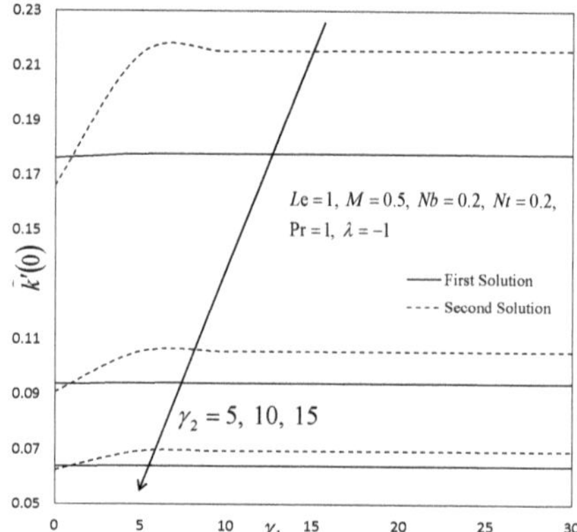

Figure 12. Variations of $k'(0)$ towards γ_1 for selected values of γ_2.

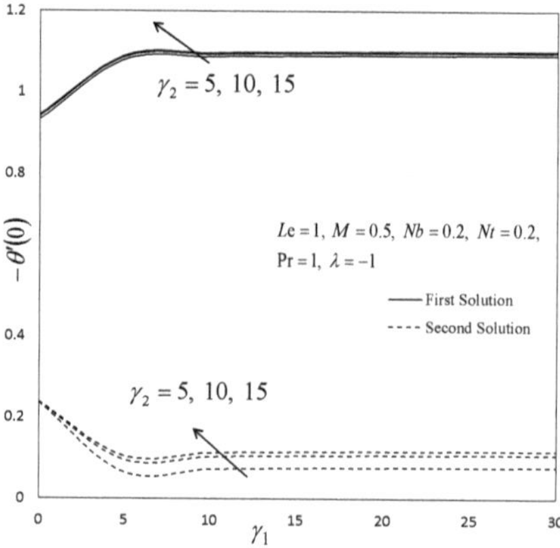

Figure 13. Variations of $-\theta'(0)$ towards γ_1 for selected values of γ_2.

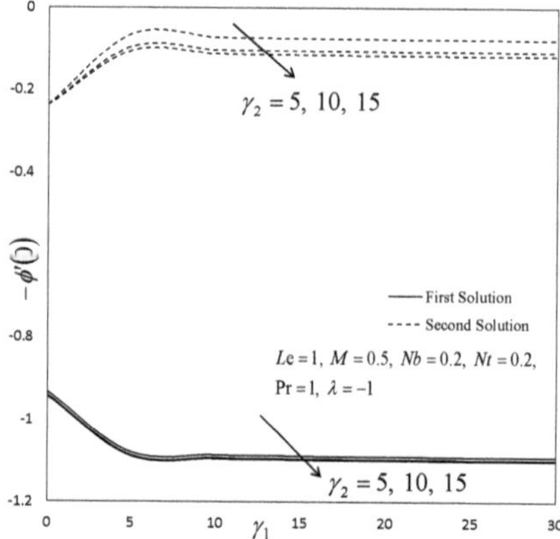

Figure 14. Variations of $-\phi'(0)$ towards γ_1 for selected values of γ_2.

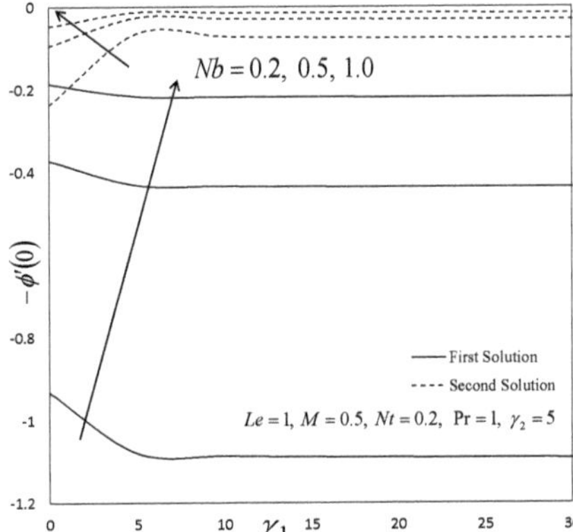

Figure 15. Variations of $-\phi'(0)$ towards γ_1 for selected values of Nb.

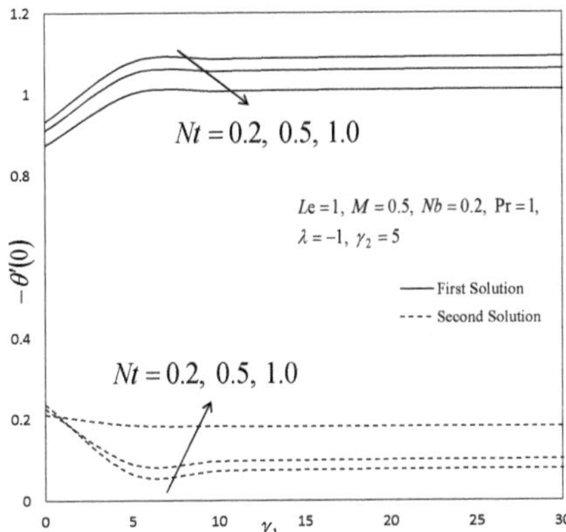

Figure 16. Variations of $-\theta'(0)$ towards γ_1 for selected values of Nt.

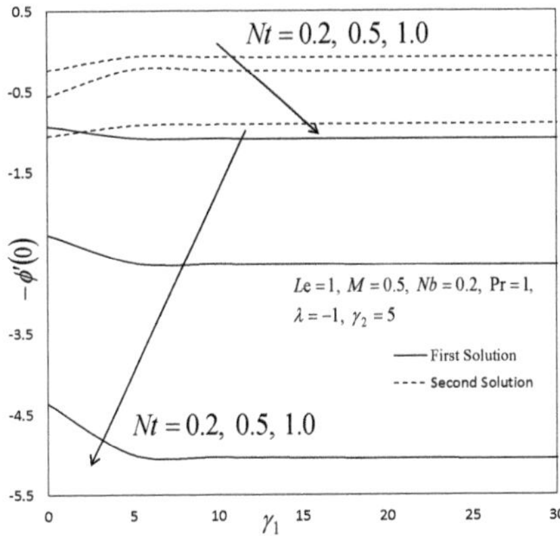

Figure 17. Variations of $-\phi'(0)$ towards γ_1 for selected values of Nt.

Table 2. σ_1 for diverse γ_1 when $\gamma_2 = 5$, $Le = 1$, $Nb = Nt = 0.2$, $M = 0.5$, $Pr = 1$ and $\lambda = -1$.

γ_1	σ_1 (First Solution)	σ_1 (Second Solution)
3	0.2051	−0.7977
5	0.3720	−0.8347
10	0.4375	−0.8626
13	0.4590	−0.8692
15	0.4690	−0.8721

5. Conclusions

A numerical investigation on three-dimensional MHD stagnation point flow due to a moving plate with the presence of anisotropic slip has been accomplished. Buongiorno's model of nanofluid is selected to integrate the thermophoresis and Brownian motion parameters. A set of transformations was used to reduce the governing model into a set of nonlinear differential equations. The similarity equations are then transformed into the bvp4c algorithm to perform the numerical computation. The numerical results are illustrated graphically on the specific physical parameters such as thermophoresis parameter Nt, slip parameters γ_1, γ_2 and Brownian motion parameter Nb to study the flow, heat and mass transfer characteristics of the nanofluid. Non-unique solutions exist boundlessly for all positive values of related parameters in the present study. The implementation of stability analysis using bvp4c code proves the reliability of the first solution. The thickness of boundary layer for the non-physical solution is larger compared to the physical solution. The rate of heat transfer is reduced with the augmentation of the anisotropic slip (difference of slip velocities), while the opposite result is obtained for the mass transfer rate. The rising values of Nb reduce the boundary layer thickness for nanoparticle concentration and increase the Sherwood number while an adverse trend is observed for the higher values of Nt.

Author Contributions: Research design, N.S.K., N.M.A., R.N. and I.P.; Formulation and methodology, N.S.K. and N.M.A.; Result analysis, N.S.K.; Validation, N.M.A., R.N., E.H.H. and N.W.; Article preparation, N.S.K.; Review and editing, N.M.A., R.N., E.H.H., N.W. and I.P.

Funding: The authors would like to express their great appreciation to the Universiti Putra Malaysia through the Putra Grant (9570600). The main author also would like to acknowledge Universiti Teknikal Malaysia Melaka and Ministry of Education (Malaysia) for the financial support through a UTEM-SLAB scholarship. The work by Ioan Pop has been supported from the grant PN-III-P4-ID-PCE-2016-0036, UEFISCDI, Romania.

Acknowledgments: The authors appreciate the valuable feedback and recommendations by the reviewers.

Conflicts of Interest: The authors declare no conflict of interest.

Abbreviations

The following abbreviations are used in this manuscript:

B_0	magnitude of the magnetic field strength	
C	nanoparticle concentration	
D_B	Brownian diffusion coefficient	
D_T	thermophoretic diffusion coefficient	
Le	Lewis number	
M	magnetic parameter	
Nb	Brownian motion parameter	
Nt	thermophoresis parameter	
Pr	Prandtl number	
Re_x	local Reynolds number	
S_1, S_2	slip coefficient in x- and y-directions, respectively	
T	nanofluid temperature	
T_w	wall temperature	
T_∞	free stream temperature	
U_e	free stream velocity along the x- axis	m/s
U_x	moving plate velocity along the x- axis	m/s
V_e	free stream velocity along the y- axis	m/s
V_y	moving plate velocity along the y- axis	m/s
b	strength of the stagnation flow	
f, g	dimensionless stream function along x- and y-directions, respectively	
h, k	slip velocity along x- and y-directions, respectively	
t	time	s
u, v, w	velocities along the x-, y-, z- directions, respectively	
α	thermal diffusivity of the fluid	
γ_1, γ_2	slip parameters in the x- and y- axis, respectively	
θ	dimensionless temperature	
λ	moving parameter	
μ	dynamic viscocity	
ν	kinematic viscocity	m^2/s
ρ	density of base fluid	kg/m^3
σ	unknown eigenvalue	
σ_M	electrical conductivity of the fluid	$\mathrm{s}^3\mathrm{A}^2/(\mathrm{kgm}^3)$
τ	dimensionless time variable	
τ_1	ratio of the heat capacity of the nanoparticles to the base fluid	
ϕ	dimensionless nanoparticle volume fraction/concentration	

References

1. Haddad, Z.; Oztop, H.F.; Abu-Nada, E.; Mataoui, A. A review on natural convective heat transfer of nanofluids. *Renew. Sust. Energy Rev.* **2012**, *16*, 5363–5378. [CrossRef]
2. Mahian, O.; Kianifar, A.; Kalogirou, S.A.; Pop, I.; Wongwises, S. A review of the applications of nanofluids in solar energy. *Int. J. Heat Mass Trans.* **2013**, *57*, 582–594. [CrossRef]
3. Mohyud-Din, S.T.; Khan, U.; Ahmed, N.; Hassan, S.M. Magnetohydrodynamic flow and heat transfer of nanofluids in stretchable convergent/divergent channels. *Appl. Sci.* **2015**, *5*, 1639–1664. [CrossRef]
4. Sidik, N.A.; Yazid, M.N.; Mamat, R. A review on the application of nanofluids in vehicle engine cooling system. *Int. Commun. Heat Mass.* **2015**, *68*, 85–90. [CrossRef]
5. Sarkar, J.; Ghosh, P.; Adil, A. A review on hybrid nanofluids: recent research, development and applications. *Renew. Sustain. Energy Rev.* **2015**, *43*, 164–177. [CrossRef]
6. Ebrahimnia-Bajestan, E.; Moghadam, M.C.; Niazmand, H.; Daungthongsuk, W.; Wongwises, S. Experimental and numerical investigation of nanofluids heat transfer characteristics for application in solar heat exchangers. *Int. J. Heat Mass Trans.* **2016**, *92*, 1041–1052. [CrossRef]

7. Al-Waeli, A.H.; Sopian, K.; Kazem, H.A.; Chaichan, M.T. Photovoltaic solar thermal (PV/T) collectors past, present and future: A review. *Int. J. Appl. Eng. Res.* **2016**, *11*, 10757–10765.

8. Fetecau, C.; Vieru, D.; Azhar, W.A. Natural convection flow of fractional nanofluids over an isothermal vertical plate with thermal radiation. *Appl. Sci.* **2017**, *7*, 247. [CrossRef]

9. Kasaeian, A.; Daneshazarian, R.; Mahian, O.; Kolsi, L.; Chamkha, A.J.; Wongwises, S.; Pop, I. Nanofluid flow and heat transfer in porous media: a review of the latest developments. *Int. J. Heat Mass Trans.* **2017**, *107*, 778–791. [CrossRef]

10. Khan, N.S.; Gul, T.; Islam, S.; Khan, I.; Alqahtani, A.M.; Alshomrani, A.S. Magnetohydrodynamic nanoliquid thin film sprayed on a stretching cylinder with heat transfer. *Appl. Sci.* **2017**, *7*, 271. [CrossRef]

11. Alobaid, M.; Hughes, B.; Heyes, A.; O'Connor, D. Determining the effect of inlet flow conditions on the thermal efficiency of a flat plate solar collector. *Fluids* **2018**, *3*, 67. [CrossRef]

12. Maleki, H.; Safaei, M.R.; Togun, H.; Dahari, M. Heat transfer and fluid flow of pseudo-plastic nanofluid over a moving permeable plate with viscous dissipation and heat absorption/generation. *J. Therm. Anal. Calorim.* **2018**, 1 –12. [CrossRef]

13. Maleki, H.; Safaei, M.R.; Alrashed, A.A.; Kasaeian, A. Flow and heat transfer in non-Newtonian nanofluids over porous surfaces. *J. Therm. Anal. Calorim.* **2018**, 1–12. [CrossRef]

14. Choi, S.U.; Zhang, Z.G.; Yu, W.; Lockwood, F.E.; Grulke, E.A. Anomalous thermal conductivity enhancement in nanotube suspensions. *Appl. Phys. Lett.* **2001**, *79*, 2252–2254. [CrossRef]

15. Buongiorno, J. Convective transport in nanofluids. *J. Heat Trans.* **2006**, *128*, 240–250. [CrossRef]

16. Nield, D.A.; Kuznetsov, A.V. The Cheng-Minkowycz problem for natural convective boundary-layer flow in a porous medium saturated by a nanofluid. *Int. J. Heat Mass Trans.* **2009**, *52*, 5792–5795. [CrossRef]

17. Buongiorno, J.; Venerus, D.C.; Prabhat, N.; McKrell, T.; Townsend, J.; Christianson, R.; Tolmachev, Y.V.; Keblinski, P.; Hu, L.W.; Alvarado, J.L.; et al. A benchmark study on the thermal conductivity of nanofluids. *J. Appl. Phys.* **2009**, *106*, 094312. [CrossRef]

18. Kuznetsov, A.V.; Nield, D.A. The Cheng–Minkowycz problem for natural convective boundary layer flow in a porous medium saturated by a nanofluid: A revised model. *Int. J. Heat Mass Trans.* **2013**, *65*, 682–685. [CrossRef]

19. Muhammad, T.; Alsaedi, A.; Shehzad, S.A.; Hayat, T. A revised model for Darcy–Forchheimer flow of Maxwell nanofluid subject to convective boundary condition. *Chi. J. Phys.* **2017**, *55*, 963–976. [CrossRef]

20. Rehman, S.U.; Haq, R.U.; Khan, Z.H.; Lee, C. Entropy generation analysis for non-Newtonian nanofluid with zero normal flux of nanoparticles at the stretching surface. *J. Taiwan Inst. Chem. Eng.* **2016**, *63*, 226–235. [CrossRef]

21. Rahman, M.M.; Al-Rashdi, M.H.; Pop, I. Convective boundary layer flow and heat transfer in a nanofluid in the presence of second order slip, constant heat flux and zero nanoparticles flux. *Nucl. Eng. Des.* **2016**, *297*, 95–103. [CrossRef]

22. Uddin, I.; Khan, M.A.; Ullah, S.; Islam, S.; Israr, M.; Hussain, F. Characteristics of buoyancy force on stagnation point flow with magneto-nanoparticles and zero mass flux condition. *Results Phys.* **2018**, *8*, 160–168. [CrossRef]

23. Ur Rahman, M.; Khan, M.; Manzur, M. Boundary layer flow and heat transfer of a modified second grade nanofluid with new mass flux condition. *Results Phys.* **2018**, *10*, 594-600. [CrossRef]

24. Jusoh, R.; Nazar, R.; Pop, I. Three-dimensional flow of a nanofluid over a permeable stretching/shrinking surface with velocity slip: A revised model. *Phus. Fluids* **2018**, *30*, 033604. [CrossRef]

25. Roşca, N.C.; Pop, I. Unsteady boundary layer flow of a nanofluid past a moving surface in an external uniform free stream using Buongiorno's model. *Comput. Fluids* **2014**, *95*, 49–55. [CrossRef]

26. Mabood, F.; Khan, W.A.; Ismail, A.M. MHD boundary layer flow and heat transfer of nanofluids over a nonlinear stretching sheet: A numerical study. *J. Magn. Magn. Mater.* **2015**, *374*, 569–576. [CrossRef]

27. Sheikholeslami, M.; Ganji, D.D.; Javed, M.Y.; Ellahi, R. Effect of thermal radiation on magnetohydrodynamics nanofluid flow and heat transfer by means of two phase model. *J. Magn. Magn. Mater.* **2015**, *374*, 36–43. [CrossRef]

28. Bakar, N.A.; Bachok, N.; Arifin, N.M. Moving plate in a nanofluid using Buongiorno model and thermophysical properties of nanoliquids. *JP J. Heat Mass Trans.* **2017**, *14*, 119. [CrossRef]

29. Bakar, N.A.; Bachok, N.; Arifin, N.M. Rotating flow over a shrinking sheet in nanofluid using Buongiorno model and thermophysical properties of nanoliquids. *J. of Nanofluids* **2017**, *6*, 1215–1226. [CrossRef]

30. Bakar, N.A.; Bachok, N.; Arifin, N.M. Stability analysis on the flow and heat transfer of nanofluid past a stretching/shrinking cylinder with suction effect. *Results Phys.* **2018**, *9*, 1335–1344. [CrossRef]
31. Othman, N.A.; Yacob, N.A.; Bachok, N.; Ishak, A.; Pop, I. Mixed convection boundary-layer stagnation point flow past a vertical stretching/shrinking surface in a nanofluid. *Appl. Therm. Eng.* **2017**, *115*, 1412–1417. [CrossRef]
32. Rashidi, M.M.; Laraqi, N.; Parsa, A.B. Analytical modeling of heat convection in magnetized micropolar fluid by using modified differential transform method. *Heat Trans. Asian Res.* **2011**, *40*, 187–204. [CrossRef]
33. Rashidi, M.M.; Abelman, S.; Mehr, N.F. Entropy generation in steady MHD flow due to a rotating porous disk in a nanofluid. *Int. J. Heat Mass Trans.* **2013**, *62*, 515–525. [CrossRef]
34. Sheikholeslami, M.; Gorji-Bandpy, M.; Ganji, D.D. MHD free convection in an eccentric semi-annulus filled with nanofluid. *J. Taiwan Inst. Chem. Eng.* **2014**, *45*, 1204–1216. [CrossRef]
35. Hayat, T.; Imtiaz, M.; Alsaedi, A.; Kutbi, M.A. MHD three-dimensional flow of nanofluid with velocity slip and nonlinear thermal radiation. *J. Magn. Magn. Mater.* **2015**, *396*, 31–37. [CrossRef]
36. Kandasamy, R.; Jeyabalan, C.; Prabhu, K.S. Nanoparticle volume fraction with heat and mass transfer on MHD mixed convection flow in a nanofluid in the presence of thermo-diffusion under convective boundary condition. *Appl. Nanosci.* **2016**, *6*, 287–300. [CrossRef]
37. Bhatti, M.M.; Abbas, T.; Rashidi, M.M. Numerical study of entropy generation with nonlinear thermal radiation on magnetohydrodynamics non-Newtonian nanofluid through a porous shrinking sheet. *J. Magn.* **2016**, *21*, 468–475. [CrossRef]
38. Bhatti, M.M.; Abbas, M.A.; Rashidi, M.M. Combine effects of Magnetohydrodynamics (MHD) and partial slip on peristaltic Blood flow of Ree–Eyring fluid with wall properties. *Eng. Sci. Technol. Int. J.* **2016**, *19*, 1497–1502. [CrossRef]
39. Makulati, N.; Kasaeipoor, A.; Rashidi, M.M. Numerical study of natural convection of a water–alumina nanofluid in inclined C-shaped enclosures under the effect of magnetic field. *Adv. Powder Technol.* **2016**, *27*, 661–672. [CrossRef]
40. Hussain, S.M.; Jain, J.; Seth, G.S.; Rashidi, M.M. Free convective heat transfer with hall effects, heat absorption and chemical reaction over an accelerated moving plate in a rotating system. *J. Magn. Magn. Mater.* **2017**, *422*, 112–123. [CrossRef]
41. Hiemenz, K. Die Grenzschicht an einem in den gleichformigen Flussigkeitsstrom eingetauchten geraden Kreiszylinder. *Dinglers Polytech. J.* **1911**, *326*, 321–324.
42. Homann, F. Der Einfluss grosser Zähigkeit bei der Strömung um den Zylinder und um die Kugel. *ZAMM J. Appl. Math. Mech./Z. Angew. Math. Mech.* **1936**, *16*, 153–164. [CrossRef]
43. Howarth, L. The boundary layer in three dimensional flow—Part II. The flow near a stagnation point. *Lond. Edinb. Dublin Philos. Mag. J. Sci.* **1951**, *42*, 1433–1440. [CrossRef]
44. Wang, C.Y. Stagnation flow on a plate with anisotropic slip. *Eur. J. Mech. B/Fluids.* **2013**, *38*, 73–77. [CrossRef]
45. Wang, C.Y. Stagnation flows with slip: Exact solutions of the Navier-Stokes equations. *Z. Angew. Math. Phys. (ZAMP)* **2003**, *54*, 184–189. [CrossRef]
46. Wang, C.Y. Stagnation slip flow and heat transfer on a moving plate. *Chem. Eng. Sci.* **2006**, *61*, 7668–7672. [CrossRef]
47. Hussain, S.; Aziz, A.; Aziz, T.; Khalique, C.M. Slip flow and heat transfer of nanofluids over a porous plate embedded in a porous medium with temperature dependent viscosity and thermal conductivity. *Appl. Sci.* **2016**, *6*, 376. [CrossRef]
48. Khan, M; Hashim; Hafeez, A. A review on slip-flow and heat transfer performance of nanofluids from a permeable shrinking surface with thermal radiation: Dual solutions. *Chem. Eng. Sci.* **2017**, *173*, 1–11. [CrossRef]
49. Belyaev, A.V.; Vinogradova, O.I. Electro-osmosis on anisotropic superhydrophobic surfaces. *Phys. Rev. Lett.* **2011**, *107*, 098301. [CrossRef]
50. Jung, T.; Choi, H.; Kim, J. Effects of the air layer of an idealized superhydrophobic surface on the slip length and skin-friction drag. *J. Fluid Mech.* **2016**, *790*. [CrossRef]
51. Pearson, J.T.; Bilodeau, D.; Maynes, D. Two-pronged jet formation caused by droplet impact on anisotropic superhydrophobic surfaces. *J. Fluids Eng.* **2016**, *138*, 074501. [CrossRef]
52. Schäffel, D.; Koynov, K.; Vollmer, D.; Butt, H.J.; Schönecker, C. Local flow field and slip length of superhydrophobic surfaces. *Phys. Rev. Lett.* **2016**, *116*, 134501. [CrossRef] [PubMed]

53. Fan, B.; Bandaru, P.R. Anisotropy in the hydrophobic and oleophilic characteristics of patterned surfaces. *Appl. Phys. Lett.* **2017**, *111*, 261603. [CrossRef]

54. Alinovi, E.; Bottaro, A. Apparent slip and drag reduction for the flow over superhydrophobic and lubricant-impregnated surfaces. *Phys. Rev. Fluids* **2018**, *3*, 124002. [CrossRef]

55. Lu, J.; Jang, H.K.; Lee, S.B.; Hwang, W.R. Characterization on the anisotropic slip for flows over unidirectional fibrous porous media for advanced composites manufacturing. *Compos. Part A Appl. Sci. Manuf.* **2017**, *100*, 9–19. [CrossRef]

56. Pasquier, S.; Quintard, M.; Davit, Y. Modeling flow in porous media with rough surfaces: Effective slip boundary conditions and application to structured packings. *Chem. Eng. Sci.* **2017**, *165*, 131–146. [CrossRef]

57. Rashad, A.M. Unsteady nanofluid flow over an inclined stretching surface with convective boundary condition and anisotropic slip impact. *Int. J. Heat Technol.* **2017**, *35*, 82–90. [CrossRef]

58. Hafidzuddin, E.H.; Nazar, R.; Arifin, N.M.; Pop, I. Effects of anisotropic slip on three-dimensional stagnation-point flow past a permeable moving surface. *Eur. J. Mech. B/Fluids.* **2017**, *65*, 515–521. [CrossRef]

59. Raees, A.; Raees-ul-Haq, M.; Xu, H.; Sun, Q. Three-dimensional stagnation flow of a nanofluid containing both nanoparticles and microorganisms on a moving surface with anisotropic slip. *Appl. Math. Model.* **2016**, *40*, 4136–4150. [CrossRef]

60. Uddin, M.J.; Khan, W.A.; Ismail, A.M.; Bég, O.A. Computational study of three-dimensional stagnation point nanofluid bioconvection flow on a moving surface with anisotropic slip and thermal jump effect. *J. Heat Trans.* **2016**, *138*, 104502. [CrossRef]

61. Al-Balushi, L.M.; Rahman, M.M.; Pop, I. Three-dimensional axisymmetric stagnation-point flow and heat transfer in a nanofluid with anisotropic slip over a striated surface in the presence of various thermal conditions and nanoparticle volume fractions. *Therm. Sci. Eng. Prog.* **2017**, *2*, 26–42. [CrossRef]

62. Sadiq, M.A. MHD Stagnation point flow of nanofluid on a plate with anisotropic slip. *Symmetry* **2019**, *11*, 132. [CrossRef]

63. Merkin, J.H. On dual solutions occurring in mixed convection in a porous medium. *J. Eng. Math.* **1986**, *20*, 171–179. [CrossRef]

64. Weidman, P.D.; Kubitschek, D.G.; Davis, A.M. The effect of transpiration on self-similar boundary layer flow over moving surfaces. *Int. J. Eng. Sci.* **2006**, *44*, 730–737. [CrossRef]

65. Harris, S.D.; Ingham, D.B.; Pop, I. Mixed convection boundary-layer flow near the stagnation point on a vertical surface in a porous medium: Brinkman model with slip. *Trans. Porous Med.* **2009**, *77*, 267–285. [CrossRef]

66. Roşca, A.V.; Pop, I. Flow and heat transfer over a vertical permeable stretching/shrinking sheet with a second order slip. *Int. J. Heat Mass Trans.* **2013**, *60*, 355–364. [CrossRef]

67. Anuar, N.; Bachok, N; Pop, I. A stability analysis of solutions in boundary layer flow and heat transfer of carbon nanotubes over a moving plate with slip effect. *Energies* **2018**, *11*, 3243. [CrossRef]

68. Salleh, S.N.A.; Bachok, N.; Arifin, N.M.; Ali, F.M.; Pop, I. Stability analysis of mixed convection flow towards a moving thin needle in nanofluid. *Appl. Sci.* **2018**, *8*, 842. [CrossRef]

69. Najib, N.; Bachok, N.; Arifin, N.M.; Ali, F.M. Stability analysis of stagnation-point flow in a nanofluid over a stretching/shrinking sheet with second-order slip, soret and dufour effects: A revised model. *Appl. Sci.* **2018**, *8*, 642. [CrossRef]

70. Abu Bakar, S.; Arifin, N.M.; Md Ali, F.; Bachok, N.; Nazar, R.; Pop, I. A stability analysis on mixed convection boundary layer flow along a permeable vertical cylinder in a porous medium filled with a nanofluid and thermal radiation. *Appl. Sci.* **2018**, *8*, 483. [CrossRef]

71. Dzulkifli, N.; Bachok, N.; Yacob, N.; Md Arifin, N.; Rosali, H. Unsteady stagnation-point flow and heat transfer over a permeable exponential stretching/shrinking sheet in nanofluid with slip velocity effect: A stability analysis. *Appl. Sci.* **2018**, *8*, 2172. [CrossRef]

72. Jamaludin, A.; Nazar, R.; Pop, I. Three-dimensional magnetohydrodynamic mixed convection flow of nanofluids over a nonlinearly permeable stretching/shrinking sheet with velocity and thermal slip. *Appl. Sci.* **2018**, *8*, 1128. [CrossRef]

73. Yahaya, R.; Md Arifin, N.; Mohamed Isa, S. Stability analysis on magnetohydrodynamic flow of casson fluid over a shrinking sheet with homogeneous-heterogeneous reactions. *Entropy* **2018**, *20*, 652. [CrossRef]

74. Jusoh, R.; Nazar, R.; Pop, I. Flow and heat transfer of magnetohydrodynamic three-dimensional Maxwell nanofluid over a permeable stretching/shrinking surface with convective boundary conditions. *Int. J. Mech. Sci.* **2017**, *124*, 166–173. [CrossRef]
75. Weidman, P.D.; Sprague, M.A. Flows induced by a plate moving normal to stagnation-point flow. *Acta Mech.* **2011**, *219*, 219–229. [CrossRef]
76. Jamaludin, A.; Nazar, R.; Pop, I. Three-dimensional mixed convection stagnation-point flow over a permeable vertical stretching/shrinking surface with a velocity slip. *Chin. J. Phys.* **2017**, *55*, 1865–1882. [CrossRef]
77. Rahman, M.M. Effects of second-order slip and magnetic field on mixed convection stagnation-point flow of a Maxwellian fluid: Multiple solutions. *J. Heat Tran.* **2016**, *138*, 122503. [CrossRef]
78. Borrelli, A.; Giantesio, G.; Patria, M.C.; Roşca, N.C.; Roşca, A.V.; Pop, I. Buoyancy effects on the 3D MHD stagnation-point flow of a Newtonian fluid. *Commun. Nonlinear Sci. Numer. Simul.* **2017**, *43*, 1–3. [CrossRef]

Article

Simulation of Vortex Heat Transfer Enhancement in the Turbulent Water Flow in the Narrow Plane-Parallel Channel with an Inclined Oval-Trench Dimple of Fixed Depth and Spot Area

Sergey Isaev [1,*], Alexandr Leontiev [2], Yaroslav Chudnovsky [3], Dmitry Nikushchenko [4], Igor Popov [5] and Alexandr Sudakov [1]

[1]   Saint-Petersburg State University of Civil Aviation, Pilots Str. 38, Saint-Petersburg 196210, Russia; sudakov-1950@mail.ru

[2]   Bauman Moscow State Technical University, 2nd Bauman Str. 5/1, Moscow 105005, Russia; nchmt@iht.mpei.ac.ru

[3]   Gas Technology Institute, 1700 S Mount Prospect Road, Des Plaines, IL 60018, USA; yaroslav.chudnovsky@gastechnology.org

[4]   Saint-Petersburg State Marine Technical University, Lotsmanskaya, 3, Saint-Petersburg 190121, Russia; ndmitry@list.ru

[5]   A. N. Tupolev Kazan National Research Technical University (Kazan Aviation Institute), Kazan 420111, Russia; popov-igor-alex@yandex.ru

*   Correspondence: isaev3612@yandex.ru; Tel.: +7-921-404-5516

Received: 1 February 2019; Accepted: 29 March 2019; Published: 4 April 2019

Abstract: This article is devoted to the development of the multiblock technique for numerical simulation of vortex heat transfer enhancement (VHTE) by inclined oval-trench dimples. Special attention is paid both to the analysis of numerical predictions of different-type boundary conditions at the wall: T = const and q = const and to the comparison of the standard and modified shear stress transport models. The article discusses the mechanism of change in the flow structure and secondary flow augmentation due to an increase in a relative length of an oval-trench dimple (at its fixed spot area, depth and orientation) where a long spiral vortex is formed.

Keywords: vortex; heat; dimple; channel; simulation

---

1. Introduction

The use of a structured surface with discrete artificial roughness is a popular method of vortex heat transfer enhancement (VHTE) in energy equipment [1]. Tubes with periodic protrusions many times enhance heat transfer in comparison to smooth tubes. However, hydraulic losses in this case grow faster, thus, an increased total pressure drop is required. Replacing protrusions by grooves allows hydraulic losses to be decreased cardinally.

Prior to the three-dimensional (3D) printing era, the type of discrete roughness was mainly determined by technology. Cylindrical cavities made by mechanical extrusion [2–4] were the simplest and well-studied technological forms. However, hydraulic losses in ducts with such cavities appeared to be rather high.

Another simple form of a surface cavity is a semi-spherical dimple formed by a sphere that was pressed into the wall. In the late 20th century, such cavities attracted the attention of researchers and engineers [5–8], although at a later date segment-spherical dimples of a relative depth of less than 0.5 (in terms of spot diameter) [9,10] were quite often considered. It is interesting to note that the self-oscillatory flow regime was for the first time revealed in a semi-circular dimple, on the sides of which vortex jets were alternatively formed [5].

One of the first researchers who drew attention to the vortex mechanism of heat transfer intensification using spherical dimples were Kiknadze et al. [11] and Ligrani et al. [12].

In the 1990s, Afanasiev et al. [13] and Chyu et al. [14] have carried out a series of pioneering experimental investigations of turbulent convective heat transfer at a dimpled plane wall in rectangular channels. The most complete complex study of hydrodynamics and heat transfer on surfaces with spherical recesses was conducted By Terekhov [10]. Results of this work were used as a basis for studying flow characteristics in dimpled channels at the University of Rostock [15].

A large number of experimental works on hydrodynamics and heat transfer in dimpled channels were performed at the beginning of the new millennium [16–21]. Later, these publications were supplemented [22–24] using new experimental methods with the implication of gradient heat flux sensors and pressure pulsation detectors. The studies on the visualization of vortex flows in the vicinity of dimples such as [25,26] should also be mentioned here.

Recent interest in laminar heat transfer enhancement was initiated by the research of heat transfer in microchannels [27]. The research in [28] analyzed the detailed maps of flow around single spherical dimples. There has also been research that [29,30] contained the important results that were used for verification and validation of numerous numerical predictions. Lastly, there has been research [31] that analyzed the technique of making longer non-spherical dimples at the wall.

The analysis of the experimental works as presented here shows that most of them are devoted to spherical dimples. As noted in [11,12], the jet-vortex nature of heat transfer enhancement by such dimples is beyond doubt; however, the intensity to form swirled flows appears to be low and the problem to select rational shapes of vortex generators with the highest thermal and hydraulic performance still remains open. In [29], it was experimentally established that heat fluxes decay on the leeward side of the spherical dimple in the zone of low-velocity separated flow in comparison to the plane wall. Heat transfer coefficients are low inside the dimple.

The objective of the present study was to apply the specialized multiblock computational technique (MCT) based on simple-topology different-scale structured grids with their partial overlapping as being realized in the VP2/3 code in order to calculate convective heat transfer in dimpled channels. The authors analyzed different interpretations of turbulent transfer, including the consideration of the flow curvature influence on turbulent characteristics. We validated the numerical predictions and verified the turbulence models when comparing predicted and experimental results.

This study analyzes the physical mechanisms of VHTE based on the control of large-scale spiral vortices in single oval-trench dimples with fixed spot area and depth at an angle of inclination to the main flow at a Reynolds number of 10^4. It is shown that increasing the dimple length changes the flow structure in the dimple. The separation zone substantially decreases, backflow enhances, which leads to an almost two-fold growth of heat transfer occurring in the separation zone and also results in an increase in the secondary flow velocity comparable to the bulk velocity in the channel.

2. Problem Statement

In this work, we considered the convective heat transfer in the channel with a cross-sectional aspect ratio of 2.5×0.33 (the dimensions are related to the dimple diameter). At the entrance to the channel, we set the velocity profile of a fully developed turbulent flow. We are looking at an incompressible fluid. On the walls, the conditions of non-slip are observed. The profiles of the longitudinal, vertical and transverse components of the velocity u, v, w, as well as the turbulence characteristics (energy k, specific scattering velocity ω, vortex viscosity μt) are found from the solution of a special problem in the development of the flow in the computational domain of the selected section of the channel under periodic boundary conditions under the selected Reynolds number 10^4. As parameters are assigned to the normalization of the volumetric velocity U, a spot diameter d of basic spherical and $10°$-truncated conical dimples, the density ρ and viscosity μ of the coolant-water (Pr = 7). The depressions of the average depth of 0.13 (according to the classification [32]) are located at some distance (about 3) from the selected channel input. The center of the Cartesian coordinates x,

y, z are in the longitudinal middle plane of the channel at the point of projection of the center of the dimple on the section coinciding with the wall of the lower plane (Figure 1). The rounding radius of dimple edges is taken as equal to 0.025. While maintaining the area of the oval dimple spot equal to the area of the spherical dimple, its width b varies from 0.731 to 0.346 in width from 1.68 to 6.78 (Figure 1 and Table 1).

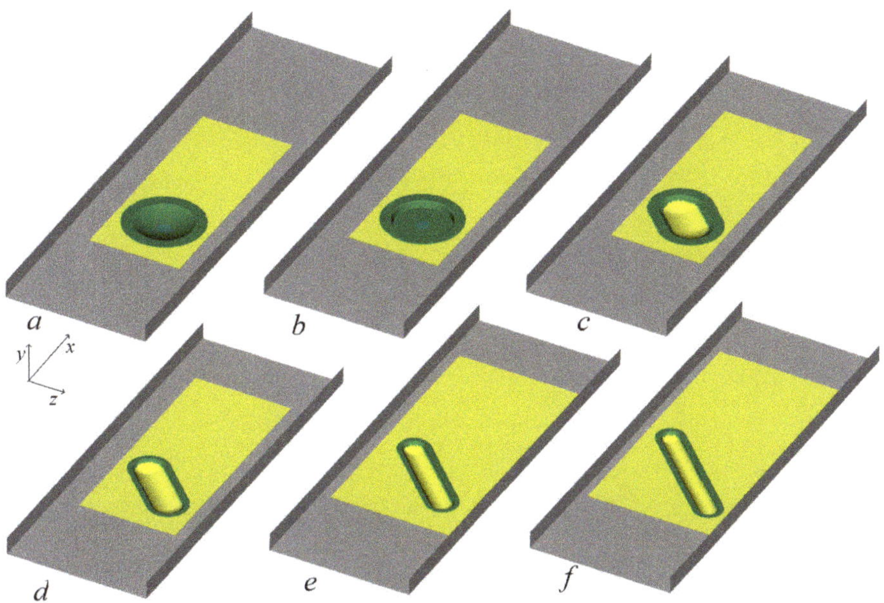

Figure 1. Channels with spherical (**a**), 10°-truncated conical (**b**) and oval dimples of width **b** = 0.731 (**c**), 0.549 (**d**), 0.429 (**e**) and 0.346 (**f**).

Table 1. The cylindrical insert length, the width of the oval dimple and its length in terms of width.

L	b	χ
0.5	0.731	1.68
0.625	0.678	1.92
0.675	0.659	2.02
0.75	0.631	2.19
0.9	0.58	2.55
1	0.549	2.82
1.25	0.482	3.59
1.5	0.429	4.50
1.75	0.383	5.57
2	0.346	6.78

The channel ends at a distance $L = 7$ after the dimple (normalized by the diameter of the dimple spot).

We solved the problems of heat transfer and hydrodynamics separately. The flow at the channel inlet is isothermal at T_{ref} = 293 K. The streamed bottom wall of the dimpled channel was heated and a supplied constant heat flux q was re-calculated in dimensionless form by the equation (of order 3.4×10^{-5}).

$$\bar{q} = \frac{q}{\lambda \mathrm{PrRe} T_{ref} / D} \tag{1}$$

Here, λ is the thermal conductivity of water. The side walls of the channel are adiabatic, while the top wall is isothermal at T_{ref} (taken as the normalization scale). The outflow conditions are predetermined for T at the channel outlet.

Since the previous studies [32,33] considered heated isothermal walls, the testing part of this work compares the boundary conditions T = const and q = const; the heated wall temperature was assigned as equal to 1.036 in terms of T_{ref}. We need to note that in [32,33], air was considered as a heat carrier and the wall temperature was set to 373 °C.

3. Models, Methods, Computational Grids

In the section below, suggestions on turbulence model and computation methodology selection are presented, multi-block mesh structure is described; also in addition, we pay attention to iteration process convergence of the problem solution.

3.1. Turbulence Models

To solve the problem of convective heat transfer in a turbulent flow of incompressible fluid in a channel with single dimple on the heated walls, a mathematical model based on a system of Reynolds-averaged stable Navier-Stokes equations (RANS) and an energy equation similar to [32,33] was used.

To close them, the standard Menter SST (Shear Stress Transport) model and the modified Menter SST model were successfully used for typical wall flows, including those with separation [34,35]. The k-ω model proposed by Menter is a generalization of two turbulence models: the Launder–Spalding k-ε model for shear flow zones far from the wall and the Saffman–Wilcox k-ω model for the near-wall region. In addition, designing the zonal shear stress transport model the ideas were taken from the Johnson–King turbulence model. As earlier mentioned, in determining eddy viscosity, the Menter 1993 model [34] uses the vorticity modulus Ω, and the Menter 2003 model [34], as the majority of semi-empirical models of differential type, includes the strain tensor modulus S into the expression for eddy viscosity. It is important to emphasize that the semi-empirical models are calibrated mainly in near-wall flows. As a result, there is a need to correct them to high-intensity separated flows. As noted in [36], the Rodi–Leschziner approach to correcting eddy viscosity within the framework of the high-Reynolds version of the Launder–Spalding dissipative two-parameter turbulence model [37] widely used in the 1980–1990s [37] is that it is affected by the correction function $f_c = 1/(1 + C_c \times R_{it})$. The constant C_c equal to 0.57 was determined analytically when calculating turbulent annular and twin parallel jets [38] and a limitation was imposed on the product $f_c \times C_\mu$: $0.02 < f_c \times C_\mu < 0.15$ (in the standard k-ε model [37], the semi-empirical constant $C_\mu = 0.09$ in the expression for eddy viscosity). Isaev generalized the Rodi–Leschziner approach (RLI approach) to the Menter 2003 model. Kharchenko, Usachov and Isaev selected the constant $C_c = 0.02$ [36] from the condition of the best agreement of numerical predictions and experimental data for numerous separated flows. Recently Smirnov and Menter [39] have proposed one more correction of the Menter 2003 model (SM correction) when they extended the Shur–Spalart correction in the Spalart–Allmares eddy viscosity model [40] to the Menter two-parameter model.

Near-wall conditions for SST models are formulated so that the normal derivative to the wall for turbulence energy is equal to zero and the specific dissipation rate of turbulence energy in the near-wall cell is determined as in [41].

3.2. Computational Methodology

It is beyond doubt that software based on solving the Reynolds-averaged Navier–Stokes equations with the use of semi-empirical models became a powerful tool to predict flow parameters and turbulence characteristics. The present study used the multiblock computational technique [42] realized in the original research VP2/3 code (velocity-pressure, two-dimensional (2D)/3D). Computational algorithms realized in this code are based both on the concept of splitting with respect to physical processes and on the application of grid methods for solution of the governing equations [42,43].

The use of the concept of splitting allows a system of partial differential equations to be divided into blocks containing momentum equations in natural variables (including Cartesian velocity components for incompressible viscous liquid flows), which replace the continuity equation, as well as the pressure correction equation (SIMPLEC (Semi-Implicit Method for Pressure Linked Equations Consistent) [44,45]) and the equations for their closure (from the chosen turbulence model).

Thus, the system of steady equations in discrete form is solved block by block at each time step during the global iteration process (about 10–20 iterations), when at each time step for one iteration in the course of solution of the momentum equation, several (about 10–15 iterations) iterations are performed in the pressure correction block and about four–six iterations in the turbulence and energy blocks. The governing equations are preliminarily linearized [46].

In the calculation algorithm we used: (1) the pressure correction procedure SIMPLEC [44,47,48]; (2) the approximation of the convective terms in the explicit hand-side of the momentum equation using Leonard's one-dimensional quadratic upwind scheme [49] to reduce the influence of numerical difference specific for a considered type of separated flows and using van Leer's scheme [50]; (3) the representation of the convective terms in the implement hand-side of transport equations using the upwind scheme with one-sided differences; (4) methods with preconditioners for solution of difference equations [51].

In this paper, we used a Rhie–Chow's generalized approach [48] to avoid difficulties in the calculation of unsteady flows. The method for solution of algebraic equations is the BiCGSHAB (biconjugate gradient stabilized method) preconditioner [51] with an AMG (Algebraic Multigrid) preconditioner from Demidov's library (amgl) [52] for pressure correction and the ILU0 preconditioner for another variables.

MCTs realized in the VP2/3 code are outlined elsewhere [42,53]. Their essence is to introduce a set of difference-scale, tier and structured overlapping grids to resolve the flow structure in the physical problem of corresponding scale. The parameters for two rows of near-boundary cells of each of the overlapping or overset grids are determined using linear interpolation [42,54] in the manner, as done in [55]. Computation from grid to grid with the use of MCTs involving linear interpolation is a source of errors; however, the test calculations [54] showed that the errors were acceptable.

3.3. Computational Grids

In this paper, we used multiblock overlapping structured grids of different scale (Figure 2). We considered grid structures similar in topology to those used in [32,33,56]; however, these are more condensed and have a greater number of cells. A rectangular channel is covered with a Cartesian grid condensing to the walls and to the area around a dimple. The total number of cells in the channel is 700,000–1,500,000.

In the case of spherical and conical dimples, a detailed grid is Cartesian and in the case of oval dimples, it is curvilinear, fitted to the streamlined bottom wall of the channel.

The area around spherical and conical grids is divided by a cylindrical grid adjusted to the dimple surface (Figure 2a,b); an additional grid close to a rectangular one ('patch') is introduced to prevent nodes from condensing in the near-axis zone. The minimum grid step near the edge is 0.002. In the case of oval dimples, a special edge grid is introduced to describe high-gradient zones. The total number of multiblock grid cells is about 1.5–3 mln cells.

Figure 3a illustrates the multiblock computational grid in the axonometric projection for an oval-trench dimple.

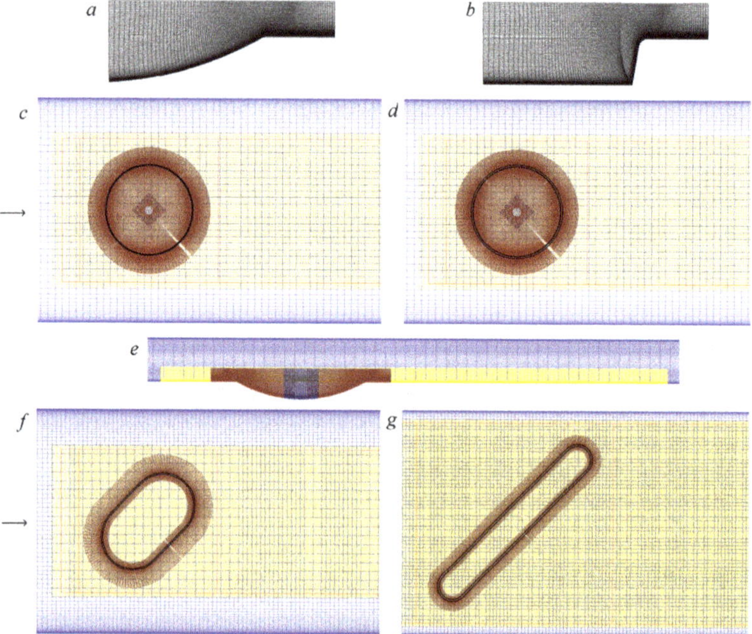

Figure 2. Spherical (**a**) and 10°-truncated conical (**b**) inserts in the multiblock structured grids in the narrow channel with single spherical (**c,e**), conical (**d**), oval (**f,g**) grids of different topology and width: (**c,d,f,g**)—view from the heated wall; **e**—middle section of the channel; (**f–b**) = 0.731; (**g**)—0.346.

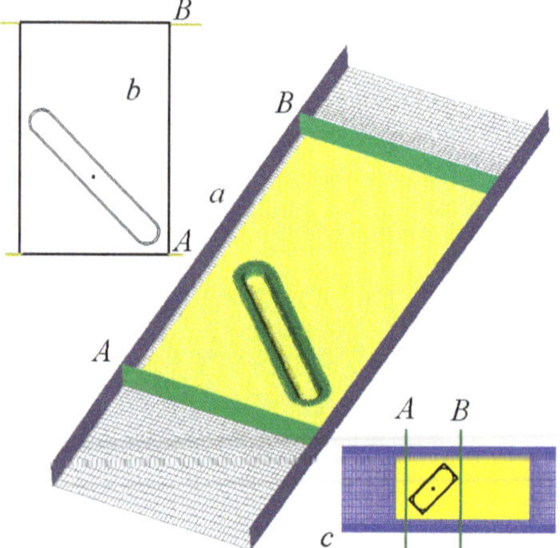

Figure 3. Multiblock grid for a narrow channel with an oval-trench dimple covered with a detailed grid in the dimple area and with edge selection (**a**), 3 × 2 dimpled section (**b**) and rectangular area of oval dimple (**c**). Channel cross sections *A-B* serve for determination of hydraulic losses.

3.4. Data Processing

The integral characteristics of flow and heat transfer in the narrow channel with single dimples were calculated in terms of the selected channel sections surrounding the dimple. As in [32,33], hydraulic losses in the channel were determined between the assigned cross sections A-B shown in Figure 3. Since we analyzed very long oval dimples, the size of the channel section (in comparison to [32,33,57]) was somewhat increased to assess relative heat transfer and to take it equal to 3×2 in the dimple center at a distance of 1 from the front boundary of the section (Figure 3b). In the present work, the thermal and hydraulic performance of the oval dimple was for the first time determined in the rectangular section surrounding the oval dimple at the $45°$ angle of orientation to the flow (Figure 3c).

We analyzed local dimensionless and relative characteristics of flow and heat transfer, including static pressure, friction, temperature and Nusselt number, as a function of longitudinal and transverse coordinates in the middle sections of the channel and the dimple at the bottom, heated and isothermal top walls. We compared Cartesian velocity components, turbulence energy and eddy viscosity normalized by the Reynolds number as profiles in the vertical coordinate in the dimple centers.

Temperature and Nusselt number fields are combined with the flow maps. We also considered vortex structures that are formed in dimples and obtained by computer visualization of labeled liquid particles.

3.5. Analysis of Convergence

Similarly [32,33] the SST model [36] with curvature correction within the RLI approach is chosen as the basic model.

Figure 4 shows the plots of maximum errors versus iteration step: convergence trajectories, transverse load and total Nusselt number determined within the dimple when the size of an additional grid is assigned (Figure 3a). Figure 4 illustrates the convergence trajectories chosen for maximum errors of the following variables u, p, T, k at each iteration step N_{it}: $Err_u, Err_p, Err_T, Err_k$ at linear and logarithmic scales.

Calculation is over when the maximum error does not exceed 10^{-5}. However, the experience with calculating separated flows [58] shows that it is not enough to control how a solution is set only in terms of errors and that it is necessary to observe the convergence in terms of integral characteristics. In the current study, transverse load R_z and heat transfer Nu_s in the dimple region are chosen as integral parameters.

The analysis of the convergence trajectories for conical and oval-trench dimples says that despite a greater number of iterations sometimes because of the use of a more detailed grid in the case of an oval dimple, the decrease in errors due to an increase in N_{it} is generally regressive in character and close to a linear one (Figure 4a,b,e,f). In the case of the conical dimple, R_z needs much time to be set and practically this takes place during the entire convergence process (about 3000 iterations).

This is mainly associated with the fact that a symmetrical vortex is formed in the dimple. At the same time, in the case of the oval dimple, R_z was set during 1500–3000 iterations, whereas the entire process occurred during more than 10,000 iterations. Approximately, the same situation was observed when total heat is transported from the dimpled region, although in the case of the conical dimple Nu_s was set slightly faster than R_z. In the case of the oval dimple, Nu_s was set to the 1000th iteration; wherein R_z still significantly changes. As a whole, the convergence for integral characteristics in the case of the oval dimple was set faster than in the case of the conical dimple.

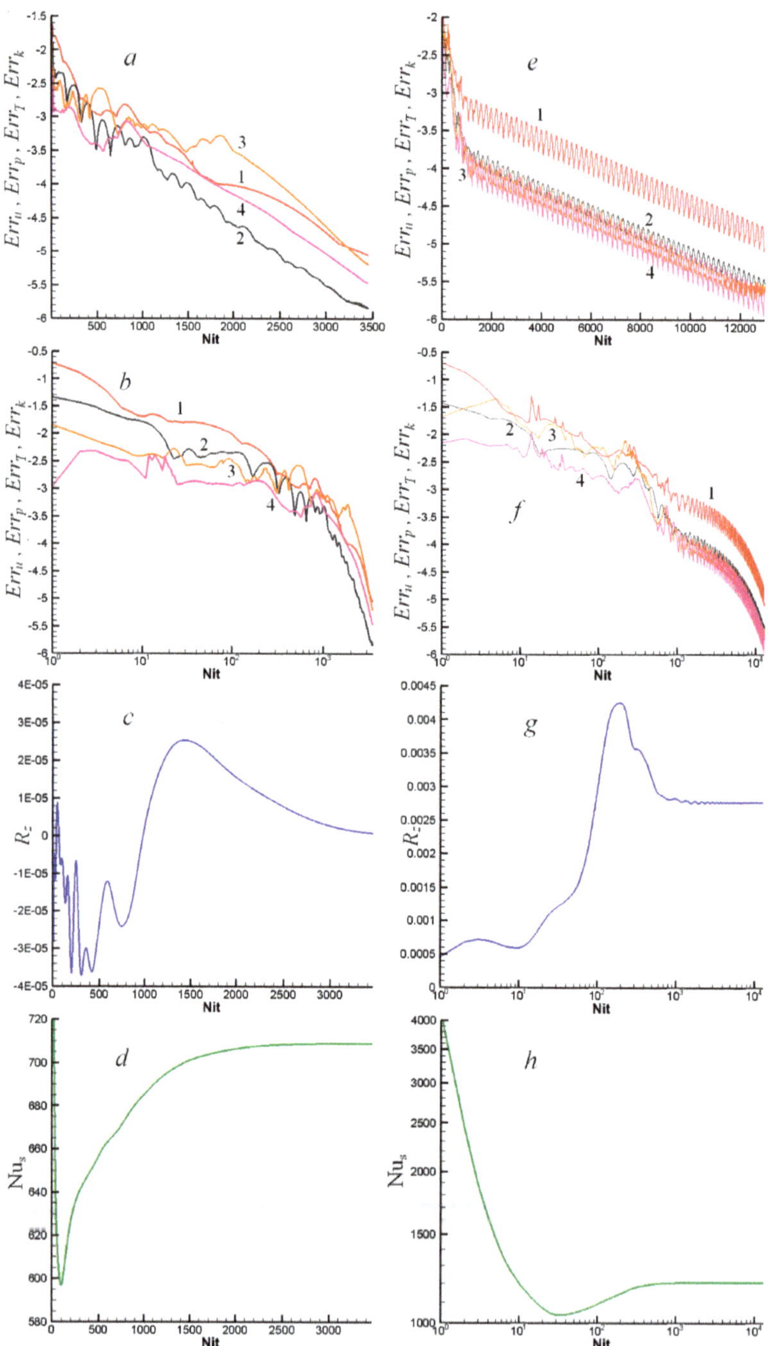

Figure 4. Convergence trajectories at linear (**a**,**e**) and logarithmic (**b**,**f**) scales, the dependences of the transverse load R_z (**c**,**g**) and the Nusselt number Nu_s (**d**,**h**) in the dimple region. Numbered curves: 1—Err_u; 2—Err_p; 3—Err_T; 4—Err_k.

4. Testing, Verification, Validation

The test block contained several series of results. First, there is the verification of methodology and turbulence model by comparison of numerical predictions with Terekhov's experiments in area of heat exchange in a spherical dimple on a thin channel wall. Second is a comparison of boundary conditions by T = const and q = const on example of flow over a spherical dimple. Third is the validation of computations in evaluation of influence of multi-block meshes on accuracy of solution of problem of convective heat exchange close to a tiled oval-trench dimple on wall of a thin channel.

4.1. Comparison of the Numerical Predictions with V.I. Terekhov's Experimental Data

Figure 5 presents some numerical and assessed calculations of flow and heat transfer in the narrow channel with a spherical dimple of depth 0.13 at the thermally insulated channel wall with a heated dimple as done in [10].

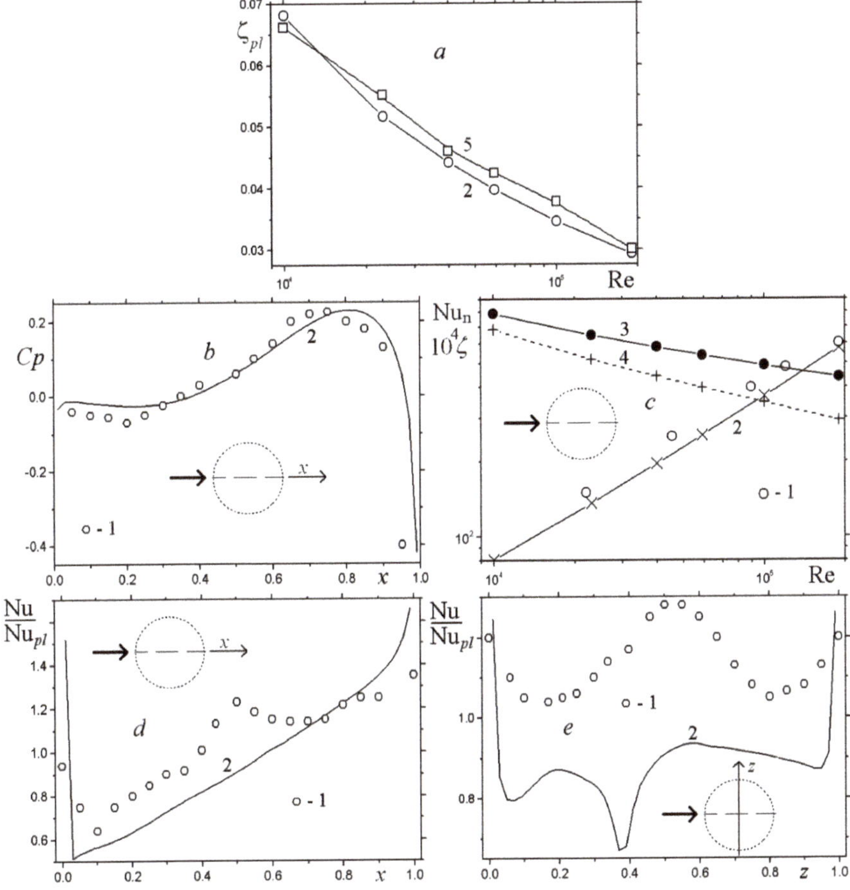

Figure 5. Verification: the comparison of the experimental data [10] with the numerical predictions of flow and heat transfer characteristics in the narrow channel with a spherical dimple (dimple spot heating at q = const): 1, 3—experiment [10]; 2, 4—calculation; 3, 4—hydraulic losses; 5—data [59]. (a) ζ_{pl}(Re); (b) $Cp(x)$; (c) $\mathrm{Nu_n}$(Re), $10^4\zeta$(Re); (d) $\mathrm{Nu}/\mathrm{Nu}_{pl}(x)$; (e) $\mathrm{Nu}/\mathrm{Nu}_{pl}(z)$

In Figure 5a, the methods for assessment of hydraulic losses ζ in the narrow plane-parallel channel [32] are verified by comparing the obtained numerical predictions with the estimates according to the data [59]. A satisfactory agreement is obtained between the results that attest to the acceptable accuracy of the methods used.

The comparison of the calculated and measured distributions of the longitudinal distribution of the pressure coefficient C_p in the middle section of the dimple (Figure 5b), the Reynolds number dependences of heat transfer on the dimple spot and hydraulic losses (Figure 5c), as well as the longitudinal and transverse relative Nusselt number distributions in the middle section of the dimple (Figure 5d,e) shows as a whole their satisfactory agreement.

4.2. Predictions of Convective Heat Transfer in the Narrow Channel with a Spherical Dimple at $T = const$ and $q = const$

Figures 6 and 7 and Table 2 demonstrate some of the comparative analysis results for the influence of boundary conditions at the heated bottom wall with a spherical dimple. As in [32,33], three sections near a dimple are chosen for analysis and are numbered as in Table 3: 10—2.5 × 1.5 section with a spherical dimple center at a distance of 1 from the front boundary of the section; 20—square section surrounding the spherical dimple; 30—2 × 1.5 section in the dimple wake.

As follows from Table 2, the integral characteristics of the thermal and hydraulic performance of the channel with a spherical dimple are practically independent of the type of boundary conditions for heat transfer within the turbulent flow regime. However, the local distributions (Figures 6 and 7) are significantly different in the near-edge zone and in the dimple center. Difference in maximum relative local heat transfer values is of the order of 1.5; at $T = const$, the loads against the near-edge zone are significantly higher than those at $q = const$.

Table 2. Thermal and hydraulic performance of three sections of the narrow channel with a spherical dimple at different boundary conditions $q = const$ and $T = const$.

Boundary Condition Type	Nu_{n10}/Nu_{np110}	ζ/ζ_{pl10}	Nu_{n20}/Nu_{npl20}	ζ/ζ_{pl20}	Nu_{n30}/Nu_{npl30}	ζ/ζ_{pl30}
$q = const$	1.098 (1.083)	1.072	1.138 (1.083)	1.16	1.085	1.010
$T = const$	1.094 (1.08)	1.071	1.17 (1.11)	1.16	1.064	1.008

Table 3. Predictions of thermal and hydraulic performance of two sections of the narrow channel with the oval dimple of width $b = 0.383$ obtained by the modified SST models and different grids.

Model	Nu_{n1}/Nu_{npl1}	ζ_1/ζ_{pl1}	$(Nu_{n1}/Nu_{npl1})/(\zeta_1/\zeta_{pl1})$	Nu_{n2}/Nu_{npl2}	ζ_2/ζ_{pl2}	$(Nu_{n2}/Nu_{npl2})/(\zeta_2/\zeta_{pl2})$
SST-model Standard [34]	1.242 (1.196)	1.079	1.151 (1.108)	1.953 (1.518)	1.150	1.698 (1.320)
SST-model Modified [35]	1.231 (1.185)	1.069	1.152 (1.109)	1.933 (1.502)	1.134	1.705 (1.325)
SST-model [35] modified within RLI approach	1.233 (1.187)	1.068	1.155 (1.111)	1.949 (1.515)	1.132	1.726 (1.338)
SST-model [35] modified within SM approach	1.228 (1.183)	1.064	1.154 (1.112)	1.950 (1.516)	1.127	1.730 (1.345)
SST-model [35] modified within RLI approach *	1.222 (1.177)	1.082	1.129 (1.088)	1.836 (1.427)	1.144	1.605 (1.247)

(* the grid contains 3 mln cells).

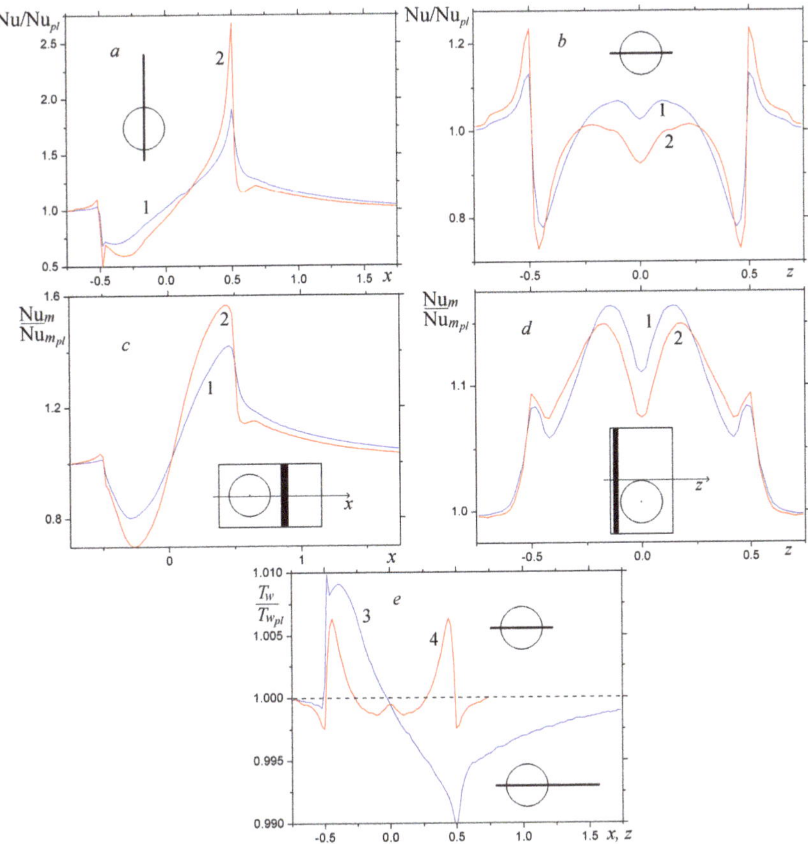

Figure 6. Calculated distributions of relative Nusselt numbers: local (**a**,**b**) and integrated over transverse (**c**) and longitudinal (**d**) strips of the selected section, as well as of relative wall temperature (**e**) in longitudinal (**a**,**c**,**3**) and transverse (**b**,**d**,**4**) middle sections of the dimple at different boundary conditions q = const (curves 1,3,4) and T = const (curves 2) at the heated wall of the narrow channel as in V.I. Terekhov's setup.

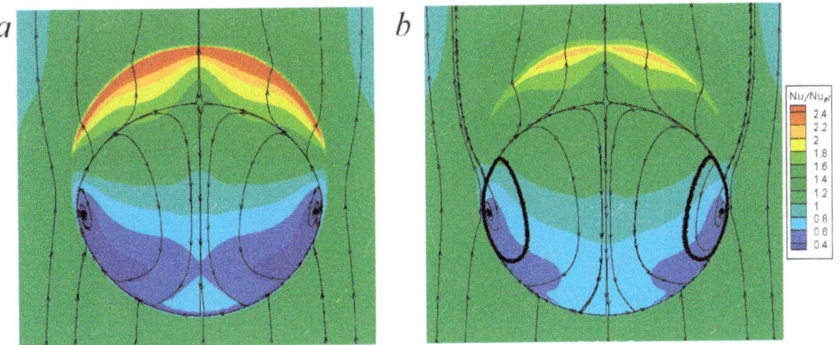

Figure 7. Calculated relative Nusselt number fields near the spherical dimple at the heated wall of the narrow channel at different boundary conditions and the streamlines: (**a**)—T = const; (**b**)—q = const.

4.3. Comparison of the Predictions Obtained by the Modified SST Models

In addition to the testing of the standard SST model [34] and the modified SST model [35] with curvature correction within the RLI and SM approaches for steady and unsteady two-dimensional separated flows [36,60–62]: circulation flow in square and circular cavities at the walls of plane-parallel and return channels and flow around a semi-circular body at a zero angle of attack, we compared the numerical predictions obtained using the modified SST models for three-dimensional steady separated flows in the narrow channel with an oval-trench dimple of width b = 0.383. Figure 8 and Table 3 show some of the obtained results.

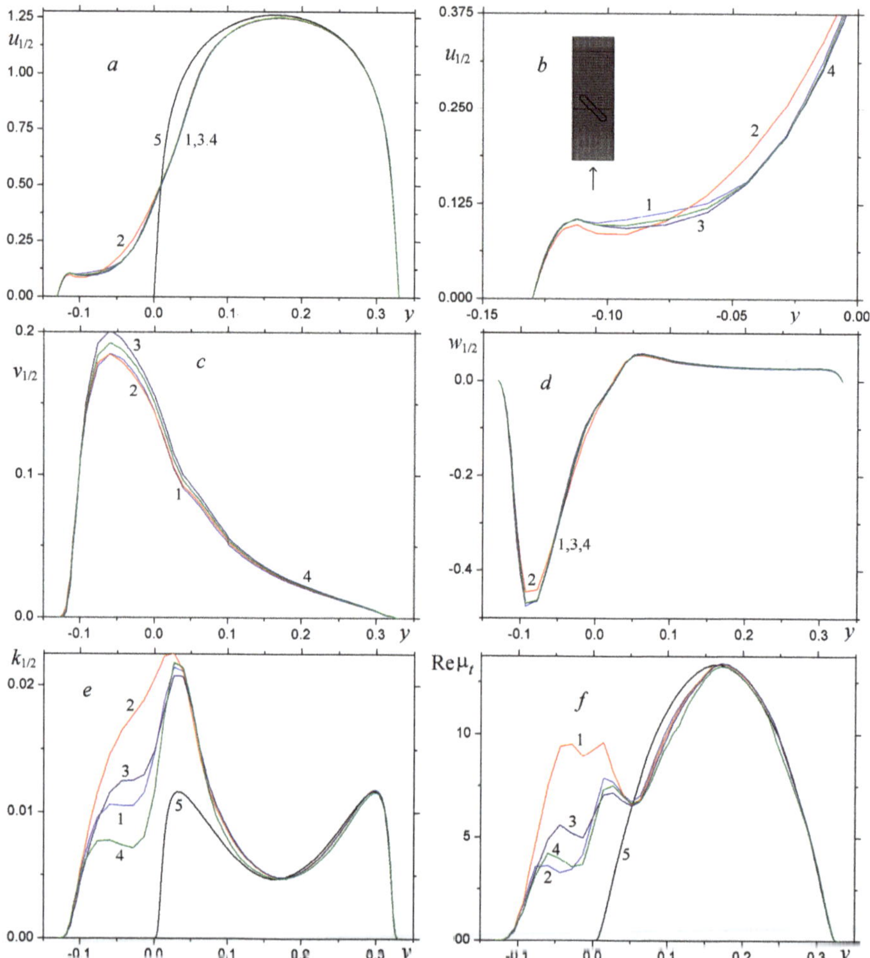

Figure 8. Distributions of longitudinal (**a**,**b**), vertical (**c**), transverse, (**d**) components of local flow velocity, turbulence energy (**e**) and normalized vortex viscosity (**f**) in the center of the oval dimple of width 0.383 obtained by the modified SST model: 1—SST model [34]; 2—SST model [35]; 3—SST model [35] with RLI correction; 4—SST model [35] with SM correction; 5—plane-parallel channel. *b*—longitudinal velocity component distribution (enlarged fragment).

We compared the local and integral characteristics of flow and heat transfer in the narrow channel with an oval dimple. We predicted about 1.6 mln cells at the wall. Table 4 is also supplemented by the refined calculations obtained by the modified SST model [35] with curvature correction within the RLI approach on the grid with 3 mln cells.

Table 4. VP2/3 code validation of the predictions of extreme flow, heat transfer and turbulence parameters for the narrow channel with the oval dimple of width $b = 0.383$.

Type	u_{min}	v_{min}	v_{max}	w_{min}	w_{max}	k_{max}	v_{tmax}	T_{wmax}
Grid A	-0.473	-0.349	0.526	-0.847	0.331	0.0412	0.00141	1.083
Grid B	-0.472	-0.337	0.508	-0.818	0.377	0.0405	0.00142	1.085

As shown by the two-dimensional tests, especially in the calculation of unsteady vortex flows, the modified SST model [35] predicts artificial turbulent viscosity in the cores of large-scale vortices. The reason for this is the strain rate tensor modulus introduced into the definition of eddy viscosity. As a result, there is a need to modify the SST model [35] in order to eliminate this non-physical viscosity. Here, two approaches were under consideration: direct viscosity correction in terms of the inverse linear dependence on turbulent Richardson number, when the semi-empirical constant $C_c = 0.02$ is added (RLI approach), and correction functions introduced into the system of equations for the modified SST model [35] (SM approach). We assumed that the correction within the RLI approach is preferable according to the test results [36,60–62].

The comparison of the integral characteristics of Table 3, as well as of the distributions of the local parameters at the streamlined surface and of the strip-integrated Nusselt numbers around and inside the dimple showed rather a good proximity of the numerical predictions obtained by all modified SST models. Some distinctions of the modified SST models [35] are not big but noticeable, especially in the case of local high heat fluxes.

The qualitative differences in the numerical predictions of local flow parameters and turbulence characteristics determined by different SST models manifested themselves in the vertical distributions in the dimple center (Figure 9). The differences in the velocity component distributions obtained by the modified SST model [35] are very noticeable, and the scatters of turbulence energy and Re-normalized eddy viscosity are particularly large. The overestimation of k and Re_{vt} is especially noticeable in the spiral vortex core. This indicates that an error is present in the modified SST model [35] in the calculation of high-intensity 3D separated flows.

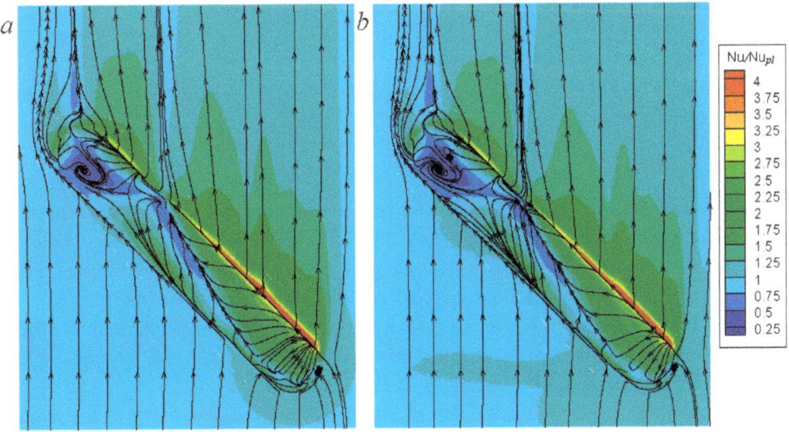

Figure 9. Relative Nusselt number distributions in the section of the narrow channel with the oval dimple of width 0.383 calculated on different grids: (**a**)—initial variant (about 1.6 mln cells); (**b**)—refinement (about 3 mln cells).

4.4. Validation

We compared the calculation results for convective heat transfer in the narrow channel wall with an oval dimple of width 0.383 on grids having 1.6 mln cells (Grid A) and about 3 mln cells (Grid B). These data are cited in Tables 3 and 4 and shown in Figures 9 and 10. The modified SST model corrected within the RLI approach [35] was used in the present study.

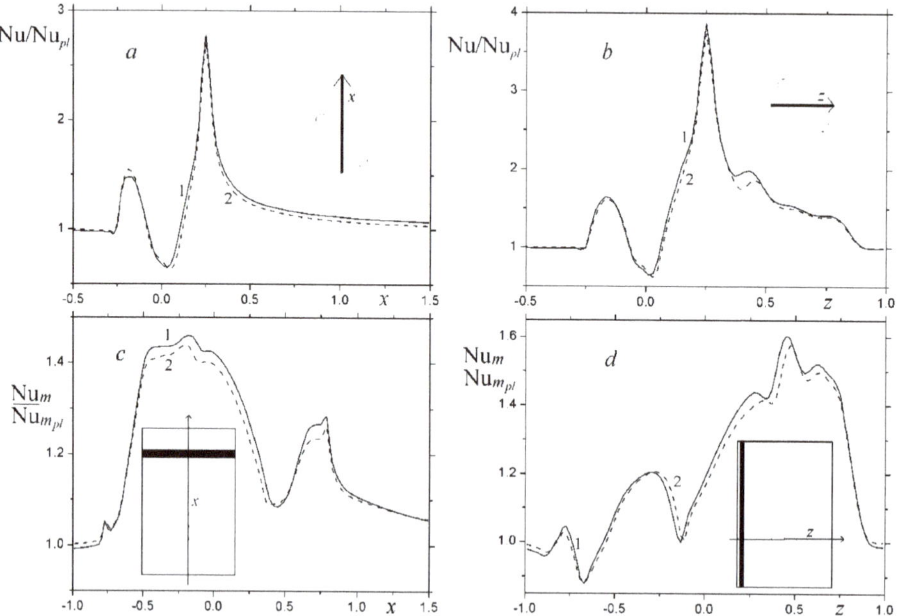

Figure 10. Distributions of relative local Nusselt numbers (**a,b**) and Nusselt numbers averaged over transverse (**c**) and longitudinal (**d**) strips of the rectangular section with the oval dimple (**c,d**) in longitudinal (**a,c**) and lateral (**b,d**) directions calculated on different grids: 1—initial variant (about 1.6 mln cells); 2—refinement (about 3 mln cells).

The comparisons of the numerical predictions of local and integral characteristics of flow and heat transfer in the channel with an oval-trench dimple on grids with about 1.6 mln cells and about 3 mln cells demonstrate their proximity. This means that the accuracy of the data for a moderate-depth grid is quite acceptable.

5. Results and Discussion

In the present study, main attention was paid both to the assessment of the influence of increase in the oval dimple length on fluid dynamics and heat transfer in the narrow channel with a dimple of fixed spot area and to the comparison of an oval dimple with spherical and conical dimples. The objective of the study was to select an oval dimple that is the best in the thermal and hydraulic performance. The work done was the embodiment of the concept of heat transfer enhancement by spiral vortices formed in oval-trench dimples in order to enhance secondary flow in the channel. The study as presented here clarifies and develops the research began in [63].

Figure 11 illustrates the temperature field variations at the heated wall of the narrow channel and the streamlines. First of all, it is of interest to note that in solving the thermal problem, where a fully developed isothermal flow at the channel inlet was assigned, heat transfer had time to stabilize in the vicinity of the dimple. As the dimple width decreased, the regions with a temperature close to characteristic and equal to 293 Kincreased. It can be seen that the flow structure near the dimple changes and the temperature field transformed. Very long decreased temperature regions appeared behind the dimple in comparison to the temperature regions at the plane wall of the narrow plane-parallel channel.

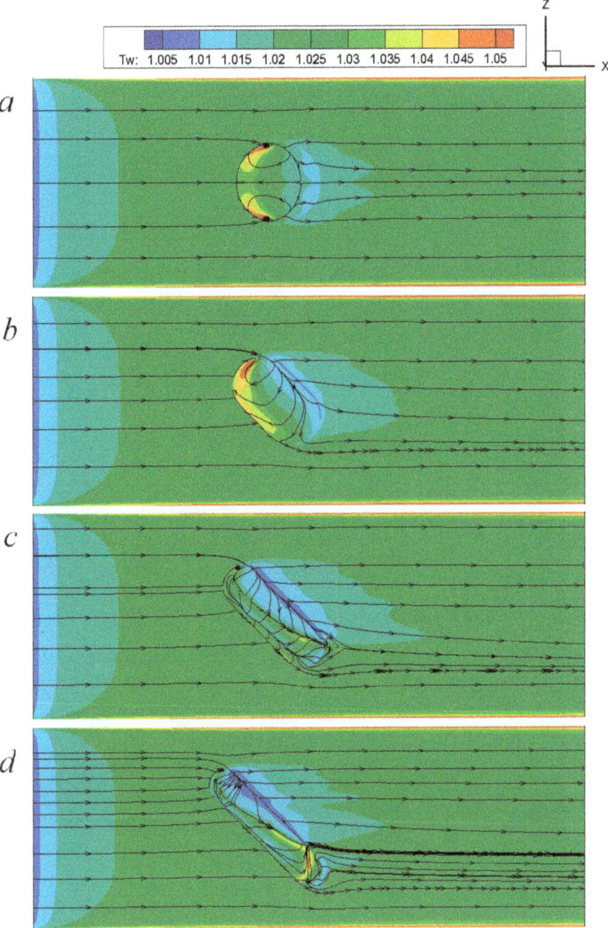

Figure 11. Temperature fields at the heated wall with oval dimples of width 1 (**a**), 0.731 (**b**), 0.549 (**c**), 0.429 (**d**) and the streamlines.

Figures 12–14 compare the temperature fields and dimple width variations for the sake of a careful analysis of the changes in the flow structure and their influence on heat transfer. It is noteworthy that when considering conical, spherical and oval dimples with cylindrical insert lengths of 0.5 and 0.625 (Figure 12), the increased wall temperature regions were related to the stagnated flow regions (leeward region of a conical dimple) and to the places, where vortices wer generated, on the sides of the spherical dimple and in the leeward edge vicinity of oval dimples.

All flow patterns demonstrated a separated flow within the entire space of dimples, although for the spherical dimple the flow was attached on the windward side rather far from the trailing edge. Varying the cylindrical insert length and the dimple width transformed the internal flow on the leeward side of the oval dimple. A line appeared that separated the separated backflow zone behind the leading edge from the flow issuing from the dimple.

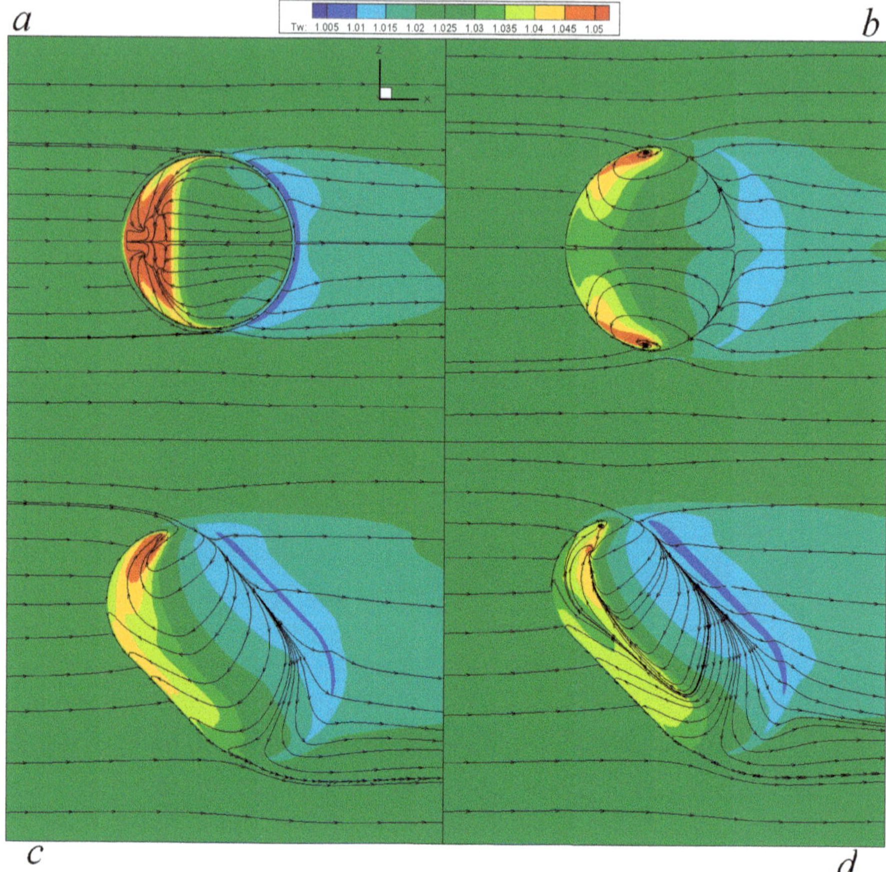

Figure 12. Temperature fields at the heated wall with conical (**a**) and oval dimples of width 1 (**b**), 0.731 (**c**), 0.678 (**d**) and the streamlines.

The oval dimple with the insert of length $L = 0.675$ (Figure 13a) is characteristic of two separating streamlines formed on the leeward side. In the dimple with the insert of length $L = 0.75$ (Figure 13b), a local vortex structure formed in the vicinity of the trailing edge and is accompanied by an increased temperature region. In the dimple with the cylindrical insert of length $L = 0.9$, a spiral vortex finally formed (Figure 13c). First, the separation zone was localized in the vicinity of the leading edge. Second, the backflow zone started forming behind the trailing edge. The dimple with the cylindrical insert of length $L = 1$ is one of the basic dimples analyzed; in fact, the flow around this dimple repeats flow around the dimple with $L = 0.9$ (Figure 13d). The dimple with the insert of length $L = 1.25$ (Figure 14a) revealed some flow instability. The flow in the vicinity of the trailing edge became separated in character. The spiral vortex formed in the dimple was ready to leave it. The flow structure changed in the dimple with the cylindrical insert of length $L = 1.5$ (Figure 14b).

It can be seen how the spiral vortex leaves the oval-trench dimple, not having reached the vicinity of the trailing edge. It is interesting to note that in this zone, a sink with an increased temperature region formed. Increasing the insert length L to 1.75 (Figure 14c) is accompanied by a further development of a secondary vortex zone in the vicinity of the trailing edge. The increased temperature region has rather large sizes. The largest oval dimple had a cylindrical insert of length $L = 2$ and a relative length of 6.78 (in terms of dimple width). As a whole, it is interesting to note that temperature fields in the vicinity of long dimples (L = 1.5, 1.75, 2) were similar, if their geometric sizes are considered in terms of dimple width. This is fairly indicative of the fact that a self-similar vortex is formed in the oval-trench dimple. The relative friction distributions $f/f_{pl}(s)$ in the central section of the oval-trench dimple at different cylindrical insert lengths L are shown in Figure 15a,b. The coordinate s is taken from the leading edge of the dimple. In Figure 12, the group of oval dimples of moderate length (L = 0.5–0.625) is characteristic of the formation of a large-scale separated flow zone covering almost the entire inner surface of dimples. A minimum relative friction value is −0.5, and a sharp friction increase (of order 2) is seen at the trailing edge of dimples.

Decreasing the width of the oval dimple, when its depth is increased and is kept constant, caused the flow structure to change in the separation zone. Backflow on the spherical portion of the dimple enhances while on the trench portion was slowed down. At L = 0.9 (dashed line), relative friction on the trench portion becomes positive, i.e., the separation zone was localized on the entrance portion of the oval-trench dimple, while $(f/f_{pl})_{min}$ decreased to −0.8.

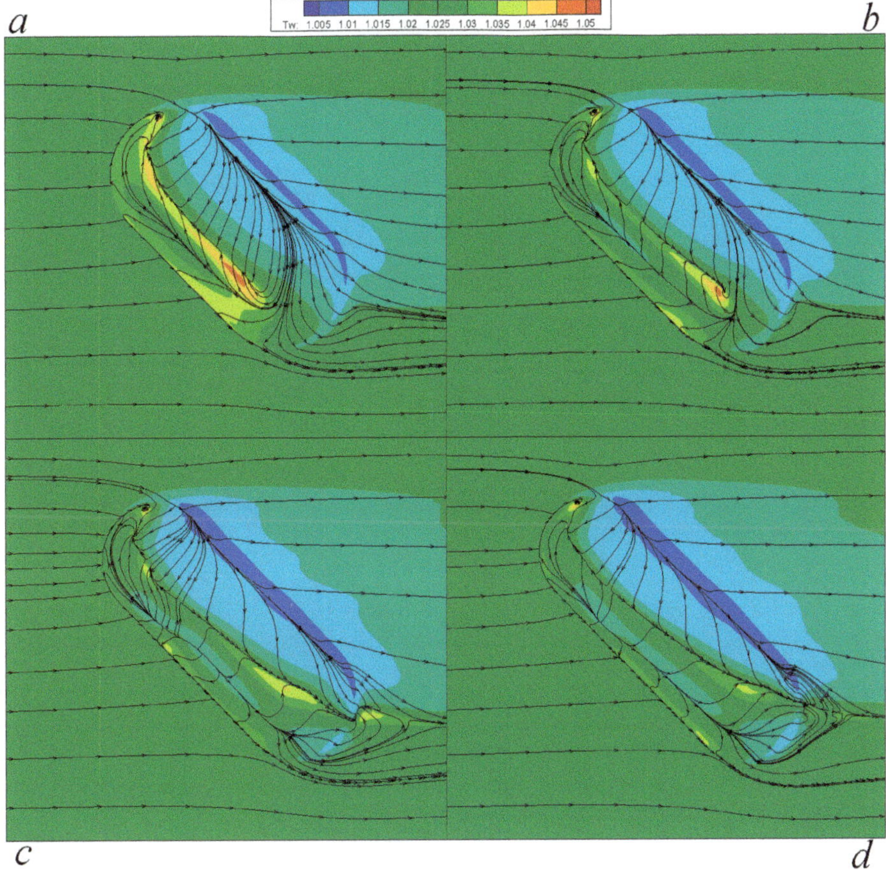

Figure 13. Temperature fields at the heated wall with oval dimples of width 0.659 (**a**), 0.631 (**b**), 0.58 (**c**) and 0.549 (**d**) and the streamlines.

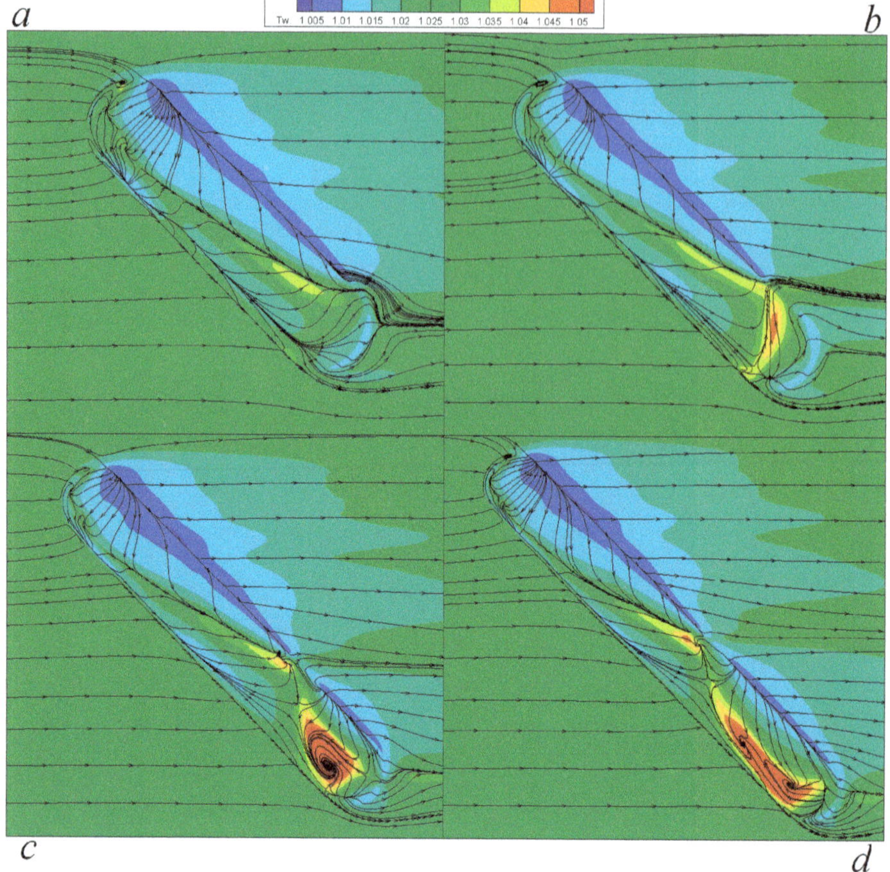

Figure 14. Temperature fields at the heated wall with oval dimples of width 0.482 (**a**), 0.429 (**b**), 0.383 (**c**) and 0.346 (**d**) and the streamlines.

As seen from Figure 15a,b, with a further increase in the dimple length, the separated flow intensity considerably enhances. At $L = 2$, $(f/f_{pl})_{min}$ decreases to -1.5. Behind the separation zone, the trench portion is characteristic of the flow acceleration zone with a local maximum relative friction value equal to 0.3–0.4.

As the width of the oval dimple is decreased, the relative Nusselt number distribution changes (Figure 16c,d). Similar to decreasing the minimum relative friction value in Figure 15a,b, the maximum value of Nu/Nu_{pl} increased in the vicinity of the spherical portion of the oval dimple from 1.2 to 1.6 when L is varied from 0.5 to 0.9. With a further increase in the dimple length, the maximum value of the Nusselt number (Nu/Nu_{pl}) in the separation zone reached 2.1 at $L = 1.5$–2.

In many ways, the behavior of the Nusselt number is defined by the wall temperature (Figure 15e,f). Thus, the mentioned minimum value of heat loads at $L = 1.75$ corresponds to a maximum value of T_w/T_{wpl} equal to 1.055. The fact is that high wall temperatures in the oval dimple with a moderate length of the cylindrical insert correlate with low heat transfer in separation zones. At the same time, the subcooling of the surface of the dimple, i.e., when the surface temperature decreases below the temperature of the plane-parallel channel wall and is accompanied by the growth of heat transfer.

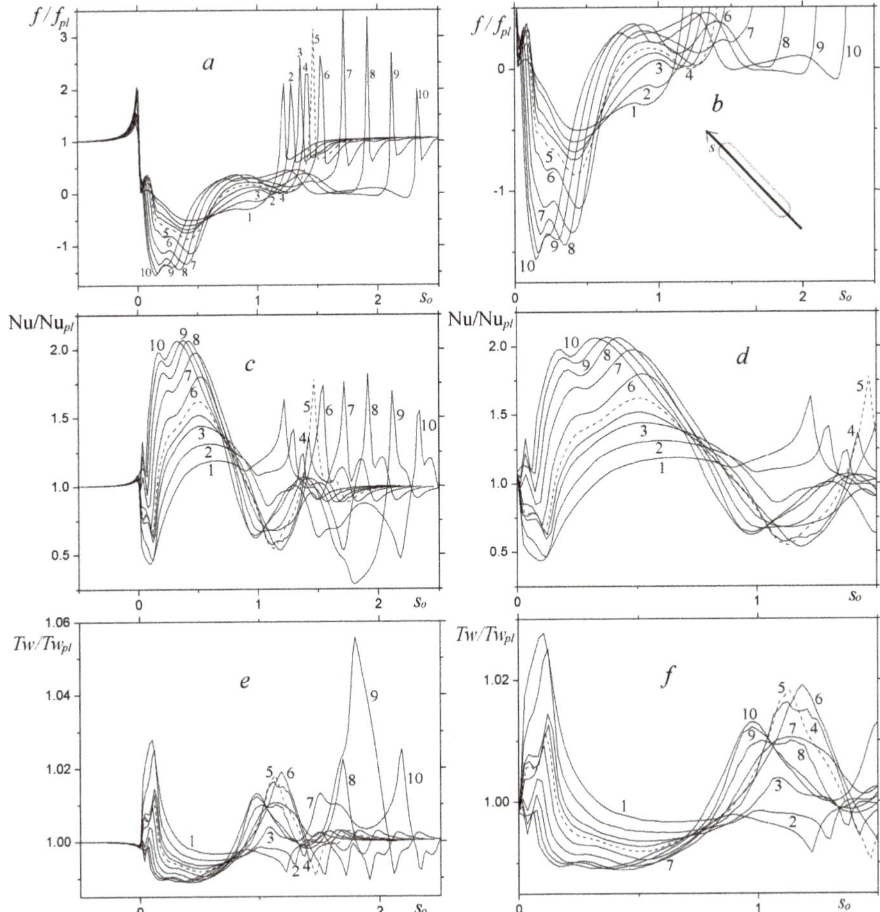

Figure 15. Comparison of the dependences of relative friction (**a,b**), Nusselt number (**c,d**), wall temperature (**e,f**) in the middle longitudinal cross sections for single oval dimples of different insert length: 1—L = 0.5; 2—0.625; 3—0.75; 4—0.825; 5—0.9; 6—1; 7—1.25; 8—1.5; 9—1.75; 10—2. b, d, f—enlarged fragments of the dependences.

As the length of the cylindrical insert of the oval dimple L is increased above 1.25, T_w/T_{wpl} becomes less than 1 in the separation zone. The region of dimple subcooling coincides with the zone of enhanced heat transfer and, vice versa, the increased temperature regions correspond to the low heat transfer zones.

To determine the influence of the oval dimple width b on the integral characteristics of flow and heat transfer is an important subject of the present study (Figure 16). With a decrease in b, when the dimple spot area is kept constant, the length of the cylindrical insert of the dimple is increased from 1 to 6.78, and therefore, the degree of influence of the dimple on flow in the near-wake. The total Nusselt number Nu_n is calculated on the control area of the rectangular section 3 in length and 2 in width (with a shift by 0.5 relative to the center downstream) around the dimple without and with regard to the increase in the curvilinear surface of the dimple. Figure 16 serves to illustrate the ratio of the total Nusselt number Nu_n in the section (in Table 3 the section is designated by 1) of the dimpled wall to the equivalent characteristic for the plane channel Nu_{npl}. Hydraulic losses are determined, as described in [32], in terms of the boundaries of the control section of the dimpled (ζ) and plane

(ζ_{pl}) channels (Figure 3). Thermal and hydraulic performance (THP) are calculated as the ratio of the thermal performance Nu_n/Nu_{npl} in the selected section to relative hydraulic losses ζ/ζ_{pl} at the boundaries of the section.

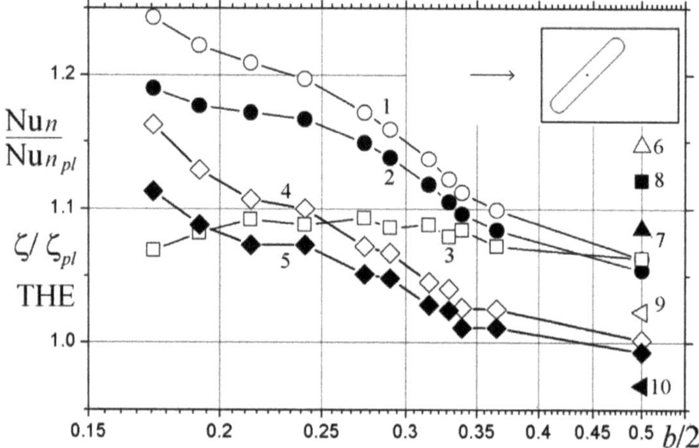

Figure 16. Influence of the oval dimple width b on thermal (1,2,6,7) and thermal-hydraulic (4,5,9,10) performances, as well as relative hydraulic losses (3,8); 6,7,8,9,10—conical dimple; 2,5,7,10—with regard to the area of the dimple inner surface in the 3 × 2 section and the shift of the dimple center by 1 from the front boundary.

With increasing the length of the cylindrical insert of the oval dimple to 6.78 (in terms of width), the thermal and hydraulic performance of the rectangular section of the dimpled channel were dramatically improved in comparison to the spherical dimple: THP = 1.17 versus 1.002; for the last THP with the consideration of increase in the area of the streamlined wall of the channel, it was less than 1.

For the rectangular section with longer oval dimples, the rate of increasing the thermal performance was considerably ahead of increase in hydraulic losses. The thermal performance of the dimple with $L = 2$ was six times higher than that of the spherical dimple with no regard to the area of the inner surface of dimples and was four times more preferred with regard to the area of the streamlined wall (Nu_n/Nu_{npl} = 1.245 versus 1.063 and 1.19 versus 1.054, respectively).

Hydraulic losses in the section with the oval dimple were maximum at the dimple width $b = 0.549$ (the cylindrical insert length was equal to 1) that exceed by a factor of 1.5 hydraulic losses in the case of the spherical dimple. Hydraulic losses in the section with a narrow dimple with $L = 2$ appeared to be the smallest and practically equal to hydraulic losses in the section with a basic spherical dimple.

6. Conclusions

1. The analysis of longer oval dimples located at a 45° angle of orientation to the flow in the channel showed that methodologically, it was important to fix a spot area of a dimple and its depth for the same channel.

2. Tasks of hydrodynamics and heat transfer were solved with the use of original MCTs on different-scale structured overlapping grids of simple topology. These technologies meant for solution of RANS—steady Reynolds-averaged Navier–Stokes equations—were implemented in the VP2/3 code and were tested in the present study using the turbulence models and boundary conditions for heat transfer.

3. Testing calculations:

a. Testing was performed on the experimental setup [9] for determining characteristics of convective heat transfer near a heated spherical dimple of depth 0.13 over the Re number range 10^4–10^5 (MCTs) with the use of the VP2/3 code and the shear stress transport (SST) model [34] with curvature correction within the Rodi–Leschziner–Isaev (RLI) approach [36]. A fair agreement between numerical predictions and measurement data was obtained.

b. The comparison of the boundary conditions T = const and q = const in the problem on heat transfer in the vicinity of a shallow spherical dimple in the narrow channel showed that integral characteristics of the thermal and hydraulic performance of the channel with a spherical dimple practically are independent of the type of boundary conditions for heat transfer within the turbulent flow regime. However, local distributions are substantially different in the near-edge zone and in the dimple center. Difference in maximum relative local heat transfer values is 1.5; at the same time, loads against the near-edge zone are much higher at T = const than at q = const.

c. The comparison of the SST models [34,35] and the SST model [36,39] with curvature correction within the Rodi–Leschziner–Isaev (RLI) approach and the Smirnov–Menter (SM) approach showed that the numerical predictions of integral characteristics are pretty close according to the standard and modified SST models. Some differences in the SST model [35] are small, but noticeable, especially in zones of extreme local heat fluxes. However, it was seen that the values of k and Re_{vt} in the spiral vortex core were too high. This is indicative of the fact that error is available in the standard SST model [34] in the calculation of high-intensity 3D separated flows.

d. The computational algorithm was validated by comparing numerical predictions for local and integral characteristics of flow and heat transfer in the channel with an oval-trench dimple that were obtained on the grids with 1.6 mln cells and about 3 mln cells. Their fair agreement shows that the data for the dimple with a moderate cylindrical insert length are quite acceptable in accuracy.

4. We revealed a series of oval-trench dimples with the cylindrical insert length 0.625–0.9, in which the separated flow structure gradually changed, the separation zone was localized behind the leading edge and backflow enhanced in it. The oval-trench dimple becomes flowing and non-separated throughout behind the separation zone.

Author Contributions: Conceptualization, S.I. and A.L.; Methodology, S.I. and A.S.; Software, S.I.; Validation, S.I., Y.C. and I P.; Formal Analysis, S.I.; Investigation, S.I. and D.N.; Resources, A.S.; Data Curation, S.I. and D.N.; Writing-Original Draft Preparation, S.I. and D.N.; Writing-Review & Editing, Y.C. and I.P.; Visualization, I.P.; Supervision, S.I. and A.L.; Project Administration, S.I.; Funding Acquisition, S.I.

Funding: This research was funded by the Russian Scientific Foundation grant number 19-19-00259.

Acknowledgments: The authors thank Professor Victor Terekhov for the discussion of the project results.

Conflicts of Interest: The authors declare no conflict of interest.

References

1. Dreitser, G.A. Problems in developing highly efficient tubular heat exchangers. *Therm. Eng.* **2006**, *53*, 279–287. [CrossRef]

2. Hagen, R.L.; Danak, A.M. Heat transfer in the field of the turbulent boundary layer separation over a dimple. *Heat Transf.* **1967**, *4*, 62–69.

3. Presser, K.H. Empirische gleichungen zur berechnung der stoff–und warmeubertragung fur den spezialfal der abgerissenen stromung. *Int. J. Heat Mass Transf.* **1972**, *15*, 2447–2471. [CrossRef]

4. Hiwada, M.; Kawamura, T.; Mabuchi, J.; Kumada, M. Some characteristics of flow pattern and heat transfer past a circular cylinder cavity. *Bull. JSME* **1983**, *26*, 1744–1758. [CrossRef]

5. Snedeker, S.; Donaldson, D.P. Observation of bistable flow in hemispherical cavity. *AIAA J.* **1966**, *4*, 735–736.

6. Gromov, P.R.; Zobnin, A.B.; Rabinovich, M.I.; Sushchik, M.M. Creation of solitary vortices in a flow around shallow spherical depressions. *Sov. Tech. Phys. Lett.* **1986**, *12*, 1323–1328.

7. Kesarev, V.S.; Kozlov, A.P. Convective heat transfer in turbulized flow past a hemispherical cavity. *Heat Transf. Sov. Res.* **1993**, *25*, 156–160.

8. Syred, N.; Khalatov, A.; Kozlov, A.; Shchukin, A.; Agachev, R. Effect of surface curvature on heat transfer and hydrodynamics within a single hemispherical dimple. *ASME J. Turbomach.* **2001**, *123*, 609–613. [CrossRef]

9. Terekhov, V.I.; Kalinina, S.V.; Mshvidobadze, Y.M. Pressure field and resistance of a single cavity with sharp and rounded edges. *J. Appl. Mech. Tech. Phys.* **1993**, *34*, 331–338. [CrossRef]

10. Terekhov, V.I.; Kalinina, S.V.; Mshvidobadze, Y.M. Heat transfer coefficient and aerodynamic resistance on a surface with a single dimple. *Enhanc. Heat Transf.* **1997**, *4*, 131–145. [CrossRef]

11. Kiknadze, G.I.; Gachechiladze, I.A.; Gorodkov, A.Y. Self-organization of tornado-like jets in flows of gases and liquids and the technologies utilizing this phenomenon. In Proceedings of the 2009 ASME Summer Heat Transfer Conference, San Francisco, CA, USA, 19–23 July 2009. Paper No. HT 2009–88644.

12. Ligrani, P.M.; Oliveira, M.M.; Blaskovich, T. Comparison of heat transfer augmentation techniques. *AIAA J.* **2003**, *41*, 337–362. [CrossRef]

13. Afanasyev, V.N.; Chudnovsky, Y.P.; Leontiev, A.I.; Roganov, P.S. Turbulent ow friction and heat transfer characteristics of spherical cavities on a plate. *Exp. Therm. Fluid Sci.* **1993**, *7*, 1–8. [CrossRef]

14. Chyu, M.K.; Yu, Y.; Ding, H. Heat transfer enhancement in rectangular channels with concavities. *Enhanc. Heat Transf.* **1999**, *6*, 429–439. [CrossRef]

15. Turnow, J. Flow Structure and Heat Transfer on Dimpled Surfaces. Ph.D. Thesis, University of Rostock, Rostock, Germany, 2011.

16. Ligrani, P.M.; Harrison, J.L.; Mahmmod, G.I.; Hill, M.L. Flow structure due to dimple depressions on a channel surface. *Phys. Fluids* **2001**, *13*, 3442–3451. [CrossRef]

17. Mahmood, G.I.; Sabbagh, M.Z.; Ligrani, P.M. Heat transfer in a channel with dimples and protrusions on opposite walls. *J. Thermophys. Heat Transf.* **2001**, *15*, 275–283. [CrossRef]

18. Ligrani, P.M.; Mahmood, G.I.; Harrison, J.L.; Clayton, C.M.; Nelson, D.I. Flow structure and local Nusselt number variations in a channel with dimples and protrusions on opposite walls. *Int. J. Heat Mass Transf.* **2001**, *45*, 2011–2020. [CrossRef]

19. Ekkad, S.V.; Nasir, H. Dimple enhanced heat transfer in high aspect ratio channels. *J. Enhanc. Heat Transf.* **2003**, *10*, 395–405. [CrossRef]

20. Mahmood, G.I.; Ligrani, P.M. Heat transfer in a dimpled channel combined influences of aspect ratio, temperature, Reynolds number and flow structure. *Int. J. Heat Mass Transf.* **2004**, *45*, 2011–2020. [CrossRef]

21. Hwang, S.D.; Kwon, H.G.; Cho, H.H. Heat transfer with dimple/protrusion arrays in a rectangular duct with a low Reynolds number range. *Int. J. Heat Fluid Flow* **2008**, *29*, 916–926. [CrossRef]

22. Mityakov, V.Y.; Mityakov, A.V.; Sapozhnikov, S.Z.; Isaev, S.A. Local heat fluxes on the surfaces of dimples. Ditches and cavities. *Therm. Eng.* **2007**, *54*, 200–203. [CrossRef]

23. Voskoboinik, A.V.; Voskoboinik, V.A.; Isaev, S.A.; Zhdanov, V.L.; Kornev, N.V.; Turnow, J. Bifurcation of vortex flow inside a spherical dimple in the narrow channel. *Appl. Hydromech.* **2011**, *13*, 3–21. (In Russian)

24. Voskoboinick, V.; Kornev, N.; Turnow, J. Study of Near Wall Coherent Flow Structures on Dimpled Surfaces Using Unsteady Pressure Measurements. *Flow Turbul. Combust.* **2013**, *90*, 709–722. [CrossRef]

25. Mitsudharmadi, H.; Tay, C.M.J.; Tsai, H.M. Effect of rounded edged dimple arrays on the boundary layer development. *Vis. Soc. Jpn. J. Vis.* **2009**, *12*, 17–25. [CrossRef]

26. Tay, C.M.; Chew, Y.T.; Khoo, B.C.; Zhao, J.B. Development of flow structures over dimples. *Exp. Therm. Fluid Sci.* **2014**, *52*, 278–287. [CrossRef]

27. Xiao, N.; Zhang, Q.; Ligrani, P.M.; Mongia, R. Thermal performance of dimpled surfaces in laminar flows. *Int. J. Heat Mass Transf.* **2009**, *52*, 2009–2017. [CrossRef]

28. Kovalenko, G.V.; Terekhov, V.I.; Khalatov, A.A. Flow regimes in a single dimple on the channel surface. *J. Appl. Mech. Tech. Phys.* **2010**, *51*, 839–848. [CrossRef]

29. Kwon, H.G.; Hwang, S.D.; Cho, H.H. Measurement of local heat/mass transfer coefficients on a dimple using naphthalene sublimation. *Int. J. Heat Mass Transf.* **2011**, *54*, 1071–1080. [CrossRef]

30. Lan, J.; Xie, Y.; Zhang, D. Effect of leading edge boundary layer thickness on dimple flow structure and separation control. *J. Mech. Sci. Technol.* **2011**, *25*, 3243–3251. [CrossRef]

31. Heo, S.-C.; Seo, Y.-H.; Ku, T.-W.; Kang, B.-S. Formability evaluation of dimple forming process based on numerical and experimental approach. *J. Mech. Sci. Technol.* **2011**, *25*, 429–439. [CrossRef]

32. Isaev, S.A.; Kornev, N.V.; Leontiev, A.I.; Hassel, E. Influence of the Reynolds number and the spherical dimple depth on the turbulent heat transfer and hydraulic loss in a narrow channel. *Int. J. Heat Mass Transf.* **2010**, *53*, 178–197. [CrossRef]

33. Isaev, S.A.; Schelchkov, A.V.; Leontiev, A.I.; Baranov, P.A.; Gulcova, M.E. Numerical simulation of the turbulent air flow in the narrow channel with a heated wall and a spherical dimple placed it for vortex heat transfer enhancement depending on the dimple depth. *Int. J. Heat Mass Transf.* **2016**, *94*, 426–448. [CrossRef]

34. Menter, F.R. Zonal two equation k–ω turbulence models for aerodynamic flows. In Proceedings of the 23rd Fluid Dynamics, Plasmadynamics, and Lasers Conference, Orlando, FL, USA, 6–9 July 1993. AIAA Paper No. 93-2906.

35. Menter, F.R.; Kuntz, M.; Langtry, R. Ten years of industrial experience with the SST turbulence model. In *Turbulence, Heat and Mass Transfer 4*; Hajalic, K., Nogano, Y., Tummers, M., Eds.; Begell House Inc.: Danbury, CT, USA, 2003; 8p.

36. Isaev, S.A.; Baranov, P.A.; Zhukova, Y.V.; Usachov, A.E.; Kharchenko, V.B. Correction of the shear-stress-transfer model with account for the curvature of streamlines in calculating separated flows of an incompressible viscous fluid. *J. Eng. Phys.* **2014**, *87*, 1002–1015. [CrossRef]

37. Launder, B.E.; Spalding, D.B. The numerical computation of turbulent flow. *Comp. Methods Appl. Mech. Eng.* **1974**, *3*, 269–289. [CrossRef]

38. Leschziner, M.; Rodi, W. Calculation of annular and twin parallel jets using various discretization schemes and turbulence-model variations. *Trans. ASME. J. Fluids Eng.* **1981**, *103*, 352–365. [CrossRef]

39. Smirnov, P.E.; Menter, F. Sensitization of the SST turbulence model to rotation and curvature by applying the Spalart-Shur correction term. *J. Turbomach.* **2009**, *131*, 041010. [CrossRef]

40. Spalart, P.R.; Shur, M.L. On the sensitization of turbulence models to rotation and curvature. *Aerosp. Sci. Technol.* **1997**, *1*, 297–302. [CrossRef]

41. Menter, F.; Ferreira, J.C.; Esch, T.; Konno, B. Turbulence model with improved wall treatment for heat transfer predictions in gas turbines. In Proceedings of the Internatinal Gas Turbine Congress, Tokyo, Japan, 2–7 November 2003.

42. Isaev, S.A.; Baranov, P.A.; Usachov, A.E. *Multiblock Computational Technologies in the VP2/3 Package on Aerothermodynamics*; LAP LAMBERT Academic Publishing: Saarbrucken, Germany, 2013; 316p.

43. Ferziger, J.H.; Peric, M. *Computational Methods for Fluid Dynamics*; Springer Science & Business Media: Berlin/Heidelberg, Germany, 1999; 389p.

44. Van Doormaal, J.P.; Raithby, G.D. Enhancement of the SIMPLE method for predicting incompressible fluid flow. *Numer. Heat Transf.* **1984**, *7*, 147–163. [CrossRef]

45. Jasak, H. Error Analysis and Estimation for the Finite Volume Method with Applications to Fluid Flows. Ph.D. Thesis, University of London and Diploma of Imperial College of Science, Technology and Medicine, London, UK, 1996.

46. Isaev, S.A.; Kudryavtsev, N.A.; Sudakov, A.G. Numerical modeling of a turbulent incompressible viscous flow along bodies of a curvilinear shape in the presence of a mobile shield. *J. Eng. Phys. Thermophys.* **1998**, *71*, 613–626.

47. Rhie, C.M.; Chow, W.L. A numerical study of the turbulent flow past an isolated airfoil with trailing edge separation. *AIAA J.* **1983**, *21*, 1525–1532. [CrossRef]

48. Pascau, A.; Garcia, N. Consistency of SIMPLEC scheme in collocated grids. In Proceedings of the V European Conference on Computational Fluid Dynamics ECCOMAS CFD 2010, Lisbon, Portugal, 14–17 June 2010.

49. Leonard, B.P. A stable and accurate convective modeling procedure based on quadratic upstream interpolation. *Comp. Methods Appl. Mech. Eng.* **1979**, *19*, 59–98. [CrossRef]

50. Van Leer, B. Towards the ultimate conservative difference scheme V. A second order sequel to Godunov's method. *J. Comp. Phys.* **1979**, *32*, 101–136. [CrossRef]

51. Saad, Y. *Iterative Methods for Sparse Linear Systems*, 2nd ed.; Society for Industrial and Applied Mathematics: Philadelphia, PA, USA, 2003; 567p.

52. Demidov, D. AMGCL: C++ Library for Solving Large Sparse Linear Systems with Algebraic Multigrid Method. Available online: http://amgcl.readthedocs.org (accessed on 3 April 2019).

53. Isaev, S.A.; Zhdanov, V.L.; Niemann, H.-J. Numerical study of the bleeding effect on the aerodynamic characteristics of a circular cylinder. *J. Wind Eng. Ind. Aerodyn.* **2002**, *90*, 1217–1226. [CrossRef]

54. Isaev, S.A.; Sudakov, A.G.; Baranov, P.A.; Zhukova, Y.V.; Usachov, A.E. Analysis of errors of multiblock computational technologies by the example of calculating a circulation flow in a square cavity with a moving cover at Re = 1000. *J. Eng. Phys.* **2013**, *86*, 1134–1150.

55. Zheng, Y.; Liou, M.-S. A novel approach of three-dimensional hybrid grid methodology: Part 1. Grid generation. *Comput. Methods Appl. Mech. Eng.* **2003**, *192*, 4147–4171. [CrossRef]

56. Isaev, S.A.; Leontiev, A.I.; Gul'tsova, M.E.; Popov, I.A. Transformation and intensification of tornado-like flow in a narrow channel during elongation of an oval dimple with constant area. *Tech. Phys. Lett.* **2015**, *41*, 606–609. [CrossRef]

57. Isaev, S.A.; Leontiev, A.I. Numerical simulation of vortex enhancement of heat transfer under conditions of turbulent floe past a spherical dimple on the wall of a narrow channel. *High Temp.* **2003**, *41*, 665–679. [CrossRef]

58. Isaev, S.A.; Baranov, P.A.; Kudryavtsev, N.A.; Lisenko, D.A.; Usachov, A.E. Complex analysis of turbulence models, algorithms, and grid structures at the computation of recirculating flow in a cavity by means of VP2/3 and FLUENT packages. Part. 1. Scheme factors influence. *Thermophys. Aeromech.* **2005**, *12*, 549–569.

59. Idelchik, I.E. *Handbook on Hydraulic Resistance*; Mashinostroyeniye: Moscow, Russia, 1992; 672p.

60. Isaev, S.A.; Baranov, P.A.; Usachov, A.E.; Zhukova, Y.V.; Vysotskaya, A.A.; Malyshkin, D.A. Simulation of the turbulent air flow over a circular cavity with a variable opening angle in an U-shaped channel. *J. Eng. Phys. Thermophys.* **2015**, *88*, 902–917. [CrossRef]

61. Isaev, S.A.; Baranov, P.A.; Zhukova, Y.V.; Kalinin, E.I.; Miau, J.J. Verification of the shear-stress transfer model and its modifications in the calculation of a turbulent flow around a semicircular airfoil with a zero angle of attack. *J. Eng. Phys. Thermophys.* **2016**, *89*, 73–89. [CrossRef]

62. Isaev, S.; Baranov, P.; Popov, I.; Sudakov, A.; Usachov, A.; Guvernyuk, S.; Sinyavin, A.; Chulyunin, A.; Mazo, A.; Demidov, D. Ensuring safe descend of reusable rocket stages–numerical simulation and experiments on subsonic turbulent air flow around a semi-circular cylinder at zero angle of attack and moderate Reynolds number. *Acta Astronaut.* **2018**, *150*, 117–136. [CrossRef]

63. Isaev, S.A.; Schelchkov, A.V.; Leontiev, A.I.; Gortyshov, Y.F.; Baranov, P.A.; Popov, I.A. Tornado-like heat transfer enhancement in the narrow plane-parallel channel with the oval-trench dimple of fixed depth and spot area. *Int. J. Heat Mass Transf.* **2017**, *109*, 40–62. [CrossRef]

Article

How Big Is an Error in the Analytical Calculation of Annular Fin Efficiency?

Mladen Bošnjaković [1,*], Simon Muhič [2], Ante Čikić [3] and Marija Živić [4]

1   Technical Department, College of Slavonski Brod, Dr. M. Budaka 1, 35000 Slavonski Brod, Croatia
2   Faculty of Mechanical Engineering, University of Novo mesto, Na Loko 2, 8000 Novo mesto, Slovenia; simon.muhic@fs-unm.si
3   Department of Mechanical Engineering, University North, 104. brigade 3, 42000 Varaždin, Croatia; acikic@unin.hr
4   Mechanical Engineering Faculty in Slavonski Brod, Josip Juraj Strossmayer University of Osijek, Trg Ivane Brlić Mažuranić 2, 35000 Slavonski Brod, Croatia; mzivic@sfsb.hr
*   Correspondence: mladen.bosnjakovic@vusb.hr

Received: 14 April 2019; Accepted: 8 May 2019; Published: 10 May 2019

Abstract: An important role in the dimensioning of heat exchange surfaces with an annular fin is the fin efficiency. The fin efficiency is usually calculated using analytical expressions developed in the last century. However, these expressions are derived with certain assumptions and simplifications that involve a certain error in the calculation. The purpose of this paper is to determine the size of the error due to the assumptions and simplifications made when performing the analytical expression and to present what has the greatest impact on the amount of error, and give a recommendation on how to reduce that error. In order to determine the error, but also to gain a more detailed insight into the physics of heat exchange processes on the fin surface, computational fluid dynamics was applied to the original definition of fin efficiency. This means that a numerical simulation was performed for the actual fin material and for the ideal fin material with infinite thermal conductivity for the selected fin geometry and *Re* numbers from 2000 to 18,000. The results show that fin efficiency determined by numerical simulations is greater by up to 12.3% than the efficiency calculated analytically. The greatest impact on the amount of error is the assumption of the same temperature of the fin base surface and the outer tube surface and the assumption of equal heat transfer coefficient on the entire fin surface area. Using a newly recommended expression for the equivalent length of the fin tip, it would be possible to calculate the fin efficiency more precisely and thus the average heat transfer coefficient on the fin surface area, which leads to a more accurate dimensioning of the heat exchanger.

Keywords: efficiency of annular fin; analytical and numerical method; computational fluid dynamics; fin base temperature

---

1. Introduction

Gas-liquid type heat exchangers are very widespread in industry. In order to increase the heat transfer, i.e., the reduction of the heat exchanger dimension, finned surfaces of different shapes have been developed. However, annular fins are most commonly used. Designing a heat exchanger with such finned surfaces is done on the basis of analytical solutions developed at the beginning of the 20th century. The first to analyse heat transfer on extended surfaces were Harper and Brown [1] in 1922. They introduced the concept of fin efficiency and provided analytical solutions for the two-dimensional model for a longitudinal and annular fin of uniform thickness. Also, they proposed the use of an adiabatic fin tip model with corrected fin height for the fin tip heat loss, which increases the fin height by half of the fin thickness.

In 1938, Murray [2] presented equations for the temperature gradient and fin effectiveness under conditions of a symmetrical temperature distribution around the radial fin base of uniform thickness. He also proposed a set of assumptions for extended surface analysis [2]. These limiting assumptions, which are referred to as the Murray–Gardner assumptions, are:

(a)   All temperatures and heat flow operate at steady-state levels.
(b)   The contact thermal resistance between the fin and the base surface is negligible.
(c)   There are no heat sources in the fin itself.
(d)   Radiation heat transfer from the fin is neglected.
(e)   The fin thickness is so small compared to its height that temperature gradients normal to the surface may be neglected.
(f)   The thermal conductivity of the fin and tube material is temperature-independent and uniform.
(g)   The temperature of the surrounding fluid is spatially uniform.
(h)   The heat-transfer coefficient is the same over all fin surfaces.
(i)   The temperature of the fin base surface is uniform and equal to the temperature of the tube base surface.

In 1945, Gardner [3] derived general equations for the temperature profile and fin efficiency for any form of extended surface for which the Murray–Gardner assumptions are applicable. Almost the same assumptions are applied by Darvishi at al. [4] in the numerical investigation for a hyperbolic annular fin with temperature-dependent thermal conductivity.

Ghai [5] investigated the validity of Gardner´s assumption of the constant heat transfer coefficient over the fin face. He has experimentally discovered large differences in the heat transfer coefficient from fin base to fin tip and along the fin surface in the air flow direction.

Based on the assumptions and the results of the research in the above-mentioned works, numerous analytical, experimental and numerical studies of heat transfer on finned surfaces have been performed for different geometries, materials and flow conditions. Heat transfer and the temperature distribution over the fin surface of the finned tube geometry with different fin perforation shape was investigated by Maki H. Zaidan et al. [6], applying a numerical analysis.

Nemati and Moghimi [7] studied the influence of different models of turbulence on the flow simulation that passes through a four-row tube heat exchanger. They found that K-Kl-ω, Transition SST and SST k-ω turbulence model show good match with experimental correlations.

Look and Krishnan [8] presented exact expressions for the correct terms for a one-dimensional fin temperature and heat loss when the insulated fin tip assumption is used for the case where the tip is maintained at a constant temperature.

Nemati and Samivand [9] have been numerically studying efficiency of the annular elliptical fin. They have proposed a simple correlation to approximate efficiency of the annular elliptical fin, applying the classic assumptions in determining the efficiency of annular fins.

Chen and Wang [10], in their trapezoid fin pattern research, apply standard expression for fin efficiency which does not take into account the temperature drop on the base surface of the fin.

H. J. Lane and P. J. Heggs [11] developed expressions for temperature profiles along the fins of different shapes: cone, trapezoid and dovetail fin, also with standard assumptions.

Cathal T. Ó Cléirigh, and William J. Smith [12] considered three shapes of spiral fins: full, partially segmented and fully segmented. The stationary state model with two equations of the turbulent model was chosen and the flow in Reynolds numbers 5000 to 30,000, respectively. The Nusselt number and the pressure drop calculated with the CFD software were compared with the correlations in the experimental literature.

The question is: what is the size of error in the calculation of the fin efficiency due to the above assumptions and simplifications applied to the physical model of heat exchange with fin surfaces? And, can this error be reduced? In this regard, the subjects of interest in this paper are the assumptions mentioned under (h) and (i) and their impact on the accuracy of the calculation of the fin efficiency.

It is clear that if fins increase the heat transfer, then heat must be transferred by conduction to the fin base surface from remote regions of the tube, and this conductive transport needs a temperature drop. Conduction through the fin cross section and convective heat transfer over the fin surface area are carried out at the same time. This is possible only if there is a temperature difference between the fin base surface and the fin tip as a driving force. Because of this, the fins have an average temperature lower than that of the unfinned part of the tube. Therefore, the efficiency of the fins is different to that of the unfinned part of the tube because the heat transfer between fin surface area and surrounding fluid takes place with a smaller temperature difference than between the unfinned part of the tube surface area and the fluid.

The assumption is that the magnitude of the error can be determined by applying numerical analysis to the original definition of the fin efficiency, which also provides a more detailed insight into the physics of the heat exchange process on finned surfaces.

The idea is to obtain quantitative data on temperature and heat exchanged on the particular surfaces of the annular fin of the constant thickness for the "ideal fin" and the real fin by numerical analysis. After that, the fin efficiency is calculated by using numerical simulation data, conventional analytical equations and Murray–Gardner assumptions. By comparing the results, it is possible to calculate the magnitude of the error resulting from the application of common expressions and assumptions. To the authors' knowledge, fin efficiency analysis has not been performed in this way so far.

2. Materials and Methods

2.1. Physical Basis of Heat Exchange on Finned Surfaces

In order to present the impact of the Murray–Gardner assumptions and simplifications, the physical model of heat exchange on the annular fin is presented and analysed in detail. Over the fin surface and the outer tube wall surface, heat is transferred by convection to the surrounding fluid. Specific consideration is given to the case with the same fin and tube material, i.e., same thermal conductivity. In theoretical consideration, the average temperature of the tube base area is T_t and the average temperature of the fin base area is T_b. The total heat transfer rate is equal to the sum of the heat transfer rate through the surface area of the fins $\dot{Q}_f$ and through the unfinned surface area of the tube $\dot{Q}_t$:

$$\dot{Q} = \dot{Q}_t + \dot{Q}_f.$$

(1)

Heat transfer rate through the fin surface consists of heat transfer rate through the fin face surfaces $\dot{Q}_{f,f}$ and through the fin tip surface $\dot{Q}_{f,t}$ as shown in Figure 1. The temperature in the fin falls from T_b at the fin base surface to $T_{f,t}$ at its tip.

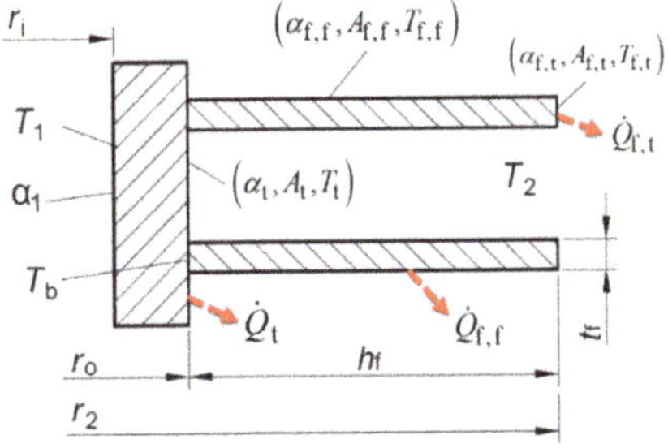

Figure 1. Characteristic dimensions of the finned surface.

The heat transfer rate $\dot{Q}_t$ from the finless tube surface A_t in the fluid is defined with Equation (2).

$$\dot{Q}_t = \alpha_t\, A_t\, (T_t - T_2),\tag{2}$$

where α_t is the associated average heat transfer coefficient. The heat transfer rate $\dot{Q}_f$ from fin surfaces A_f to the fluid is defined with Equations (3) to (5) (Figure 1).

$$\dot{Q}_f = \dot{Q}_{f,f} + \dot{Q}_{f,t},\tag{3}$$

$$\dot{Q}_{f,f} = \alpha_{f,f}\, A_{f,f}\, (T_{f,f} - T_2),\tag{4}$$

$$\dot{Q}_{f,t} = \alpha_{f,t}\, A_{f,t}\, (T_{f,t} - T_2).\tag{5}$$

The total heat transfer rate from fin surfaces to the fluid is defined with Equation (6).

$$\dot{Q}_f = \alpha_{f,f}\, A_{f,f}\, (T_{f,f} - T_2) + \alpha_{f,t}\, A_{f,t}\, (T_{f,t} - T_2).\tag{6}$$

The total fin surface area is the sum of the fin face surface area and the fin tip surface area:

$$A_f = A_{f,f} + A_{f,t}\,.\tag{7}$$

The average heat transfer coefficient for all fin surfaces can be defined as:

$$\alpha_f = \frac{\alpha_{f,f} \cdot A_{f,f} + \alpha_{f,t} \cdot A_{f,t}}{A_{f,f} + A_{f,t}}.\tag{8}$$

The average temperature for all fin surfaces can be defined as:

$$T_f = \frac{T_{f,f} \cdot A_{f,f} + T_{f,t} \cdot A_{f,t}}{A_{f,f} + A_{f,t}}.\tag{9}$$

The heat transfer rate from fin to the fluid applying Equations (8) and (9) will be:

$$\dot{Q}_f = \alpha_f\, A_f\, (T_f - T_2).\tag{10}$$

The fin's thermal performance is expressed by the fin efficiency. The maximum driving potential for convection is the difference between the fin base temperature T_b and the bulk fluid temperature T_2. Hence, the fin efficiency is defined as the ratio of the actual fin heat flux to its heat flux if the entire surface of ideal nonexisting fin were at the same temperature as its base surface area, where the fin material has an infinite thermal conductivity. However, since any fin has a finite conductivity resistance, a temperature gradient must exist along the fin. Definition of the fin efficiency therefore can be expressed as:

$$\eta_f = \frac{\dot{Q}_f}{\dot{Q}_{f,max}}.\tag{11}$$

The fin efficiency is an idealisation and has physically no meaning. An ideal fin has infinite thermal conductivity λ_f of the fin material and the same geometry and operating conditions as the actual fin. Under these assumptions, the ideal fin is at the uniform base surface temperature T_b so $T_b \approx T_t = T_{f,f} = T_{f,t}$ and the heat transfer coefficient $\alpha_{t,\lambda=\infty} \approx \alpha_t$, $\alpha_{f,\lambda=\infty} \approx \alpha_f$, $\alpha_{f,f,\lambda=\infty} \approx \alpha_{f,f}$, $\alpha_{f,t,\lambda=\infty} \approx \alpha_{f,t}$. We could write:

$$\dot{Q}_{f,max} = \alpha_{f,f}\, A_{f,f}\, (T_t - T_2) + \alpha_{f,t}\, A_{f,t}\, (T_t - T_2) = (\alpha_{f,f}\, A_{f,f} + \alpha_{f,t}\, A_{f,t})(T_t - T_2).\tag{12}$$

Applying Equation(8) in Equation(12) follows that:

$$\dot{Q}_{f,\max} = \alpha_f A_f (T_t - T_2),$$ (13)

$$\eta_f = \frac{\dot{Q}_f}{\dot{Q}_{f,\max}} = \frac{\alpha_f A_f (T_f - T_2)}{\alpha_f A_f (T_t - T_2)} = \frac{T_f - T_2}{T_t - T_2}.$$ (14)

Equation (14) shows that η_f actually corrects the mean temperature difference between the fin surface and the surrounding fluid, which is necessary because the temperature along the fin surface is not constant. From Equation (14), it follows that:

$$\dot{Q}_f = \eta_f \, \alpha_f A_f (T_t - T_2).$$ (15)

Fin efficiency is always less than one. It depends on both the heat conduction processes in the fin and the convective heat transfer, which influence each other.

In reality, the average temperature of the fin base surface T_b is different from the unfinned tube wall temperature T_t. The heat transfer rate through the fin base surface into the fin is significantly higher than the heat transfer rate from the unfinned part of the tube wall into the fluid. The temperature drop, which appears underneath the fin, causes a periodic temperature profile to develop in the tube wall. This is shown schematically in Figure 2. The temperature drop is higher as the temperature conductivity of the material is lower.

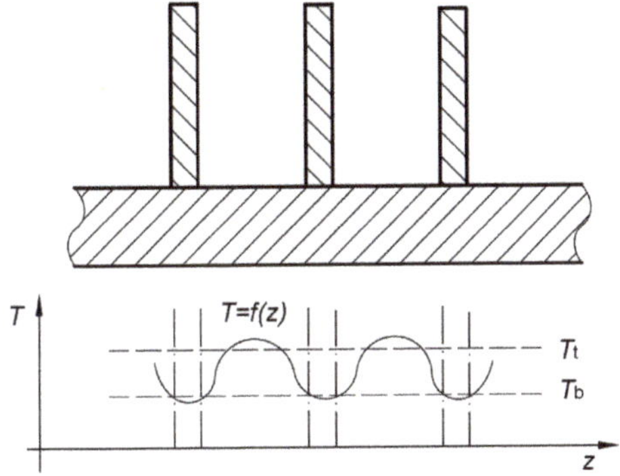

Figure 2. The periodic temperature profile in the tube wall.

In the literature [13,14], it is often assumed that:

$$T_b = T_t.$$ (16)

This simplification leads to an overestimation of the heat transfer rate. The calculated heat transfer rate can be as much as 25% bigger than it actually is, as shown by Sparrow and Hennecke [15] as well as Sparrow and L. Lee [16]. Using Equations (15) and (16) leads to:

$$\dot{Q} = \dot{Q}_t + \dot{Q}_f = \alpha_t A_t (T_t - T_2) + \eta_f \, \alpha_f A_f (T_t - T_2).$$ (17)

For thin fins densely spaced $A_f \gg A_t$, so in the Equation (17) the second term is more important, although $\eta_f < 1$. Therefore, as $\alpha_t \approx \alpha_f$ in set in the literature, we could write:

$$\dot{Q} = \alpha_f (A_t + \eta_f \, A_f)(T_t - T_2).$$ (18)

The overall surface efficiency that includes fins and unfinned part of the tube is defined with Equation (19).

$$\eta_o = \frac{\dot{Q}}{\dot{Q}_{max}}. \tag{19}$$

$\dot{Q}_{max}$ is the maximum heat transfer rate for ideal fin, which is obtained by setting $\eta_f = 1$ in Equation (18):

$$\dot{Q}_{max} = \alpha_f (A_t + A_f)(T_t - T_2). \tag{20}$$

With this we could write:

$$\eta_o = \frac{\dot{Q}}{\alpha_f (A_t + A_f)(T_t - T_2)}. \tag{21}$$

Substituting Equations (18) and (20) into Equation(21) yields:

$$\eta_o = 1 - \frac{A_f}{A_t + A_f}(1 - \eta_f). \tag{22}$$

2.2. Analytical Solution of Fin Efficiency

In general, theoretical fin efficiency is defined as follows:

$$\eta_f = \frac{\text{The actual heat transfer rate}}{\text{Maximum heat transfer rate when the entire fin}} . \tag{23}$$
$$\text{was at the base surface temperature}$$

The theoretical efficiency of an annular fin is calculated according to [13,14]:

$$\eta_{f,th} = C_2 \frac{K_1(m \cdot r_o) \cdot I_1(m \cdot r_{2c}) - I_1(m \cdot r_o) \cdot K_1(m \cdot r_{2c})}{I_0(m \cdot r_o) \cdot K_1(m \cdot r_{2c}) + K_0(m \cdot r_o) \cdot I_1(m \cdot r_{2c})}. \tag{24}$$

K_0, I_0 are modified Bessel's functions of the zero order, first and second types, a K_1, I_1 are the modified *Bessel'* function of the first order, first and the second type. With the aim of simplifying calculations, the analytical expression for the corrected fin radius, r_{2c} is introduced (Figure 3). It is based on the assumption of equivalence between the heat transferred from the actual fin with tip convection and the heat transferred from a longer, hypothetical fin with an adiabatic tip [1,13]. The corrected fin radius is defined with:

$$r_{2c} = r_2 + \frac{t_f}{2}. \tag{25}$$

This assumption increases fin surface area in the corresponding amount but does not include a greater heat transfer coefficient at the fin tip in regards to the heat transfer coefficient at the fin face surface. This enters also a certain error in the fin efficiency calculation. Table 1 shows the proportion of individual surfaces in the total tube finned surface area.

Table 1. The share of individual surfaces in the total surface area.

Surface	Area (mm^2)	Share (%)
Unfinned tube surface	628.32	11.4
Fin face surface	4712.39	85.7
Fin tip surface	157.08	2.9
Total surface area	5497.78	100.0

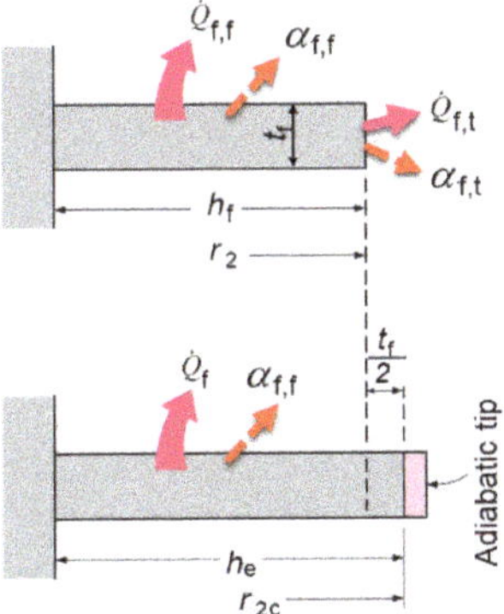

Figure 3. Adiabatic fin tip assumption.

Parameters m and C_2 are defined with:

$$m = \sqrt{\frac{2\,\alpha_0}{\lambda_f \cdot t_f}},\tag{26}$$

$$C_2 = \frac{2 \cdot r_o}{m \cdot (r_{2c}^2 - r_o^2)}.\tag{27}$$

Comparing experimentally obtained fin efficiencies with the analytical solutions, an empirical correction factor was obtained. For annular fin this correction factor has the form [17]:

$$E = 0.76 + 0.24 \cdot \eta_{f,th},\tag{28}$$

$$\eta_f = E \cdot \eta_{f,th}.\tag{29}$$

The actual average gas-side convective heat transfer coefficient α_0 is calculated applying fin efficiency:

$$\alpha_0 = \frac{\alpha_e \cdot (A_t + A_f)}{(A_t + \eta_f \cdot A_f)}.\tag{30}$$

The effective/apparent airside convective heat transfer coefficient α_e is:

$$\alpha_e = \frac{1}{\left(\dfrac{1}{U} - (A_t + A_f)\dfrac{\ln \frac{r_o}{r_i}}{2\,\pi\,L\,N_t\,\lambda_t} - \dfrac{(A_t + A_f)}{A_{t,i}\,\alpha_i}\right)}.\tag{31}$$

L is the tube length, N_t is the total number of tubes, $A_{t,i}$ is the tube side surface area and α_i is the tube side convective heat transfer coefficient. The overall heat transfer coefficient U referenced to (or based on) the outside tube heat transfer area is calculated according to Equation (32):

$$U = \frac{\dot{Q}}{(A_t + A_f) \cdot \Delta T_{\ln}}.\tag{32}$$

It is important to note that the mean logarithmic temperature difference ($\Delta T_{\ln}$) in the concept of fin efficiency is calculated by applying the (maximum) temperature of the outer surface of the tube.

In our case, the logarithmic mean temperature difference for counter-current flow acc. to Equation (33) is used:

$$\Delta T_{\text{ln}} = \frac{T_{\text{in}} - T_{\text{out}}}{\ln \frac{T_{\text{in}} - T_{\text{t}}}{T_{\text{out}} - T_{\text{t}}}} \tag{33}$$

T_{in} and T_{out} is the air temperature at the inlet and outlet of the heat exchanger.

3. Results

3.1. Numerical Analysis

In this paper, for research implementation, the tube Ø20×1.5 is selected. As a reference geometry, an annular fin Ø40/Ø20 with a thickness of 0.5 mm was selected. The fin is used as a cooling fin. The fin and tube material is stainless steel. Characteristic dimensions of the tube and fins are presented in Table 2.

Table 2. Characteristic dimensions of tube and fins.

Tube Outside Diameter	d_o	mm	20
Tube inside diameter	d_i	mm	17
Tube rows configuration	-	-	staggered
Transverse tube pitch	s_t	mm	50
Longitudinal tube pitch	s_l	mm	40
Fin height	h_f	mm	10
Fin thickness	t_f	mm	0.5
Fin pitch	s_f	mm	4.5
Number of tube rows	N_l	-	5

3.1.1. Mathematical Model

The numerical domain was chosen to perform the numerical analysis of the tube heat exchanger. For a heat exchanger with a uniform fluid flow field at the inlet, a typical repeating section in the heat exchanger is selected for the domain (Figure 4), and the solutions obtained are assumed valid for the entire heat exchanger.

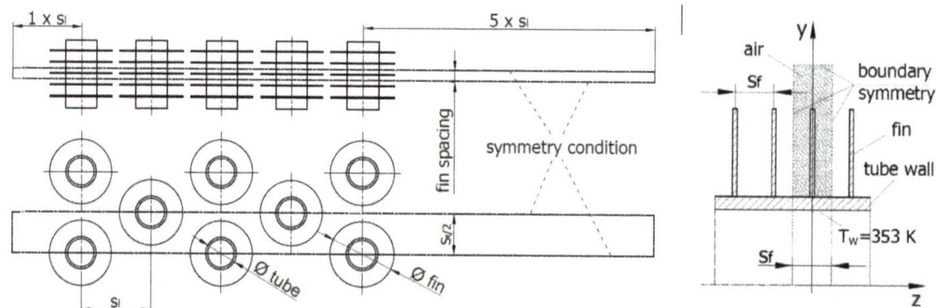

Figure 4. Schematic view of the computational domain.

Discretization of the geometrical model form the final volume and the set of all finite volumes makes a geometry mesh. The computational mesh was created in ANSYS Meshing software by using a hybrid mesh where most of the volume is structured mesh and the smaller part around the fins is unstructured mesh. Among different tested turbulence models, the k-ω SST model of turbulence with steady-state fluid flow and heat transfer was selected [18] because the results of numerical experiments show that the k-ω SST engineering turbulence model accurately captures the turbulent flow behaviour in the inner and outer regions of the boundary layer for wide range of engineering problems [19]. Second order solvers were used with residuals convergence limit 10^{-4} for all solved equations except the energy equation for which the criterion was set to 10^{-9}. The final mesh, which gives results independent of the numeric mesh consists of approximately 11 million control volumes, where dimensionless wall distance $y+ < 1$ (Figures 5–7).

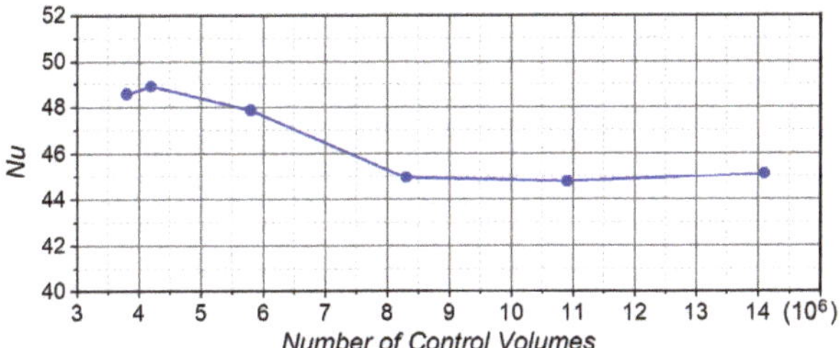

Figure 5. *Nu* dependence on the number of control volumes.

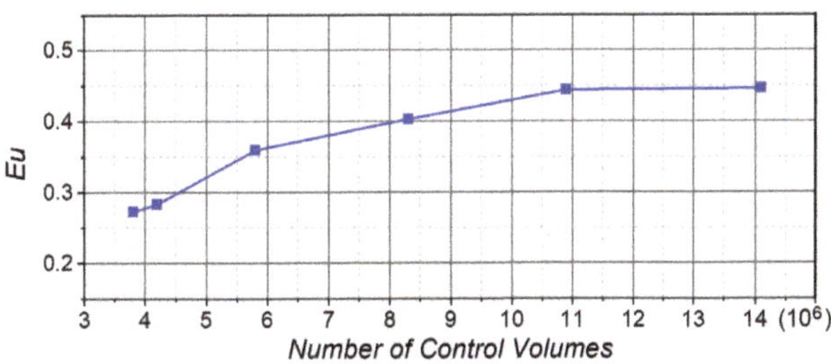

Figure 6. *Eu* dependence on the number of control volumes.

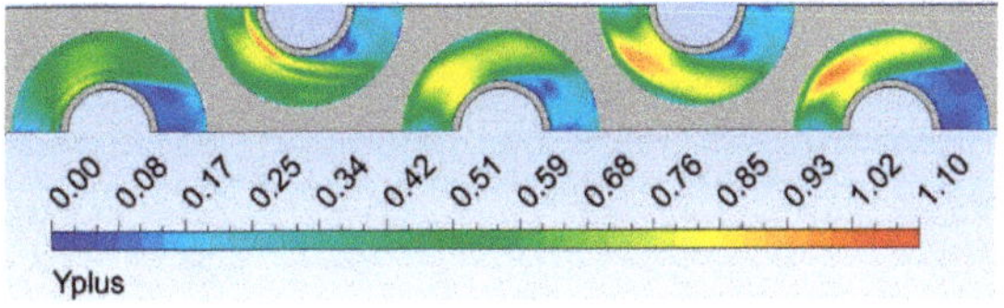

Figure 7. Dimensionless wall distance $y+$.

Details related to the convergence criteria, numerical simulation parameters and the validation of results obtained are presented in [20]. The boundary conditions were defined according to Table 3.

Table 3. Boundary conditions for Computational fluid dynamics simulations.

Inlet Air Temperature	K	288
Air velocity at the inlet of the heat exchanger	m/s	1, 2, 3, 5, 7
Inlet air turbulence intensity	%	5
Internal tube wall temperature	K	353
Outlet air pressure	Pa	101,325
Wall condition		Hydraulically smooth wall

3.1.2. Numerical Simulation Results

With the aim to simulate an ideal fin with infinite thermal conductivity, the influence of the thermal conductivity to the average heat flux was investigated. This influence was numerically analysed by setting five values of thermal conductivity (16 W/(mK), 160 W/(mK), 1600 W/(mK), 16,000W/(mK), 160,000 W/(mK)) at an air velocity of 5 m/s (*Re* = 11,218).

It can be seen from Figure 8 that by taking a value of 16,000 W/(mK) for thermal conductivity of the fin material, it is possible to simulate the infinite fin heat conductivity with sufficient accuracy because the difference in heat flux compared to the thermal conductivity of 160,000 W/(mK) is less than 0.06%. To make the calculation error even smaller, the thermal conductivity of 160,000 W/(mK) is chosen for further calculation. A numerical simulation for the case of infinite thermal conductivity was performed for both types of fins. Five airflow rates were chosen (see Table 2). In order to be able to analyse in detail the heat transfer on particular fin surfaces, the named selection for face surfaces of the fin, the fin tip surface and the face surface of the unfinned part of tubes are defined (Figure 9).

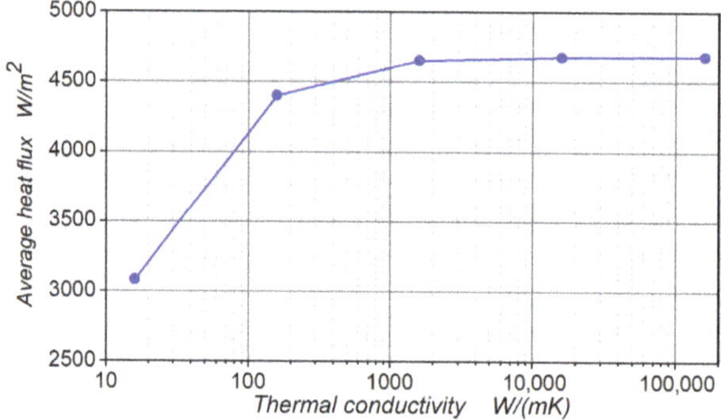

Figure 8. Heat flux versus fin thermal conductivity.

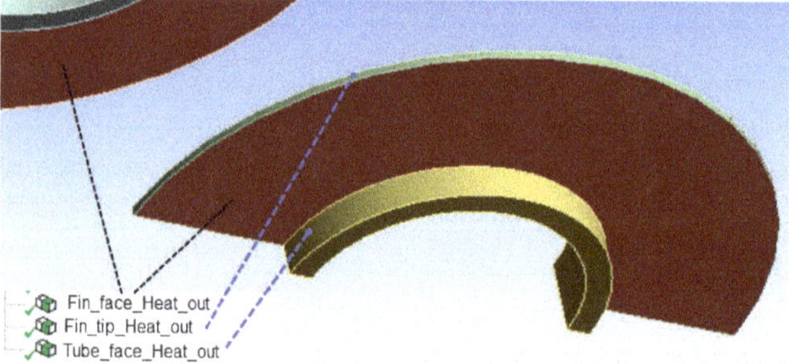

Figure 9. The named selection of fin.

In the case of infinite thermal conductivity (Figure 10), the entire fin surface temperature is close to the temperature of the fin base surface.

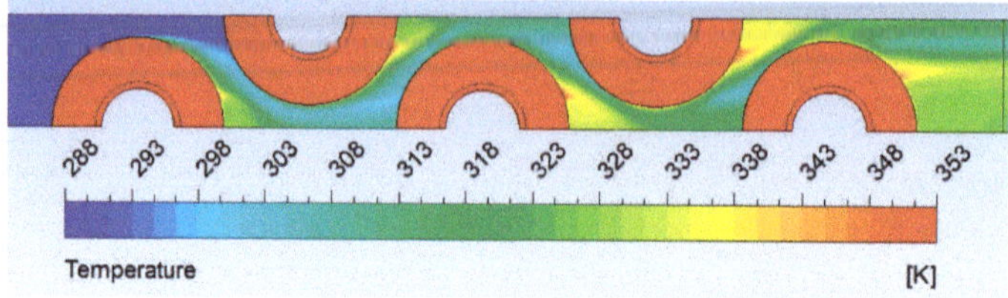

Figure 10. Temperature contours of ideal annular fins.

After that, numerical simulation for stainless steel (thermal conductivity of 16 W/(mK)) was performed for defined airflow rates. For a finite conduction resistance, a temperature gradient must exist along the fin. For stainless steel, the temperature variation in the longitudinal direction (Figure 11, left) is important. The temperature difference in the transverse section of the fin is small compared to the difference between the temperature of the fin surface and the environment as it is initially assumed for the thin fins (Figure 11, right). This can be checked by Biot (Bi) number which is in this case 0.0009 ($<< 0.1$). In addition, the temperature distribution across a fin surface area is not uniform (Figure 11, left).

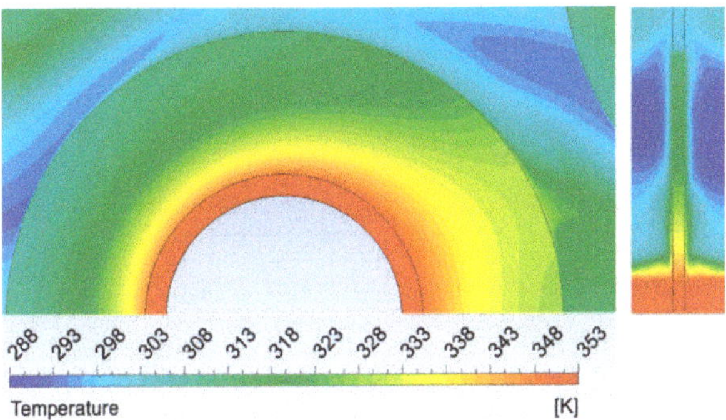

Figure 11. Temperature contours over fins in the 3rd row for air velocity 5 m/s. Left: Finned tube cross section, Right: Fin cross-section.

Therefore, due to the turbulent flow conditions in the fin sections, the heat transfer coefficient at the surface area is not constant. The assumption of a uniform heat transfer coefficient along the surface area of the fins can result in a fin efficiency difference relative to actual values up to 40% [21]. The local characteristics of the temperature field according to Figure 12b show that in the area of the fin base surface, the temperature is lower than in the tube area between the fins, so for the first row of tubes, it is 348 K.

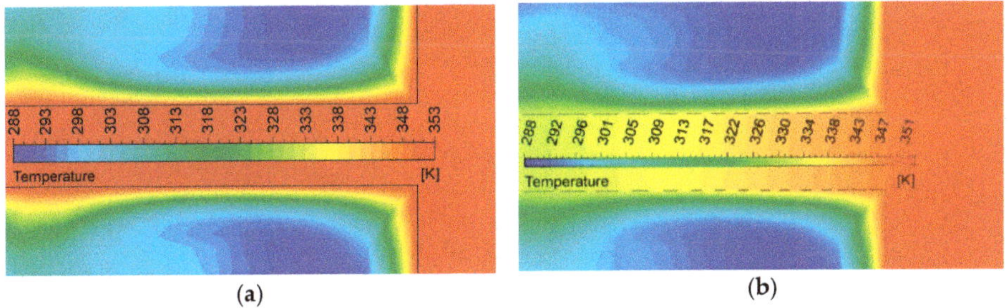

Figure 12. Fin temperature contours: (**a**) Temperature contour at $\lambda = 16{,}000$ W/(mK). (**b**) Temperature contour at $\lambda = 16$ W/(mK).

The amount of temperature drop in the fin base surface depends on the ratio h/t_f [15]. The presence of the fin, in addition to depressing its own base temperature, also depresses the temperature of the adjacent tube material, thereby slightly reducing the heat transfer from the unfinned tube wall. It can be seen that the assumption of a uniform base fin temperature equal to the temperature of the unfinned tube surface area can lead to significant errors in the calculated heat transfer rates from the fins [22]. An increase in the wall thickness increases the error. Heat flux is different on each surface. Figure 13 shows the heat flux on annular fin surfaces. The heat flux on the fin tip surface is much bigger than on the fin face surface. In addition, this indicates that the convective heat transfer on the fin tip surface is much bigger than on the fin face surface.

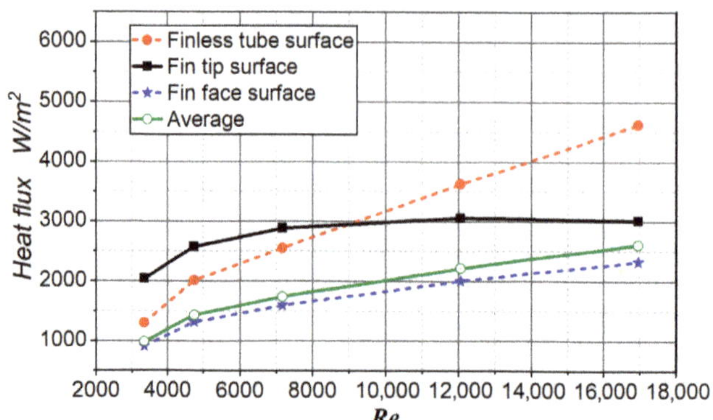

Figure 13. Heat flux on annular fin surfaces.

Figure 14 shows the share of the heat rate on particular surfaces in the total heat rate for annular fins for different *Re* numbers.

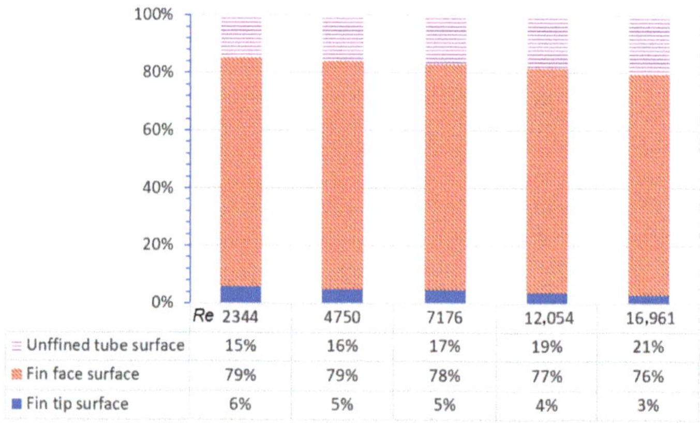

Figure 14. The share of heat rate on particular surfaces of the annular fin.

The mass average is used to calculate individual airflow variables at certain boundaries (e.g., at the inlet and outlet of the heat exchanger). In ANSYS CFD-Post, the appropriate function is massFlowAve. This function is defined by the expression:

$$massFlowAve(\phi) = \frac{\sum (m\phi)}{\sum m}.$$ (34)

The variable ϕ represents the fluid flow variable that is being observed and $\dot{m}$ is the local mass flow rate. Each member of the sum is taken at the node level of the observed section. The average air temperature in the heat exchanger (the reference temperature) is defined by the expression:

$$T_2 = \frac{T_{in} - T_{out}}{2}.$$ (35)

The average temperatures of the named surfaces were obtained from ANSYS Fluent using the *areaAve* function (Table 4). The area-weighted average of a quantity is computed by dividing the summation of the product of the selected field variable and facet area by the total area of the surface:

$$areaAve(\phi) = \frac{\sum \phi_i A_i}{A}. \tag{36}$$

Table 4. Average temperature of fin surfaces.

Air Velocity (m/s)	Fin Face Surface Temperature (K)	Fin Tip Surface Temperature (K)	Total Fin Surface Temperature (K)	Outer Unfinned Tube Surface Temp. (K)
1	338.8	333.8	338.6	352.4
2	333.4	326.8	333.2	352.1
3	330.0	322.7	329.8	351.9
5	325.3	316.9	325.0	351.5
7	321.9	313.0	321.6	351.3

Figure 15 shows the convective heat transfer coefficient on particular fin surfaces.

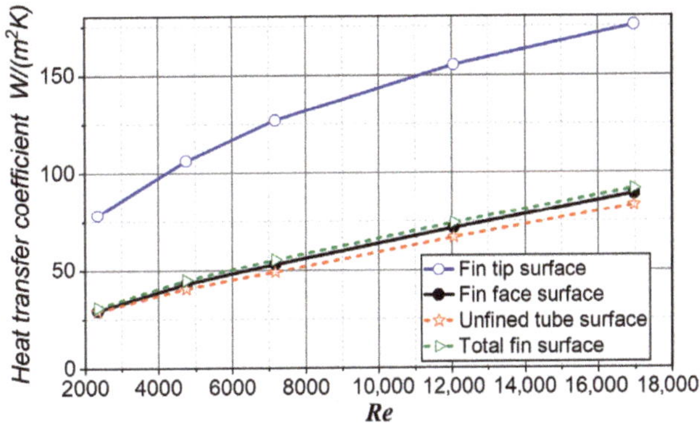

Figure 15. Convective heat transfer coefficient.

The convective heat transfer coefficient was calculated on the basis of numerical simulation data. Data of the average temperature of a given surface area, the heat rate on this surface and the amount of surface area were taken. For example, the convective heat transfer coefficient for the fin face surface is calculated according to the Equation (37):

$$\alpha_{f,f} = \frac{\dot{Q}_{f,f}}{A_{f,f}(T_{f,f} - T_2)}. \tag{37}$$

4. Discussion

Based on the numerically obtained results for the heat transfer rate $\dot{Q}_f$ and $\dot{Q}_{f,max}$ using Equation (11), the efficiency of the fin η_f is calculated. An analytically calculated theoretical value of the fin efficiency $\eta_{f,th}$ is calculated according to Equation (24) and then the corrected theoretical value of the fin efficiency η_f is calculated according to Equation (29). Figure 16 shows that the analytically calculated fin efficiency is smaller than the numerically obtained one. This is the consequence of introducing simplifications in the analytical derivation of the equation for the fin efficiency. The difference in efficiency varies from 10.7% to 12.5% for lower and higher *Re* numbers, respectively. The overall surface efficiency η_o is calculated according to Equation (19) based on the numerically obtained results for the heat transfer rate $\dot{Q}$ and $\dot{Q}_{max}$. Analytically, η_o is calculated according to Equation (22), where η_f is also analytically obtained.

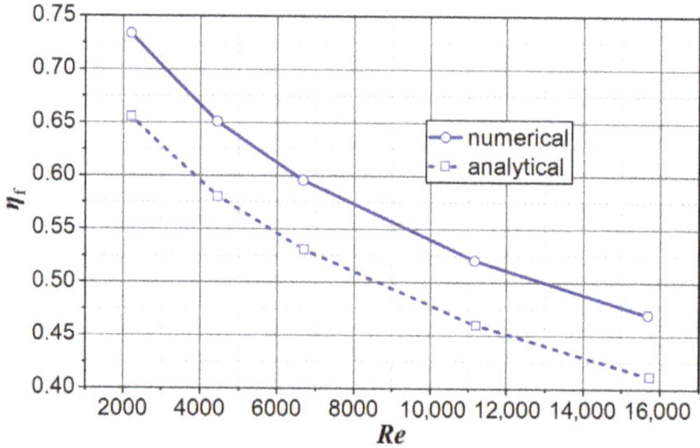

Figure 16. The efficiency of fins determined numerically and analytically.

Figure 17 shows that the analytically calculated overall surface efficiency is smaller than that obtained numerically. This is related to the error in the calculation of the analytical fin efficiency and the application of Equation (18) in which the assumption $\alpha_t \approx \alpha_f$ is embedded. The difference in numerical and analytical efficiencies for annular fins is 11.6% to 10.9% for lower and higher Re numbers, respectively.

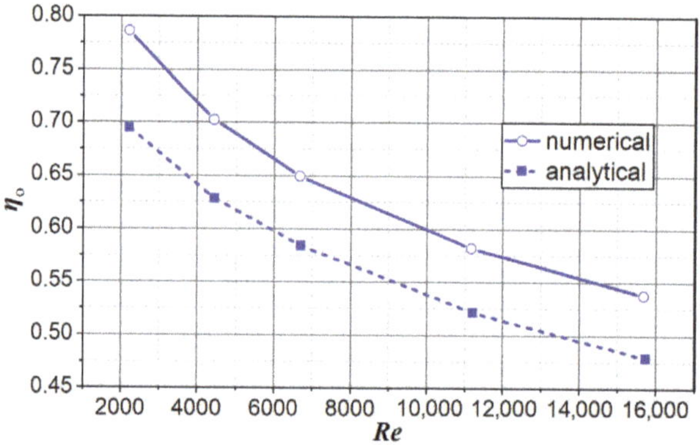

Figure 17. The overall surface efficiency η_o determined numerically and analytically.

The convective heat transfer coefficient can also be calculated based on data of numerical simulation according to Equation (38):

$$\alpha_{f,n} = \frac{\dot{Q}}{\eta_o(A_t + A_f)(T_f - T_2)}.\tag{38}$$

Figure 18 shows that the convective heat transfer coefficient based on the data obtained by numerical analysis is smaller than that calculated according to Equation (30). This is in accordance with the experimental results. The difference varies from 23.2% to 18.5% for lower and higher Re numbers, respectively.

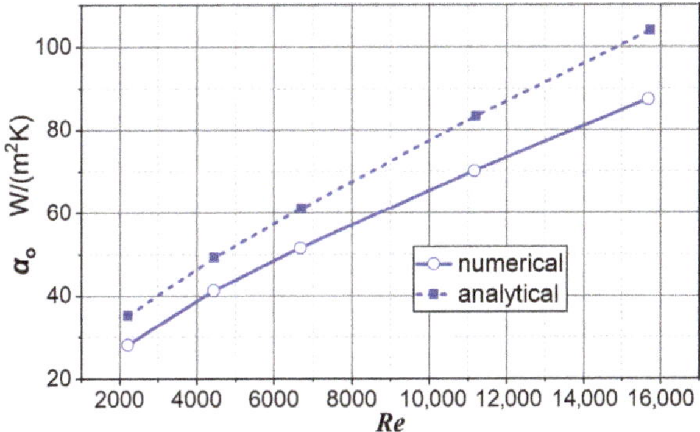

Figure 18. Convective heat transfer coefficient based on numerical analysis data and analytical calculation.

As mentioned above, the concept of the adiabatic fin tip introduces some error in calculation. The amount of error can be estimated according to:

$$Error = \frac{(Q_f - Q_{f,a})}{Q_f} \cdot 100\%, \tag{39}$$

$$Q_{f,a} = \Delta A_f \cdot q_{f,f} + Q_{f,f}, \tag{40}$$

$$\Delta A_f = \left(r_{2c}^2 - r_o^2\right)\frac{\pi}{2} - \left(r_2^2 - r_o^2\right)\frac{\pi}{2}. \tag{41}$$

Figure 19 shows that the amount of error ranges from 2.5% to 5.4% for the ratio $t_f/h_f = 0.05$ and $2000 < Re < 17{,}000$. In order to take into account the higher convection at the fin tip and reduce error, our recommendation is to take higher amounts than $t_f/2$, which is recommended in the literature.

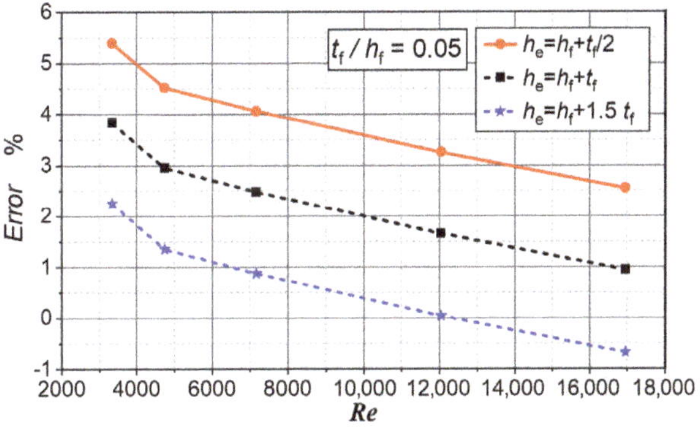

Figure 19. Error for adiabatic fin tip concept.

Good results can be obtained by for *Re* numbers up to 16,000:

$$h_e = h_f + 1.5 \cdot t_f, \tag{42}$$

and for *Re* numbers over 16,000:

$$h_e = h_f + t_f. \tag{43}$$

The increase in h_e reduces the fin efficiency to some extent.

5. Conclusions

Fin efficiency analysis was carried out with the aim of determining the error occurring when applying the commonly used equation in the literature for the efficiency of the annular fin. Stainless steel was selected as the fin material because it has lower thermal conductivity than other applied materials, thus the error is more pronounced. For the analysed interval of *Re* numbers from 2000 to 17,000, it was found that the application of analytical expressions for annular fins gives a fin efficiency coefficient of less than 10.7% to 12.6% compared to numerical analysis results. Related to these, the results for the efficiency coefficient of the entire surface are 11.6% to 10.9% lower compared to the results obtained by numerical analysis. At the same time, the convective heat transfer coefficient determined by numerical analysis is 23.2% to 18.5% lower than that calculated by using analytical terms.

The reasons for these deviations are the assumptions and simplifications introduced when deriving analytic expressions. The greatest impact on the amount of error is the assumption of equal temperature of the fin base surface and the outer surface of the tube, which neglects the temperature drop on the base surface of the fin. Also, the model of the adiabatic fin tip does not consider the higher heat flux at the fin tip compared to the fin face surface. The above-mentioned errors have opposing signs, so the overall effect is about 20% for annular fins. This should be considered when sizing the heat exchanger and one should increase the heat exchanger surface by the appropriate amount. In addition, corrected fin height is recommended for the analysed range of *Re* numbers.

Author Contributions: Conceptualization, A.Č.; data curation, formal analysis, investigation, methodology, software M.B. and S.M.; original draft preparation, M.B.; writing—review and editing, M.Ž. All authors have read and approved the final version of the manuscript.

Funding: This research received no external funding.

Conflicts of Interest: The authors declare no conflict of interest.

Nomenclature

A	surface area on the air side	[m^2]
E	empirical correction factor to the theoretical fin efficiency for fins	[mm]
h	fin high	[mm]
m	fin effectiveness parameter	[-]
L	tube length	[m]
N_t	total number of tubes	[-]
$\dot{Q}$	heat transfer rate	[W]
r	radius	[mm]
t	thickness	[mm]
ΔT_{ln}	the logarithmic mean temperature difference	[K]
T	the average temperature	[K]
U	the overall heat transfer coefficient	[W/(m^2·K)]
λ	the thermal conductivity	[W/(m·K)]
α	average convective heat transfer coefficient	[W/(m^2·K)]
η	efficiency	[-]
I_0, K_0	modified Bessel functions of the first and second kinds and zero order	[-]
I_1, K_1	modified Bessel functions of the first and second kinds and first order	[ ]

	Subscripts	
b	fin base	
e	effective/equivalent	
i	tube inside	
in	air inlet	
f	fin	
f,f	fin face	
f,t	fin tip	

f,n	fin, numerically
max	maximal (for ideal fin)
o	tube outside/overall
out	air outlet
t	tube/finless tube surface area
th	theoretical
tot	total
0	actual
1	fluid properties inside tube
2	outside fin radius/bulk fluid properties outside tube and fins
2c	outside corrected fin radius

References

1. Harper, D.R.; Brown, W.B. Mathematical equations for heat conduction in the fins of air cooled engines. *NACA R* **1922**, *158*, 32.
2. Murray, W.M. Heat transfer through an annular disk or fin of uniform thickness. *Trans. ASME J. Appl. Mech.* **1938**, *60*, A78–A81.
3. Gardner, K.A. Efficiency of extended surface. *Trans. ASME* **1945**, *67*, 621.
4. Darvishi, M.T.; Khani, F.; Aziz, A. Numerical investigation for a hyperbolic annular fin with temperature dependent thermal conductivity. *Propuls. Power Res.* **2016**, *5*, 55–62. [CrossRef]
5. Ghai, M.L. Heat transfer in straight fins. In Proceedings of the General Discussion on Heat Transfer, London, UK, 11–13 September 1951.
6. Zaidan, M.H.; Alkumait, A.A.R.; Ibrahim, T.K. Assessment of heat transfer and fluid flow characteristics within finned flat tube. *Case Stud. Therm. Eng.* **2018**, *12*, 557–562. [CrossRef]
7. Nemati, H.; Moghimi, M. Numerical study of flow over annular-finned tube heat exchangers by different turbulent models. *CFD Lett.* **2014**, *6*, 101–112.
8. Look, D.C., Jr.; Krishnan, A. One-dimensional fin-tip boundary condition correction II. *Heat Transf. Eng.* **2001**, *22*, 35–40. [CrossRef]
9. Nemati, H.; Samivand, S. Performance optimization of annular elliptical fin based on thermo-geometric parameters. *Alexandria Eng. J.* **2015**, *54*, 1037–1042. [CrossRef]
10. Chen, H.L.; Wang, C.C. Analytical analysis and experimental verification of trapezoidal fin for assessment of heat sink performance and material saving. *Appl. Therm. Eng.* **2016**, *98*, 203–212. [CrossRef]
11. Lane, H.J.; Heggs, P.J. Extended surface heat transfer-the dovetail fin. *Appl. Therm. Eng.* **2005**, *25*, 2555–2565. [CrossRef]
12. Cléirigh, C.T.Ó.; Smith, W.J. Can CFD accurately predict the heat-transfer and pressure-drop performance of finned-tube bundles? *Appl. Therm. Eng.* **2014**, *73*, 679–688. [CrossRef]
13. Incropera, F.P.; DeWitt, D.P.; Bergman, T.L.; Lavine, A.S. *Fundamentals of Heat and Mass Transfer*; John Wiley & Sons, Inc.: Hoboken, NJ, USA, 2007.
14. Kraus, A.D.; Aziz, A.; Welty, J. *Extended Surface Heat Transfer*; John Wiley & Sons, Inc.: Hoboken, NJ, USA, 2001.
15. Sparrow, E.M.; Hennecke, D.K. Temperature depression at the base of a fin. *Trans. ASME* **1970**, *92*, 204–206. [CrossRef]
16. Sparrow, E.M.; Lee, L. Effects of fin base temperature depression in a multifin array. *J. Heat Transf.* **1975**, *97*, 463–465. [CrossRef]
17. Hashizume, K.; Morikawa, R.; Koyama, T.; Matsue, T. Fin efficiency of serrated fins. *Heat Transf. Eng.* **2002**, *23*, 7–14. [CrossRef]
18. Taler, D. *Numerical Modelling and Experimental Testing of Heat Exchangers*; Springer: Berlin, Germany, 2019; Volume 161.
19. Könözsy, L. Volume I: Theoretical background and development of an anisotropic hybrid k-omega shear-stress transport/stochastic turbulence model. In *A New Hypothesis on the Anisotropic Reynolds Stress Tensor for Turbulent Flows*; Springer: Berlin, Germany, 2019; Volume 120.

20. Bošnjaković, M.; Čikić, A.; Muhič, S.; Stojkov, M. Development of a new type of finned heat exchanger. *Tehnical Gazette* **2017**, *24*, 1785–1796. [CrossRef]
21. Huang, L.J.; Shah, R.K. Assessment of calculation methods for efficiency of straight fins of rectangular profile. *Int. J. Heat Fluid Flow* **1992**, *13*, 282–293. [CrossRef]
22. Suryanarayana, N.W. Two-dimensional effects on heat transfer rates from an array of straight fins. *J. Heat Transf.* **1977**, *99*, 129–132. [CrossRef]

Article

Natural Convection of Non-Newtonian Power-Law Fluid in a Square Cavity with a Heat-Generating Element

Darya S. Loenko [1],*, Aroon Shenoy [2] and Mikhail A. Sheremet [1]

[1]   Laboratory on Convective Heat and Mass Transfer, Tomsk State University, 634050 Tomsk, Russia;
      sheremet@math.tsu.ru
[2]   S.A.I.C.O., 1111 Arlington Blvd, Arlington, VA 22209, USA; draroonshenoy@hotmail.com
*   Correspondence: whiteink@bk.ru

Received: 1 May 2019; Accepted: 1 June 2019; Published: 5 June 2019

Abstract: Development of modern technology in microelectronics and power engineering necessitates the creation of effective cooling systems. This is made possible by the use of the special fins technology within the cavity or special heat transfer liquids in order to intensify the heat removal from the heat-generating elements. The present work is devoted to the mathematical modeling of thermogravitational convection of a non-Newtonian fluid in a closed square cavity with a local source of internal volumetric heat generation. The behavior of the fluid is described by the Ostwald-de Waele power law model. The defining Navier–Stokes equations written using the dimensionless stream function, vorticity and temperature are solved using the finite difference method. The effects of the Rayleigh number, power-law index, and thermal conductivity ratio on heat transfer and the flow structure are studied. The obtained results are presented in the form of isolines of the stream function and temperature, as well as the dependences of the average Nusselt number and average temperature on the governing parameters.

Keywords: non-Newtonian fluid; natural convection; heat source of volumetric heat generation; finite difference method

---

1. Introduction

The use of liquids in various technical applications is well known, but there is a greater special interest of researchers in the study of non-Newtonian fluids because of the realization that they are now very common in everyday life [1–3]. For example, most polymeric liquids, including melts and solutions, foams, emulsions, suspensions, worm-like micelles, antifreezes, porcelain clay, sewage sludge, pharmaceutical formulations, cosmetics and toiletries, paints and synthetic lubricants, biological fluids (blood, synovial fluid, sage) and food products (butter, jams, jellies, soups, marmalade) show non-Newtonian properties according to flow characteristics [4–8].

The main feature of non-Newtonian fluids is the nonlinear dependence of shear stress on shear rate. Since dynamic viscosity changes with shear stress, the fluidity of such a fluid, and thus heat transfer, becomes more complex. Therefore, the analysis of non-Newtonian fluids behavior warrants special attention [1–9].

Many papers have been devoted to the analysis of non-Newtonian fluids flow in various configurations. The peristaltic transport of non-Newtonian fluids in a diverging tube with various forms of near-wall waves was studied by Hariharan et al. [10]. Comparison of ink exhibiting Newtonian and non-Newtonian character for laser printing was carried out by Kalaitzis et al. [11]. The rheological properties of the $TiO_2/ZnO/EG$ nanofluid were studied by Nafchi et al. [12]. Hundertmark-Zausková and Lukácová-Medvid'ová [13] considered blood flow as a pseudoplastic fluid circulating inside elastic vessels. A three-dimensional simulation of the human cardiovascular system was performed by Janela et al. [14], where the authors compared the Newtonian and non-Newtonian models for blood.

Along with the hydrodynamics of non-Newtonian fluid, the process of natural convection is being actively studied, since it is the most common method of electronic components cooling. This technique is of great interest due to its simplicity, minimal cost, low noise, smaller size and reliability [15]. In addition, cooling by natural convection in enclosures is very popular due to various areas of applicability, such as cooling systems for electronic gadgets, high-performance insulation of buildings, multi-layer structures, various furnaces, solar heat collectors, polymer processing, oil product recovery, transportation of suspensions, production of chemical substances and fertilizers, medicine and food industry, paper making, production and packaging of paints and emulsions, etc. [16,17]. The process of natural convection is also considered in various cavities and with various heat exchange agents, for example, natural convection cooling of a heat source by a nanofluid was studied by Aminossadati and Ghasemi [18]. Heat transfer using natural convection during the melting of a material with a phase transition in a closed rectangular cavity with three heaters on the left vertical wall was studied by El Qarnia et al. [19]. Laminar natural convection in a square cavity heated through side walls at small Prandtl numbers with large differences in density has been numerically studied by Pesso and Piva [20]. Aly and Raizahan [21] proposed a hydrodynamic method for incompressible particles for natural convection in a cavity filled with nanofluid, including numerous solid structures. Laminar unsteady free convection in a closed region with a heat source of various shapes was numerically studied by Gibanov and Sheremet [22]. A numerical study of natural convection in a closed cavity, in the center of which a heat source is located, was carried out by Mahalakshmi et al. [23].

Despite the active use of various nanofluids, water and air in such processes, non-Newtonian power law fluids are also assigned a significant role. For example, Sasmal et al. [24] studied the laminar natural convection of a non-Newtonian fluid in a closed square cavity with an isothermal rotating cylinder. Kiyasatfar [25] performed the analysis of convective heat exchange of a sliding flow of non-Newtonian fluid through a parallel plate and round microchannels. A numerical study of the natural convection of power law fluids in a vertical open finite channel was carried out by Zhou and Bayazitoglu [26]. Double-diffusive natural convection of a Bingham fluid was studied by Kefayati [27] taking into account the Soret and Dufour effects. Laminar natural convection in a closed trapezoidal cavity filled with a non-Newtonian nanofluid was studied numerically by Alsabery et al. [28]. Kefayati and Tang [29] investigated the natural convection of a non-Newtonian nanofluid in a cavity with a uniform magnetic field. An experimental study of the convective heat transfer of a stream of a non-Newtonian solution of Xanthan gum in a micropipe was carried out by Shojaeian et al. [30].

Based on the review of literature, it can be concluded that natural convection of a non-Newtonian fluid in a closed cavity has a great deal of significance in modern research. In this regard, the aim of the present work is the mathematical modeling of the natural convection of a power-law fluid in a closed square cavity with a heat-generating and heat-conducting element.

2. Governing Equations and Numerical Technique

In this paper, we study the process of thermogravitational convection in a closed square cavity as shown in Figure 1. Horizontal walls are heat insulated. Vertical walls are maintained at low temperature T_c. The energy source is located in the center of the bottom wall and this source has a constant internal volumetric heat generation Q. Gravity force is directed vertically down.

The cavity under investigation is filled with a non-Newtonian fluid that follows the Ostwald-de Waele power law model:

$$\tau_{ij} = 2\mu_{eff}D_{ij} = 2K(2D_{kl}D_{kl})^{\frac{n-1}{2}}D_{ij} \tag{1}$$

Here τ_{ij}-is the shear stress, μ_{eff}-is the effective viscosity, D_{ij}-is the component of the strain rate tensor, K-is the flow consistency index, n-is the power law index.

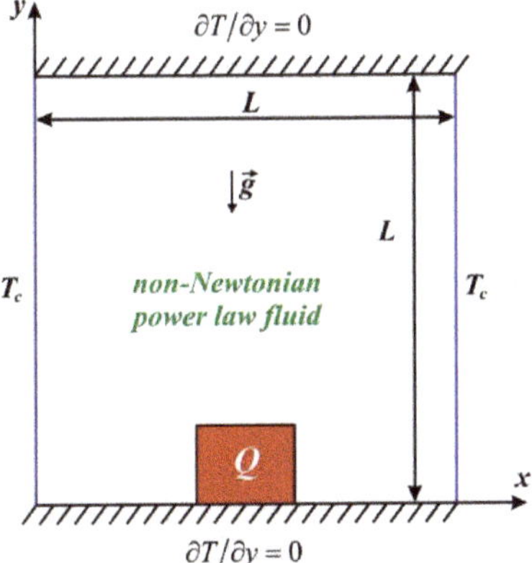

Figure 1. A schematic of the system.

The process of natural convection is described by the following time-dependent system of partial differential equations:

$$\frac{\partial u}{\partial x} + \frac{\partial v}{\partial y} = 0 \tag{2}$$

$$\frac{\partial u}{\partial t} + u\frac{\partial u}{\partial x} + v\frac{\partial u}{\partial y} = -\frac{1}{\rho}\frac{\partial p}{\partial x} + \frac{1}{\rho}\left(\frac{\partial \tau_{xx}}{\partial x} + \frac{\partial \tau_{xy}}{\partial y}\right) \tag{3}$$

$$\frac{\partial v}{\partial t} + u\frac{\partial v}{\partial x} + v\frac{\partial v}{\partial y} = -\frac{1}{\rho}\frac{\partial p}{\partial y} + \frac{1}{\rho}\left(\frac{\partial \tau_{xy}}{\partial x} + \frac{\partial \tau_{yy}}{\partial y}\right) + g\beta(T - T_c) \tag{4}$$

$$\frac{\partial T}{\partial t} + u\frac{\partial T}{\partial x} + v\frac{\partial T}{\partial y} = \alpha\left(\frac{\partial^2 T}{\partial x^2} + \frac{\partial^2 T}{\partial y^2}\right) \tag{5}$$

The heat conduction equation for the local heater is

$$(\rho c)\frac{\partial T}{\partial t} = k_{hs}\left(\frac{\partial^2 T}{\partial x^2} + \frac{\partial^2 T}{\partial y^2}\right) + Q \tag{6}$$

For ease of calculation, the problem was reduced to a transformed dimensionless form. The non-primitive variables such as the stream function $u = \frac{\partial \psi}{\partial y}$, $v = -\frac{\partial \psi}{\partial x}$ and vorticity $\omega = \frac{\partial v}{\partial x} - \frac{\partial u}{\partial y}$ were introduced. Further, dimensionless variables were included into the equations using the following parameters L is the length of the cavity; $X = x/L$, $Y = y/L$ are the coordinates; $\tau = t\sqrt{g\beta\Delta T/L}$ is the dimensionless time; $\Psi = \psi/\sqrt{g\beta\Delta T L^3}$ is the dimensionless stream function; $\Omega = \omega/\sqrt{L/g\beta\Delta T}$ is the dimensionless vorticity; $\Theta = (T - T_c)/\Delta T$ is the dimensionless temperature; $\Delta T = QL^2/k_{hs}$ is the temperature difference.

The basic equations in non-dimensional form using the stream function, vorticity and temperature variables take a new form (7)–(10):

$$\frac{\partial^2 \Psi}{\partial X^2} + \frac{\partial^2 \Psi}{\partial Y^2} = -\Omega \tag{7}$$

$$\frac{\partial \Omega}{\partial \tau} + \frac{\partial \Psi}{\partial Y}\frac{\partial \Omega}{\partial X} - \frac{\partial \Psi}{\partial X}\frac{\partial \Omega}{\partial Y} = \left(\frac{Ra}{Pr}\right)^{\frac{n-2}{2}}\left[\nabla^2(\widetilde{M}\Omega) + S_\Omega\right] + \frac{\partial \Theta}{\partial X} \tag{8}$$

$$\frac{\partial \Theta}{\partial \tau} + \frac{\partial \Psi}{\partial Y}\frac{\partial \Theta}{\partial X} - \frac{\partial \Psi}{\partial X}\frac{\partial \Theta}{\partial Y} = \frac{1}{\sqrt{Ra \cdot Pr}}\left(\frac{\partial^2 \Theta}{\partial X^2} + \frac{\partial^2 \Theta}{\partial Y^2}\right) \tag{9}$$

In the case of internal heat-generating element we have the following heat conduction equation

$$\frac{\partial \Theta_{hs}}{\partial \tau} = \frac{Ar}{\sqrt{Ra \cdot Pr}}\left(\frac{\partial^2 \Theta_{hs}}{\partial X^2} + \frac{\partial^2 \Theta_{hs}}{\partial Y^2} + 1\right) \tag{10}$$

Here $\widetilde{M} = \left[4\left(\frac{\partial^2 \Psi}{\partial X \partial Y}\right)^2 + \left(\frac{\partial^2 \Psi}{\partial Y^2} - \frac{\partial^2 \Psi}{\partial X^2}\right)^2\right]^{\frac{n-1}{2}}$, $S_\Omega = 2\left[\frac{\partial^2 \widetilde{M}}{\partial X^2}\frac{\partial^2 \Psi}{\partial Y^2} + \frac{\partial^2 \widetilde{M}}{\partial Y^2}\frac{\partial^2 \Psi}{\partial X^2} - 2\frac{\partial^2 \widetilde{M}}{\partial X \partial Y}\frac{\partial^2 \Psi}{\partial X \partial Y}\right]$, $Pr = \frac{\widetilde{\nu}}{\alpha}$ is the Prandtl number; $Ra = \frac{g\beta \Delta T L^3}{\widetilde{\nu}\alpha}$ is the Rayleigh number, $\widetilde{\nu} = \left(\frac{K}{\rho}\right)^{\frac{1}{2-n}} L^{\frac{2(1-n)}{2-n}}$ is the effective kinematic viscosity.

Initial conditions for the considered problem are $\Omega = \Psi = \Theta = 0$.
Boundary conditions:

$$X = 0 \text{ and } X = 1,\ 0 \leq Y \leq 1,\ \Psi = 0,\ \frac{\partial \Psi}{\partial X} = 0,\ \Theta = 0;$$
$$Y = 0 \text{ and } Y = 1,\ 0 \leq X \leq 1,\ \Psi = 0,\ \frac{\partial \Psi}{\partial Y} = 0,\ \frac{\partial \Theta}{\partial Y} = 0;$$

at the heat source surface: $\Psi = 0$, $\Omega = -\frac{\partial^2 \Psi}{\partial n^2}$, $\begin{cases} \Theta_{hs} = \Theta_f \\ \frac{k_{hs}}{k_f}\frac{\partial \Theta_{hs}}{\partial n} = \frac{\partial \Theta_f}{\partial n} \end{cases}$.

The main numerical technique used to solve the considered problem is the finite difference method. The differential Equation (7) for the stream function was discretized using the central differences and the obtained difference equation was solved by the successive over-relaxation method [31,32]. The optimal value of the relaxation parameter was selected on the basis of numerical experiments. The convective terms in parabolic Equations (8) and (9) were approximated using the "donor cell" difference scheme. It should be noted that this scheme is known also as the second upstream scheme. This scheme is based on one-sided differences in spatial variables and has the property of transport and conservatism. The point is that when the velocities are positive, the difference is used backwards, and vice versa. The diffusion terms were discretized by central differences. The numerical solution of parabolic Equations (8)–(10) was performed using the locally one-dimensional Samarskii scheme [31,32]. This scheme allows transformation of a two-dimensional problem to a one-dimensional problem. The obtained linear algebraic equations were solved by the Thomas algorithm [31,32].

The developed algorithm for solving the problem was implemented by a computational code using the C++ programming language. Before carrying out the basic calculations, the developed in-house code was tested in detail using some problems for verification. The first benchmark problem was the natural convection of a non-Newtonian fluid in a differentially heated cavity. The horizontal walls were heat insulated. The vertical left wall was maintained at a constant temperature T_h, while the vertical right wall had a constant temperature T_c ($T_h > T_c$). The diagram of the test problem solution area is shown in Figure 2.

Figure 3 shows the comparison of streamlines at $Ra = 10^5$, $Pr = 100$, $n = 1.0$ and 1.4, obtained using the developed in-house computational code and numerical data of Khezzar et al. [33]. As a result of this validation study, an analysis was obtained for the values of the average Nusselt number at the left isothermal wall of the cavity in comparison with the results [33,34]. Table 1 shows the obtained comparison for the average Nusselt number. Thus, the results of testing the computational code showed that the data obtained are in good agreement with the data of other authors. The numerical algorithm and developed computational code are operable and may be applicable in further studies.

In the case of thermogravitational convection of a power law fluid in a cavity with an energy source, preliminary estimation of the influence of the grid parameters and the time step on the process under study was carried out. Figure 4 shows the time dependences of the average Nusselt number (a) and average heater temperature (b). Dependencies are plotted at $n = 0.6$, $Ra = 10^5$, $k = k_{hs}/k_f = 1$, $Pr = 10^2$ with the following grid parameters: 50×50, 100×100, 150×150.

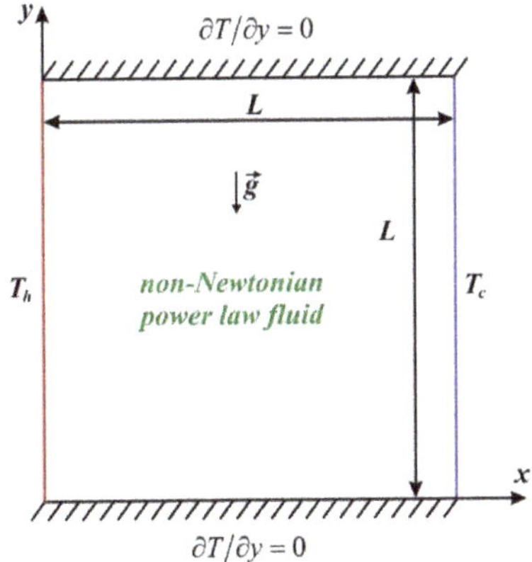

Figure 2. A schematic of the test system.

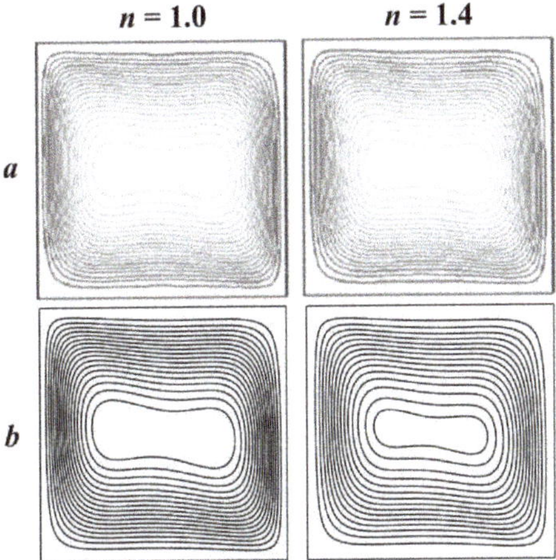

Figure 3. Streamlines: Data from [33] (**a**) and obtained results (**b**).

Table 1. The value of the average Nusselt number depending on the power law index.

n	Nu_{avg} in This Work	Nu_{avg} in [33]	Nu_{avg} in [34]
0.6	7.3823	6.9345	7.020
0.8	5.6201	5.5127	–
1.0	4.7662	4.6993	4.741
1.2	4.2227	4.1709	–
1.4	3.8464	3.7869	3.770

Based on this dependence, we chose a grid of 100×100 nodes, since it does not lead to strong discrepancies and it does not affect the process under study. In addition, in the course of the study it was revealed that the final value of dimensionless time $\tau = 200$ is not enough, so the calculations were carried out at $\tau = 2000$.

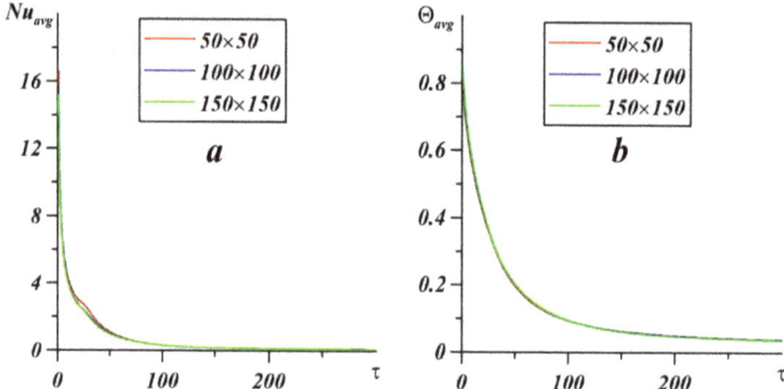

Figure 4. The time dependences of the average Nusselt number (**a**) and average temperature (**b**) at $n = 0.6$, $Ra = 10^5$, $k = 1$, $Pr = 10^2$.

3. Results

As a result of numerical simulation of thermogravitational convection of a non-Newtonian power law fluid in a closed square cavity with a heat source, an analysis was made to understand the effects of governing parameters on the process. The Rayleigh number Ra is varied in the range 10^4–10^5, the power law index n is changed from 0.8 to 1.4, the thermal conductivity ratio is $k = 1$, 10, 10^2, 10^3. The Prandtl number was fixed at $Pr = 10^2$. Streamlines, isotherms and distributions of the average Nusselt number and mean temperature within the heater are shown in Figures 5–10.

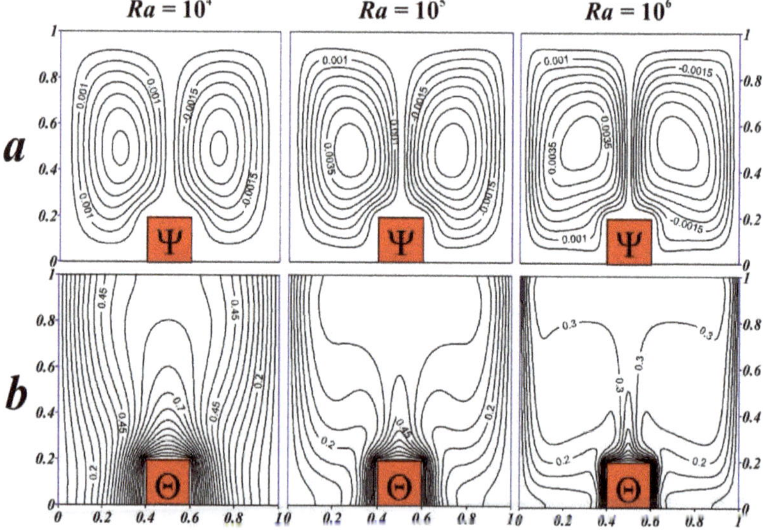

Figure 5. Streamlines Ψ (**a**) and isotherms Θ (**b**) at $n = 0.8$, $Pr = 10^2$, $k = 10^2$ and different Rayleigh numbers.

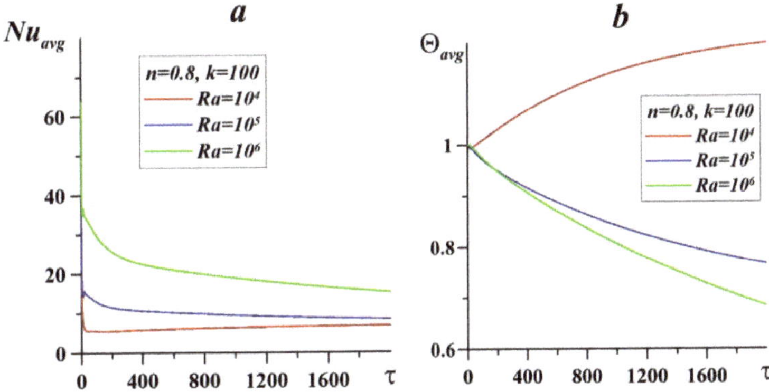

Figure 6. The time dependences of the average Nusselt number (**a**) and average temperature (**b**) at $n = 0.8$, $k = 100$, $Pr = 10^2$ and different Rayleigh numbers.

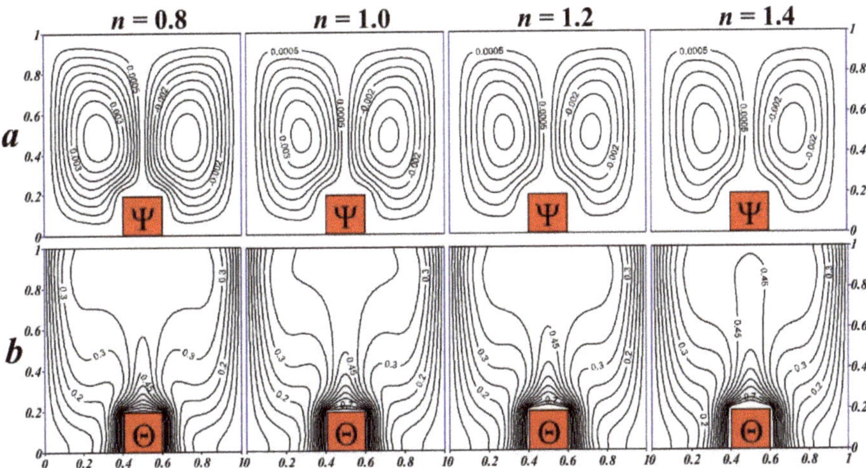

Figure 7. Streamlines Ψ (**a**) and isotherms Θ (**b**) at $Pr = 10^2$, $k = 10^2$, $Ra = 10^5$ and different n.

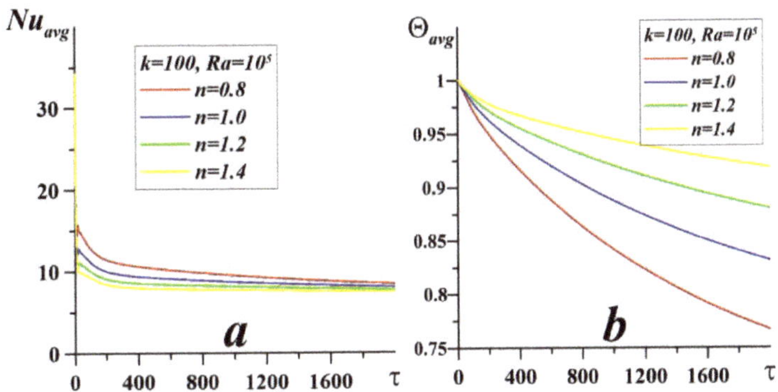

Figure 8. The time dependences of the average Nusselt number (**a**) and average heater temperature (**b**) at $k = 100$, $Pr = 100$, $Ra = 10^5$ and different n.

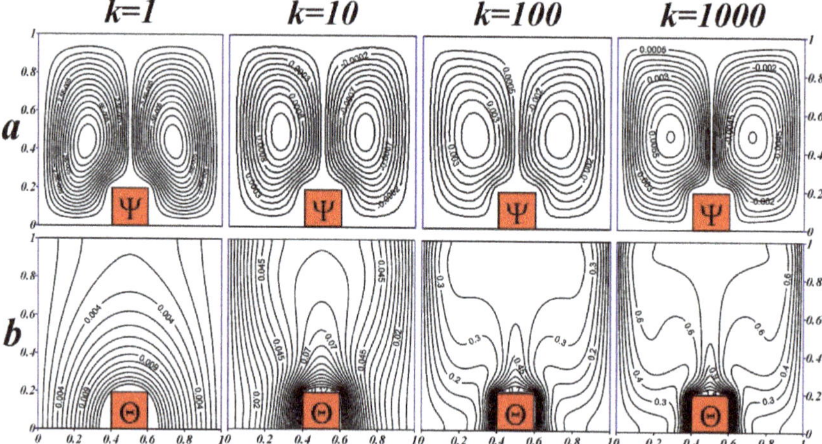

Figure 9. Streamlines Ψ (**a**) and isotherms Θ (**b**) at $n = 0.8$, $Pr = 10^2$, $Ra = 10^5$ and different k.

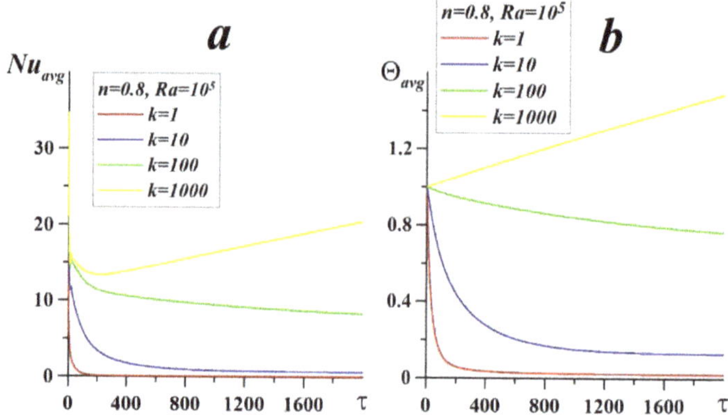

Figure 10. The time dependences of the average Nusselt number (**a**) and average temperature (**b**) at $n = 0.8$, $Pr = 100$, $Ra = 10^5$ and different k.

Figure 5 shows the effect of the Rayleigh number on the distribution of streamlines Ψ (a) and isotherms Θ (b) for $n = 0.8$, $Pr = 10^2$, $k = 10^2$. In Figure 5a, the distribution of the streamlines shows that with the increasing of the Rayleigh number, the flow structure does not change. Namely, two convective cells are formed in the cavity, and these cells are symmetrical to each other relative to the central axis. The shape of the cells varies slightly from elliptical to more rectangular with a pronounced flow around the source. Moreover, the convective cells cores are displaced to the central vertical axis and the streamlines density increases in this central part. Such behavior can be explained by a growth of convective flow intensity due to high values of the buoyancy force magnitude and as a result, the thickness of the central thermal plume decreases that reflects intensive liquid motion in this narrow zone. The temperature field in Figure 5b with $Ra = 10^4$ indicates the predominance of the conductive mechanism of heat transfer in the cavity, since the isotherms are located almost parallel to the cooling walls. Consequently, the heat transfer in the cavity is very weak. With an increase in Ra, the convective heat exchange is enhanced, which corresponds to the formation of a two-dimensional heat plume above the heat source. It is seen that an upward flow is formed above the source due to the temperature difference between the heater and the liquid near this element. The hot ascending flow reaches the upper adiabatic wall. After that, this flow separates and descends along the cooling vertical walls, which corresponds to the expansion of the thermal plume cap and an appearance of cold temperature wave in the bottom part of the cavity from the left and right sides of the heater. Similar relationship between streamlines and isotherms can be observed for all considered cases.

Figure 6 shows the time dependence of the average Nusselt number Nu_{avg} (a) and mean heater temperature Θ_{avg} (b) for different Rayleigh numbers. Dependencies are plotted for $n = 0.8$, $Pr = 10^2$, $k = 10^2$. The average Nusselt number at the heat source surface was calculated as follows $Nu_{avg} = \frac{1}{0.6} \int_0^{0.6} \left(-\frac{\partial \Theta}{\partial n}\right) d\varsigma$. Figure 6a shows that as the Rayleigh number increases, the average Nusselt number also rises. This suggests that convective heat transfer is enhanced in the cavity. In addition, it can be seen that the considered modes are unsteady, because Nu_{avg} continues to change with time. It is worth noting that for $Ra = 10^4$ the mean Nusselt number decreases during the conduction regime and increases during the weak convective regime, while for $Ra = 10^5$ and 10^6 Nu_{avg} reduces. Figure 6b shows the dependence of the average temperature within the source on the time and Rayleigh number. One can see that this dependence fully corresponds to the dependence of the average Nusselt number on the Rayleigh number, namely, the mean temperature decreases with time for $Ra = 10^5$ and 10^6, while it increases with time for $Ra = 10^4$.

An analysis of the effect of the power law index on the liquid circulation and thermal transmission can be observed in Figures 7 and 8. It should be noted that n describes the character of the relationship between the components of the stress tensor and the strain rate tensor. The case $n < 1$ describes a pseudoplastic fluid, whose viscosity reduces with increasing strain rate. The case $n = 1$ describes a Newtonian fluid. The case $n > 1$ characterizes the behavior of the dilatant fluid, the viscosity of which rises with increasing strain rate.

The weakening of the convective heat transfer is well reflected with the distribution of streamlines and isotherms, shown in Figure 7a,b, respectively. It should be noted that with an increase in the power law index, the number of streamlines decreases, which also corresponds to a slowdown of the convective flow. The structure of the distribution of streamlines does not tolerate changes. One can find also a growth of the dynamic boundary layer thickness near the vertical walls.

The distribution of isotherms is also not subject to significant changes (see Figure 7b). The two-dimensional heat plume is located above the heat source, which reflects the formation of temperature stratification zones to the left and right of the energy source. As a result, it is possible to conclude that a growth of the power law index characterizes less intensive cooling of the cavity from the vertical cooled walls. Therefore, the temperature gradient within the cavity diminishes with increasing n and liquid circulation becomes weaker.

Moreover, velocity field behavior is related to the temperature field considering the natural convection problem where there is no any external dynamic influence. This relation can be described as follows; the presence of the non-slip effect at solid walls characterizes an appearance of high velocity gradients in these zones and as a result a growth of the power law index leads to a rise of the effective viscosity. The latter illustrates the attenuation of convective flow circulation and a formation of heat conduction dominated mode near the walls. Therefore, for high n one can find less intensive cooling of the cavity from the vertical walls.

In Figure 8a, the average Nusselt number decreases slightly with increasing n at $k = 100$, $Pr = 100$, $Ra = 10^5$. This suggests that convective heat transfer slows down with an increase in the power law index, which corresponds to a growth in temperature inside the source, as illustrated in Figure 8b. Moreover, the average Nusselt number reaches the constant value for high values of the power law index, while low values require more time for reaching steady state. At the same time, the mean heater temperature rises with n, which reflects less intensive heat removal from the heater for the case of dilatant liquid, while in the case of the pseudoplastic liquid it is possible to intensify the heater cooling.

For the considered problem, a key parameter is the thermal conductivity ratio. Figures 9 and 10 demonstrate the impacts of the thermal conductivity ratio on the flow structures and heat transfer patterns. Thus, streamlines and isotherms are presented in Figure 9 for $n = 0.8$, $Pr = 10^2$, $Ra = 10^5$.

It should be noted that the considered thermal conductivity ratio is a relation between the thermal conductivity of the heater material and liquid thermal conductivity. Therefore, growth of k illustrates an increase in the heater material thermal conductivity. A growth of this parameter reflects more intensive heating of the cavity and a growth of the circulation rate due to a formation of high temperature gradient. It is worth noting that a growth of the heater material thermal conductivity illustrates more intensive heating of this local element considering the internal volumetric heat generation and as a result the liquid near the heater warms rapidly for high k. Therefore, high k illustrates less cooling, while for low values of k heat conduction is a dominating heat transfer mechanism.

Figure 10 shows the average Nusselt number at the heater surface and the mean heater temperature depending on the thermal conductivity ratio and dimensionless time. It follows that an increase in k indicates a more heat-conducting material of the energy source or a less heat-conducting non-Newtonian medium. Therefore, with a growth of k, the average temperature in the source increases, which leads to a rise in the temperature gradient inside the cavity, and thus, one can find an augmentation of the average Nusselt number. It should be noted that for $k = 1$ and 10 the considered time range is enough for reaching the steady state, while for $k \geq 100$ the time range should be increased for the steady state. As a result, one can conclude that more cooling of the local heater can be organized by the passive cooling system (from the cooled vertical walls) in the case of low values of the thermal conductivity ratio $k \leq 100$.

4. Conclusions

In the present work, the mathematical simulation of the thermogravitational convection of a non-Newtonian fluid within a closed square cavity in the presence of a local heat-generating element was carried out. Governing partial differential equations formulated using the non-primitive variables and Ostwald-de Waele power law were solved by the finite difference method. An analysis of the key parameters influence on the process under investigation was performed based on the obtained distributions of streamlines and isotherms, as well as dependences of the average Nusselt number and average temperature inside the energy source. Taking into account the obtained results the following conclusions can be formulated:

- The influence of the Rayleigh number was considered in the range between 10^4 and 10^6. It has been established that with increasing Ra the heat transport mechanism changes from conductive to convective, and the average Nusselt number increases, which corresponds to the intensification of heat removal from the heater surface. At the same time, a rise of Ra characterizes a growth of time for reaching the steady state. Moreover, a rise of the Rayleigh number characterizes a reduction of the thermal boundary layers thickness not only near the vertical walls but also for the central formed thermal plume.
- The power law index is varied within the limits 0.8 and 1.4. It has been found that the growth of n slows down the flow and heat transfer in the cavity, therefore, for a pseudoplastic fluid, heat removal from the energy source occurs more intensively. It should be noted that at a large n faster stationary mode is achieved.
- Analysis of the flow and heat transfer as a result of an increase in the thermal conductivity ratio from 1 to 1000 showed that as the flow rate and time for reaching the steady state are increased, the cavity warms up more intensively, and therefore the heat removal from the heater surface is weakened.

Author Contributions: D.S.L. and M.A.S. conceived the main concept. D.S.L., A.S. and M.A.S. contributed to the investigation and data analysis. D.S.L., A.S. and M.A.S. wrote the manuscript. All authors contributed in writing the final manuscript.

Funding: This work was supported by the Grants Council (under the President of the Russian Federation), Grant No. MD-821.2019.8.

Conflicts of Interest: The authors declare no conflict of interest.

References

1. Coussot, P. Yield stress fluid flows: A review of experimental data. *J. Non-Newton. Fluid Mech.* **2014**, *211*, 31–49. [CrossRef]
2. Lu, G.; Wang, X.-D.; Duan, Y.-Y. A critical review of dynamic wetting by complex fluids: From Newtonian fluids to non-Newtonian fluids and nanofluids. *Adv. Colloid Interface Sci.* **2016**, *236*, 43–62. [CrossRef] [PubMed]
3. Gangawane, K.M.; Manikandan, B. Laminar natural convection characteristics in an enclosure with heated hexagonal block for non-Newtonian power law fluids. *Chin. J. Chem. Eng.* **2017**, *25*, 555–571. [CrossRef]
4. Bird, R.; Armstrong, R.; Hassger, O. *Dynamics of Polymeric Liquids, Volume 1: Fluid Mechanics*, 2nd ed.; Wiley InterScience: Hoboken, NJ, USA, 1987; p. 672.
5. Shenoy, A.V.; Saini, D.R. *Thermoplastic Melt Rheology and Processing*; CRC Press: New York, NY, USA, 1996.
6. Shenoy, A.V. *Rheology of Filled Polymer Systems*; Kluwer Academic Publishers: Dordrecht, The Netherlands, 1999.
7. Chhabra, R.; Richardson, J. *Non-Newtonian Flow and Applied Rheology: Engineering Applications*, 2nd ed.; Butterworth-Heinemann: Oxford, UK, 2008; p. 536.
8. Shenoy, A. *Heat Transfer to Non-Newtonian Fluids: Fundamentals and Analytical Expressions*; Wiley-VCH: Weinheim, Germany, 2018.
9. Xu, H.; Liao, S.-J. Laminar flow and heat transfer in the boundary-layer of non-Newtonian fluids over a stretching flat sheet. *Comput. Math. Appl.* **2009**, *579*, 1425–1431. [CrossRef]
10. Hariharan, P.; Seshadri, V.; Banerjee, R.K. Peristaltic transport of non-Newtonian fluid in a diverging tube with different wave forms. *Math. Comput. Model.* **2008**, *48*, 998–1017. [CrossRef]
11. Kalaitzis, A.; Makrygianni, M.; Theodorakos, I.; Hatziapostolou, A.; Melamed, S.; Kabla, A.; de la Vega, F.; Zergioti, I. Jetting dynamics of Newtonian and non-Newtonian fluids via laser-induced forward transfer: Experimental and simulation studies. *Appl. Surf. Sci.* **2019**, *465*, 136–142. [CrossRef]
12. Nafchi, P.M.; Karimipour, A.; Afrand, M. The evaluation on a new non-Newtonian hybrid mixture composed of $TiO_2/ZnO/EG$ to present a statistical approach of power law for its rheological and thermal properties. *Phys. A Stat. Mech. Its Appl.* **2019**, *516*, 1–18. [CrossRef]
13. Hundertmark-Zausková, A.; Lukáčová-Medvid'ová, M. Numerical study of shear-dependent non-Newtonian fluids in compliant vessels. *Comput. Math. Appl.* **2010**, *60*, 572–590. [CrossRef]
14. Janela, J.; Moura, A.; Sequeria, A. A 3D non-Newtonian fluid structure interaction model for blood flow in arteries. *J. Comput. Appl. Math.* **2010**, *234*, 2783–2791. [CrossRef]
15. Ostrach, S. Natural convection in enclosures. *Adv. Heat Transf.* **1972**, *8*, 161–227.
16. Haque, A.; Nayak, A.K.; Soni, B.; Majhi, M. Thermosolutal hydromagnetic convection of power law fluids in an enclosure with periodic active zones. *Int. J. Heat Mass Transf.* **2018**, *127*, 622–642. [CrossRef]
17. Shabany, Y. *Heat Transfer: Thermal Management of Electronics*; CRC Press: Boca Raton, FL, USA, 2009; p. 523.
18. Aminossadati, S.M.; Ghasemi, B. Natural convection cooling of a localised heat source at the bottom of a nanofluid-filled enclosure. *Eur. J. Mech. B-Fluid* **2009**, *28*, 630–640. [CrossRef]
19. El Qarnia, H.; Draoui, A.; Lakhal, E.K. Computation of melting with natural convection inside a rectangular enclosure heated by discrete protruding heat sources. *Appl. Math. Model.* **2013**, *37*, 3968–3981. [CrossRef]
20. Pesso, T.; Piva, S. Laminar natural convection in a square cavity: Low Prandtl numbers and large density differences. *Int. J. Heat Mass Transf.* **2009**, *152*, 1036–1043. [CrossRef]
21. Aly, A.M.; Raizahan, Z.A.S. Incompressible smoothed particle hydrodynamics (ISPH) method for natural convection in a nanofluid-filled cavity including rotating solid structures. *Int. J. Mech. Sci.* **2018**, *146–147*, 125–140. [CrossRef]
22. Gibanov, N.S.; Sheremet, M.A. Natural convection in a cubical cavity with different heat source configurations. *TSEP* **2018**, *7*, 138–145. [CrossRef]
23. Mahalakshmi, T.; Nithyadevi, N.; Oztop, H.F.; Abu-Hamdeh, N. Natural convective heat transfer of Ag-water nanofluid flow inside enclosure with center heater and bottom heat source. *Chin. J. Phys.* **2018**, *56*, 1497–1507. [CrossRef]
24. Sasmal, C.; Gupta, A.K.; Chhabra, R.P. Natural convection heat transfer in a power-law fluid from a heated rotating cylinder in a square duct. *Int. J. Heat Mass Transf.* **2019**, *129*, 975–996. [CrossRef]

25. Kiyasatfar, M. Convective heat transfer and entropy generation analysis of non-Newtonian power-law fluid flows in parallel-plate and circular microchannels under slip boundary conditions. *Int. J. Therm. Sci.* **2018**, *128*, 15–27. [CrossRef]
26. Zhou, E.; Bayazitoglu, Y. Developing laminar natural convection of power law fluids in vertical open ended channel. *Int. J. Heat Mass Transf.* **2019**, *128*, 354–362. [CrossRef]
27. Kefayati, G.H.R. Double-diffusive natural convection and entropy generation of Bingham fluid in an inclined cavity. *Int. J. Heat Mass Transf.* **2018**, *116*, 762–812. [CrossRef]
28. Alsabery, A.I.; Chamkha, A.J.; Saleh, H.; Hashima, I. Transient natural convective heat transfer in a trapezoidal cavity filled with non-Newtonian nanofluid with sinusoidal boundary conditions on both sidewalls. *Powder Technol.* **2017**, *308*, 214–234. [CrossRef]
29. Kefayati, G.H.R.; Tang, H. Simulation of natural convection and entropy generation of MHD non-Newtonian nanofluid in a cavity using Buongiorno's mathematical model. *Int. J. Hydrog. Energy* **2017**, *42*, 17284–17327. [CrossRef]
30. Shojaeian, M.; Karimzadehkhouei, M.; Koşar, A. Experimental investigation on convective heat transfer of non-Newtonian flows of Xanthan gum solutions in microtubes. *Exp. Therm. Fluid Sci.* **2017**, *85*, 305–312. [CrossRef]
31. Shenoy, A.; Sheremet, M.; Pop, I. *Convective Flow and Heat Transfer from Wavy Surfaces: Viscous Fluids, Porous Media and Nanofluids*; CRC Press: Boca Raton, FL, USA, 2016; p. 306.
32. Bondareva, N.S.; Sheremet, M.A.; Oztop, H.F.; Abu-Hamdeh, N. Heatline visualization of natural convection in a thick walled open cavity filled with a nanofluid. *Int. J. Heat Mass Transf.* **2017**, *109*, 175–186. [CrossRef]
33. Khezzar, L.; Siginer, D.; Vinogarov, I. Natural convection of power law fluids in inclined cavities. *Int. J. Therm. Sci.* **2012**, *53*, 8–17. [CrossRef]
34. Sojoudi, A.; Saha, S.C.; Gu, Y.T.; Hossian, M.A. Steady natural convection of non-Newtonian power law fluid in a trapezoidal enclosure. *Adv. Mech. Eng.* **2003**, *5*, 8. [CrossRef]

Article

Domestic Hot Water Storage Tank Utilizing Phase Change Materials (PCMs): Numerical Approach

Ayman Bayomy [1,*], Stephen Davies [2] and Ziad Saghir [3]

[1]   Mechanical and Industrial Engineering Department, Ryerson University, Toronto, ON M5B 2K3, Canada
[2]   Ecologix Heating Technologies Inc., Cambridge, ON N1R 7L2, Canada; sdavies@ecologix.ca
[3]   Mechanical and Industrial Engineering Department, Ryerson University, Toronto, ON M5B 2K3, Canada; zsaghir@ryerson.ca
*   Correspondence: ayman.bayomy@ryerson.ca

Received: 25 April 2019; Accepted: 4 June 2019; Published: 6 June 2019

Abstract: Thermal energy storage (TES) is an essential part of a solar thermal/hot water system. It was shown that TES significantly enhances the efficiency and cost effectiveness of solar thermal systems by fulfilling the gap/mismatch between the solar radiation supply during the day and peak demand/load when sun is not available. In the present paper, a three-dimensional numerical model of a water-based thermal storage tank to provide domestic hot water demand is conducted. Phase change material (PCM) was used in the tank as a thermal storage medium and was connected to a photovoltaic thermal collector. The present paper shows the effectiveness of utilizing PCMs in a commercial 30-gallon domestic hot water tank used in buildings. The storage efficiency and the outlet water temperature were predicted to evaluate the storage system performance for different charging flow rates and different numbers of families demands. The results revealed that increases in the hot water supply coming from the solar collector caused increases in the outlet water temperature during the discharge period for one family demand. In such a case, it was observed that the storage efficiency was relatively low. Due to low demand (only one family), the PCMs were not completely crystallized at the end of the discharge period. The results showed that the increases in the family's demand improve the thermal storage efficiency due to the increases in the portion of the energy that is recovered during the nighttime.

Keywords: PCMs; storage tank; photovoltaic; computational fluid dynamics (CFD); finite elements

---

1. Introduction

Phase change materials (PCMs) are an important topic in research and industry. Due to the intermittent supply characteristics of solar energy, energy storage represents a potential solution by storing the energy during the daytime to use it during the nighttime. PCMs can store large amounts of heat at constant temperature while the material changes phase or state. When PCMs are coupled with solar technology, efficiency of traditional heating, ventilation, and air conditioning (HVAC) systems may potentially increase. The majority of applications for PCMs are for space heating/cooling and providing domestic hot water for buildings. Also, PCMs have high energy density and latent heat, and they are cost-effective. There are several applications in which PCMs were implemented as shown in Table 1.

Phase change material (PCM) is an environmentally friendly material used to improve building energy consumption and indoor thermal comfort [1,2].

An experimental study investigated a heat pump utilizing a thermal energy storage (TES) tank [3]. In their research, it was found that a PCM storage tank has 14.5% better performance. Moreover, the PCM storage tank improved indoor temperature stability within comfort 20.65% longer in time compared to the conventional water tank. Also, a thermal storage system was installed in a one family house [4], where sodium acetate trihydrate (SAT) was used as a PCM.

Table 1. Phase change material (PCM) applications.

Application	References
Thermal storage of solar energy	[5–7]
Heating and sanitary hot water	[8,9]
Cooling	[10–16]
Thermal comfort in vehicles	[17]
Solar power plants	[18–21]
Cooling of engines	[22]
Thermal protection of electronic devices	[23]
Spacecraft thermal system	[24]

The performance of thermal storage systems was analyzed [25]. A south-oriented wall was used as thermal storage with phase change materials embedded in the wall. The thermal storage system can store solar radiation up to 6–8 h after solar irradiation; this has effects on the stability of the daily temperature swings (up to 10 °C).

Zalba et al. [26] conducted an experimental study to store outdoor cold during nighttime and to release it indoors during daytime using PCMs. The results revealed that the system was successfully tested for 1000 cycles without any degradation in the performance of the system.

Kenneth [27] developed a solar system that utilizes a PCM in domestic houses in the United Kingdom (UK). The system consisted of solar flat plate collectors, which put the energy into a storage tank and PCM-filled panels. Calcium chloride was used as the PCM with a melting point of 29 °C. The results revealed that the use of PCMs reduced energy consumption by 18–32%.

The literature shows a limited number of investigations on PCM thermal storage tanks coupled with a solar thermal collector to provide domestic hot water or building heating. In the present paper, a numerical model was developed for a 30-gallon hot water tank utilizing *n*-eicosane PCM as the thermal store medium. The solar energy was used to charge the storage tank during the daytime (9 h) and then the thermal energy was recovered to provide hot domestic water during the discharge period (15 h).

2. Numerical Model Description

In this study, a three-dimensional numerical model was created using finite element techniques (COMSOL Multiphysics) for a 30-gallon domestic hot water thermal storage tank connected to a photovoltaic thermal collector. The tank height and diameter were 1.15 m and 0.46 m, respectively, as shown in Figure 1. PCMs are present in small cylindrical containers inside the tank as shown in Figure 2a. The tank contains a spiral heat exchanger to exchange the heat from the circulating water to the thermal storage medium as shown in Figure 2b. The diameter of the heat exchanger pipe was 0.02 m, and the length of the heat exchanger was 1.1 m with a spiral diameter of 0.28 m. The length and diameter of the PCM containers were 0.2 m and 0.05 m, respectively. The tank consisted of 32 PCM cylindrical containers per row with a total of five rows. The properties of the PCM used in the model are presented in Table 2.

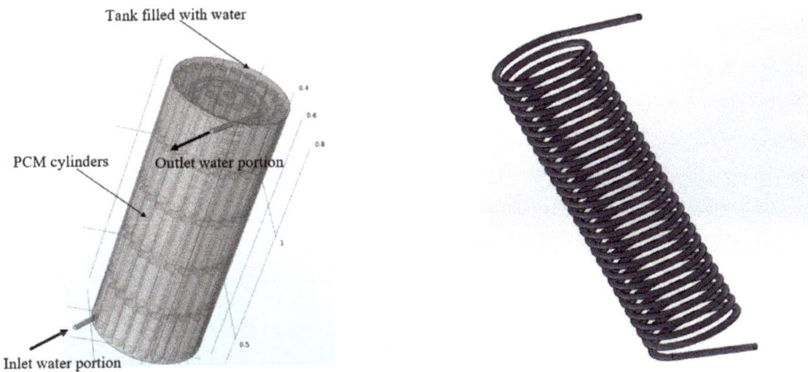

Figure 1. System schematic diagram.

(**a**) Model description   (**b**) Spiral heat exchanger

Figure 2. Numerical model description.

Table 2. PCM properties.

n-Eicosane	Solid Phase	Liquid Phase
Density (kg/m^3)	856	778
Specific heat (kJ/(kg·K))	2.136	2.1336
Thermal conductivity (W/m·K)	0.35	0.15
Melting point	36.4 °C	
Latent heat (kJ/kg)	247.3	

2.1. Governing Equations

The assumptions on which the governing equations were based prior to the formulation of the model were as follows:

(1) The fluid flow is Newtonian and incompressible;
(2) No heat is generated inside tank solid domains;
(3) The variation of thermo-physical properties of the PCM can be neglected.

When these assumptions were taken into consideration, a system of governing equations to describe the heat transfer and fluid flow were solved and coupled. In the present study, an attempt was made to solve the Navier–Stokes equations for the fluid flow through the internal spiral heat exchanger and a free convection fluid flow inside the tank. In addition, the energy equations for all domains including the tank solid domain, the PCM domain, and the fluid domain were solved. It is important to note that the Navier–Stokes equations and the energy equations were coupled. The coupling was established for each time step using the velocity field obtained from Navier–Stokes equations as an input to evaluate the convective heat transfer term in the energy equation.

The Navier–Stokes equation and the energy equation were solved numerically using COMSOL Multiphysics [3] as follows:

Momentum equation along x-direction:

$$\rho\left(u\frac{\partial u}{\partial x} + v\frac{\partial u}{\partial y} + w\frac{\partial u}{\partial z}\right) = -\frac{\partial p}{\partial x} + \mu\left(\frac{\partial^2 u}{\partial x^2} + \frac{\partial^2 u}{\partial y^2} + \frac{\partial^2 u}{\partial z^2}\right); \tag{1}$$

Momentum equation along y-direction:

$$\rho\left(u\frac{\partial v}{\partial x} + v\frac{\partial v}{\partial y} + w\frac{\partial v}{\partial z}\right) = -\frac{\partial p}{\partial y} + \mu\left(\frac{\partial^2 v}{\partial x^2} + \frac{\partial^2 v}{\partial y^2} + \frac{\partial^2 v}{\partial z^2}\right); \tag{2}$$

Momentum equation along z-direction:

$$\rho\left(u\frac{\partial w}{\partial x} + v\frac{\partial w}{\partial y} + w\frac{\partial w}{\partial z}\right) = -\frac{\partial p}{\partial z} + \mu\left(\frac{\partial^2 w}{\partial x^2} + \frac{\partial^2 w}{\partial y^2} + \frac{\partial^2 w}{\partial z^2}\right) + \rho g. \tag{3}$$

The continuity equation for this simulation can be expressed as

$$\left(\frac{\partial u}{\partial x} + \frac{\partial v}{\partial y} + \frac{\partial w}{\partial z}\right) = 0. \tag{4}$$

The energy equation was as follows:

$$\left(\rho c_p\right)\left(u\frac{\partial T}{\partial x} + v\frac{\partial T}{\partial y} + w\frac{\partial T}{\partial z}\right) = K\left(\frac{\partial^2 T}{\partial x^2} + \frac{\partial^2 T}{\partial y^2} + \frac{\partial^2 T}{\partial z^2}\right), \tag{5}$$

where ρ represents the water density, c_p represents the fluid specific heat, p represents the pressure, u, v, and w represent the coordinates of a velocity field vector, T represents the temperature, μ represents the dynamic viscosity of the fluid, and K represents the thermal conductivity.

For the phase change material,

$$c_p = \frac{1}{\rho}\left(\theta \rho_{phase1} c_{p\,phase1} + (1-\theta)\rho_{phase2} c_{p\,phase2}\right) | L\,\frac{\partial \alpha_m}{\partial T}, \tag{6}$$

$$\alpha_m = \frac{1}{2}\frac{(1-\theta)\rho_{phase2} - \theta\rho_{phase1}}{\theta\rho_{phase1} + (1-\theta)\rho_{phase2}}, \tag{7}$$

where θ represents the PCM solid fraction, L represents the latent heat of the phase change material, phases 1 and 2 represent the solid and liquid phases, respectively, and $\frac{\partial \alpha_m}{\partial T}$ represents the melting fraction per degree of temperature.

2.2. Boundary Conditions

Boundary conditions can be divided into two types: heat transfer boundary conditions and fluid flow boundary conditions. The thermal boundary conditions involve the inlet water temperature to the heat exchange, which was 50 °C during the charging period over the daytime (9 h) and 20 °C during the discharge period over the nighttime (15 h). Moreover, the thermal boundary conditions included the insulated wall of the tank outer surface ($\nabla.T = 0$).

The fluid flow boundary conditions included the inlet velocity (U) at the inlet portion (assuming a flat profile), with no pressure constraints at the outlet portion and walls (e.g., no slip condition) at the remainder of the surface (see Figure 2a). It is important to note that, in the present study, the flow rates during the charging period were set to 2 L/m, 3 L/m, and 4 L/m. In addition, the flow rate of the demand of hot water during the discharge period was set to match the typical hot water demand for one, two, three, and four families, as shown in Figure 3 [28]. A typical family consists of two adults and two children [28]. As shown in Figure 3, the demand was low over the first 5 h of the day and increased dramatically over the rest of the day.

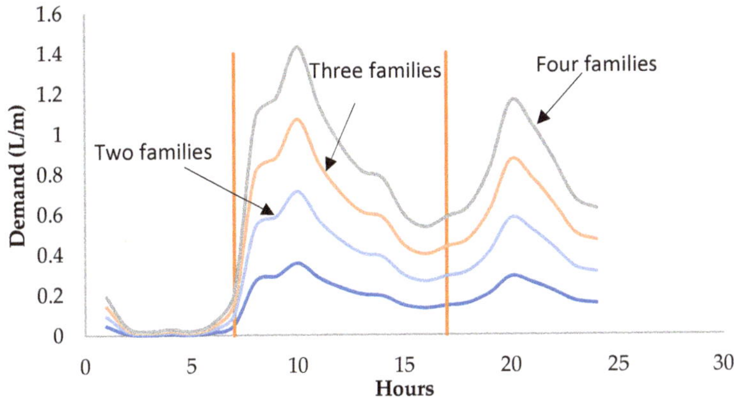

Figure 3. Typical domestic hot water demand for different numbers of families.

3. Results and Discussion

3.1. Effect of Charging Flow Rates

In this section, the flow rates over the charging period (9 h) were set to 2 L/m, 3 L/m, and 4 L/m. The demand of domestic hot water flow rate over the discharge period (during the remaining 15 h) was fixed to match the typical hot water demand for one family, as illustrated in Figure 3. Figure 4 shows the storage tank inlet and outlet temperatures versus time.

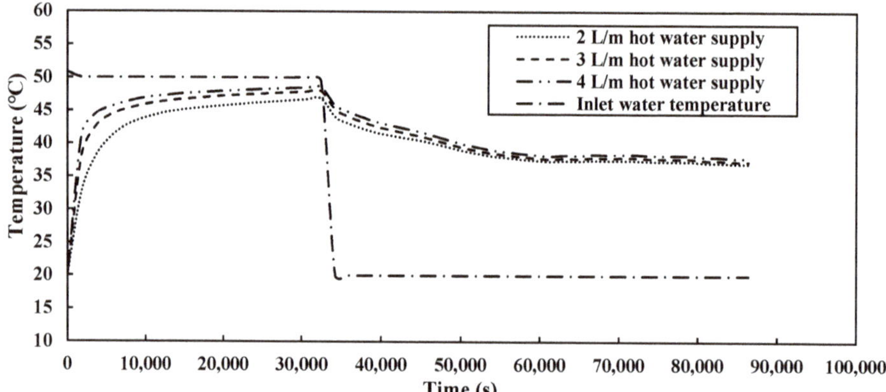

Figure 4. Storage tank inlet and outlet water temperatures.

As shown in Figure 4, the inlet temperature was 50 °C during the charging period (9 h), which came from the photovoltaic thermal (BIPVT/T) collector. Then, the inlet water temperature decreased to 20 °C during the discharge period. The amount of thermal energy was calculated as follows:

$$q_{stored} = m \cdot c_p (T_{in} - T_{out}) \Delta t; \tag{8}$$

$$q_{discharged} = m \cdot c_p (T_{out} - T_{in}) \Delta t. \tag{9}$$

It was noted that, over the charging period, the amount of heat stored (q_{stored}) increased upon increasing the hot water supply flow rate at given demand.

As a result, the average outlet water temperatures during the discharge period increased with the increase in the hot water flow rate during the 9-h charging period, as shown in Figure 4.

Table 3 shows the amount of heat being stored and extracted during charging and discharging periods, respectively. It was observed that the maximum storage efficiency was 39%. The thermal storage efficiency is defined as a ratio between the energy extracted and the energy injected into the tank. This means that only 35–39% of stored energy was successfully extracted during the discharge period. Figures 5 and 6 explain the reason for such low storage efficiency. Figure 5 shows the melting fraction of the PCM cylinders; when it is equal to one, it means that the PCMs are completely melted. As shown in this figure, PCMs were not melted completely after the 9-h charging period. This means that there was a capability of the storage tank to absorb more heat during the charging period.

Table 3. Tank storage efficiency.

Hot Water Supply Flow Rate	Heat Input (kJ)	Heat Extracted (kJ)	Storage Efficiency (%)
2 L/m	22,307.5	7869	35%
3 L/m	21,417	8253	38.5%
4 L/m	21,738	8511	39%

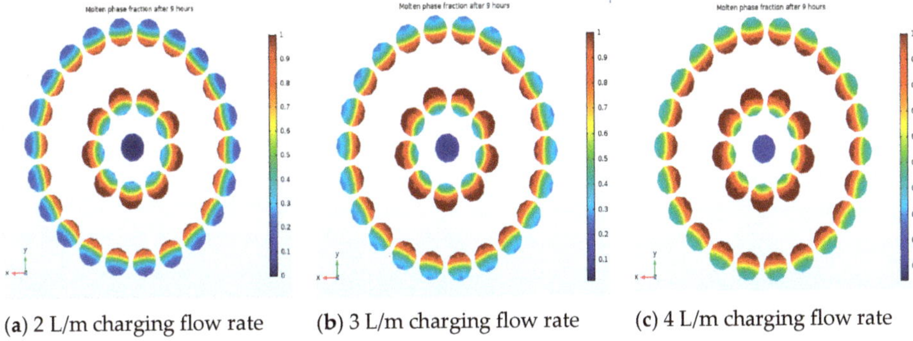

(**a**) 2 L/m charging flow rate   (**b**) 3 L/m charging flow rate   (**c**) 4 L/m charging flow rate

Figure 5. Melting fraction of phase change material (PCM) after charging period (9 h).

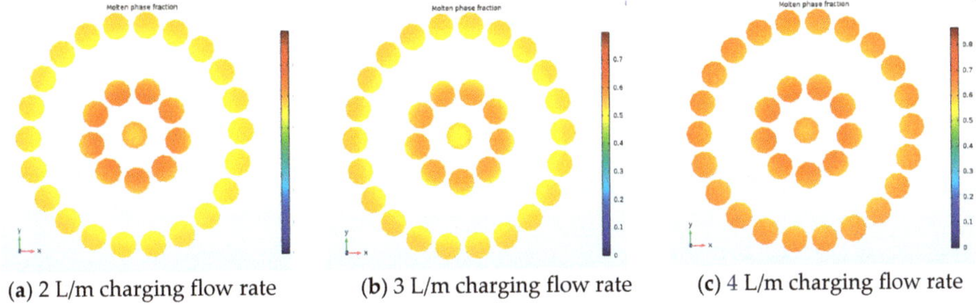

(**a**) 2 L/m charging flow rate   (**b**) 3 L/m charging flow rate   (**c**) 4 L/m charging flow rate

Figure 6. Melting fraction of phase change material after discharge period (24 h).

Moreover, Figure 6 shows that the PCMs were not solidified/frozen completely after the discharge period (24 h). It was observed that about 50% of PCMs were not completely solidified/frozen. This means that the tank still had some available heat to be used. To reach better efficiency, the hot water demand should be increased in order to absorb the remaining stored heat, allowing the PCM to be completely solidified/frozen.

In order to better understand the melting phenomenon, Figure 7 shows the melting fraction of a single PCM cylinder. As shown in this figure, around half of the cylinder was melted completely after the charging period. Moreover, after completing one cycle (e.g., 24 h), the PCM cylinder still had a liquid phase up to 0.7, as shown in Figure 7b.

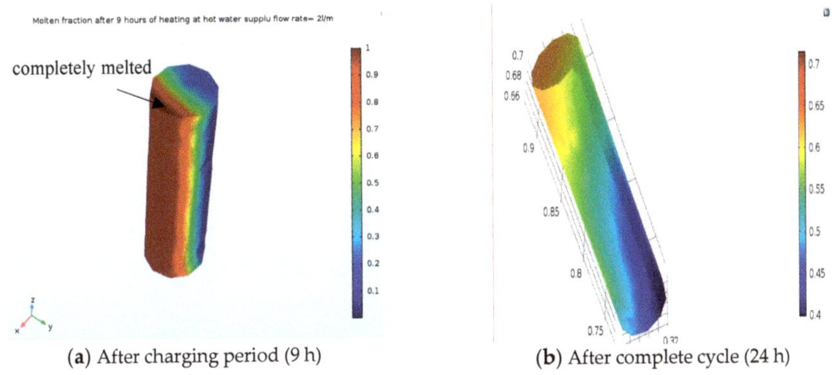

(**a**) After charging period (9 h)   (**b**) After complete cycle (24 h)

Figure 7. Melting fraction over charging and discharging periods for a single PCM cylinder.

Figure 8a shows the mid-plane temperature contours at the end of charging period. As shown in this figure, the temperature of the circulating water coming from the solar collector was about 50 °C, and the average temperature of the tank was about 34 °C. Also, Figure 8b shows the temperature contours after 24 h, where the average temperature was about 30 °C.

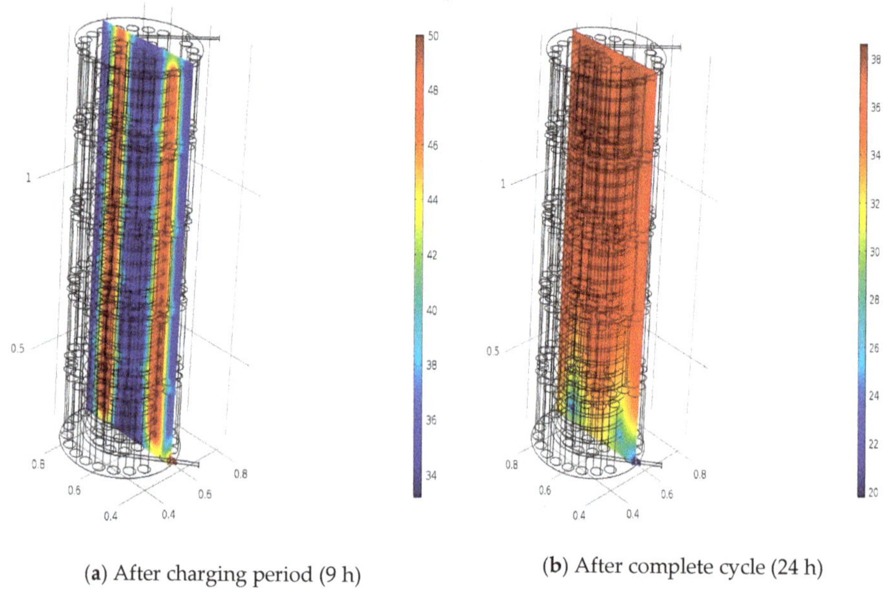

(**a**) After charging period (9 h)   (**b**) After complete cycle (24 h)

Figure 8. Mid-plane temperature contours after charging and discharging periods.

3.2. Effect of Number of Families

In order to enhance the storage efficiency, domestic hot water demand was increased to match the typical demand of two, three, and four families. Figure 9a shows the outlet temperatures from the tank versus time for the demand of one, two, three, and four families and 2 L/m of hot water supply during the charging period. It was noted that the outlet temperature decreased with the increase in the number of families. The increase in hot water demand during the discharge periods (number of families) meant increases in the thermal energy to be extracted/recovered from the thermal storage tank, as shown in Figure 9b. However, for the demand of four families, it was observed that the domestic hot water temperature over the discharge period decreased dramatically until it reached below 30 °C. In such a case, a supplement back-up system is needed to maintain the outlet domestic water above 30 °C.

In order to quantify the thermal storage efficiency, Table 4 shows the amount of heat that was injected into the thermal storage tank during the charging period (9 h) and the amount of heat that was absorbed/recovered during the discharge periods. It was observed that, for a given hot water supply, increasing the number of families increased the efficiency from 35% for one family to 82% for four families. The four families used the whole store of energy and, because of the high demand rate, a big portion of the stored energy was recovered. As a result, the storage efficiency was increased up to 82%.

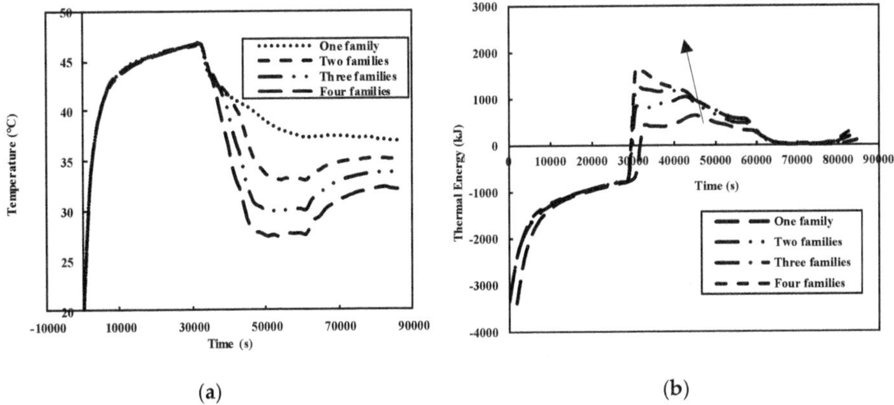

(a)   (b)

Figure 9. Outlet water temperatures and thermal energy over one day of operation. (**a**) Outlet temperature variation versus number of families; (**b**) Thermal energy stored and recovered during the day for the demands of one, two, three, and four families.

Table 4. Thermal storage efficiencies for multiple families.

Hot Water Supply Flow Rate	Number of Families	Heat Input (kJ)	Heat Extracted During Night (kJ)	Storage Efficiency (%)
2 L/m	1	22,307.5	7869	35%
2 L/m	2	22,307.5	13,415.96	60%
2 L/m	3	22,307.5	16,643.48	74%
2 L/m	4	22,307.5	18,288.89	82%

4. Conclusions

This paper presented a CFD numerical code of a domestic hot water tank utilizing a phase change material as a storage medium. The following conclusions were made:

- The increases in the hot water supply during the charging periods increased the storage efficiency from 35% to 39%.
- At given hot water supply, increasing the number of families increased the efficiency from 35% for one family to 82% for four families.
- At given hot water supply, the heat extracted over the nighttime increased from 7869 kJ to 18,288.89 kJ upon increasing the demand from one family to four families.

Further developments and future work on the present topic involve the introduction of nanofluid to enhance the thermal storage efficiency of the tank, the use of different PCMs and melting temperatures, and the calculation of the electricity consumption to determine the energy efficiency. Moreover, the effect of inlet water profile (cycling temperature) will be investigated as an extension of the present work.

Author Contributions: Conceptualization, methodology, validation, and original draft preparation, A.B.; review and editing, S.D. and Z.S.

Funding: This research was funded by Natural Sciences and Engineering Research (NSERC).

Acknowledgments: The authors acknowledge the full financial support of the National Science and Engineering Research Council (NSERC) and Ecologix Heating Technologies Inc.

Conflicts of Interest: The authors declare no conflict of interest

References

1. Baetens, R.; Jelle, B.; Gustavsen, A. Phase change materials for building applications: A state-of-the-art review. *Energy Build.* **2010**, *42*, 1361–1368. [CrossRef]
2. Lin, W. Development and evaluation of a ceiling ventilation system enhanced by solar photovoltaic thermal collectors and phase change materials. *Energy Convers. Manag.* **2014**, *88*, 218–230. [CrossRef]
3. Moreno, P. PCM thermal energy storage tanks in heat pump system for space cooling. *Energy Build.* **2014**, *82*, 399–405. [CrossRef]
4. Johansen, J. Laboratory Testing of Solar Combi System with Compact Long Term PCM Heat Storage. *Energy Procedia* **2016**, *91*, 330–337. [CrossRef]
5. Enibe, S. Performance of a natural circulation solar air heating system with phase change material energy storage. *Renew. Energy* **2002**, *27*, 69–86. [CrossRef]
6. Enibe, S. Parametric effects on the performance of a passive solar air heater with storage. In Proceedings of the World Renewable Energy Congress WII, Cologne, Germany, 19–26 October 2002.
7. Buddhi, D.; Sahoo, L. Solar cooker with latent heat storage: Design and experimental testing. *Energy Convers. Manag.* **1997**, *38*, 493–498. [CrossRef]
8. Mehling, H.; Cabeza, L.; Hippel, S.; Hiebler, S. Improvement of stratified hot water heat stores using a PCMmodule. In Proceedings of the EuroSun, Bologna, Italy, 23–26 June 2002.
9. Mehling, H. PCM-module to improve hot water heat stores with stratification. *Renew. Energy* **2003**, *28*, 699–711. [CrossRef]
10. Safarik, M.; Gramlich, K.; Schammler, G. Solar absorption cooling system with 90 _C-latent heat storage. In Proceedings of the World Renewable Energy Congress WII, Cologne, Germany, 19–26 October 2002.
11. Buick, T.; O'Callaghan, P.; Probert, S. Short-term thermal energy storage as a means of reducing the heat pump capacity required for domestic central heating systems. *Int. J. Energy Res.* **1987**, *11*, 583–592. [CrossRef]
12. Charters, W.L.; Aye, L.; Chaichana, C.; MacDonald, R. Phase change storage systems for enhanced heat pump performance. In Proceedings of the 20th International Congress of Refrigeration, IIR/IIF, Sydney, Australia, 16–17 June 1999.
13. Lorsch, H. *Improving Thermal and Flow Properties of Chilled Water. Part 2: Facility Construction and Flow Tests*; American Society of Heating, Refrigerating and Air-Conditioning Engineers, Inc.: Atlanta, GA, USA, 1997.
14. Velraj, R.; Anbudurai, K.; Nallusamy, N.; Cheralathan, M. PCM based thermal storage system for building airconditioning—Tidal Park, Chennai. In Proceedings of the World Renewable Energy Congress WII, Cologne, Germany, 29 June–5 July 2002.
15. Ismail, K. *Ice-Banks: Fundamentals and Modelling*; State University of Campinas: Campinas-SP-Brazil, Brazil, 1998.
16. Hasnain, S. Review on sustainable thermal energy storage technologies, Part II: Cool themal storage. *Energy Convers. Mgmt.* **1998**, *39*, 1139–1153. [CrossRef]
17. Bl€uher, P. Latent€wärmespeicher erh€oht den Fahrkomfort und die Fahrsicherheit. *ATZ Automob. Z.* **1911**, *93*, 3–8.
18. Hunold, D.; Ratzesberger, R.; Tamme, R. Heat transfer measurements in alkali metal nitrates used for PCM storage applications. In Proceedings of the Eurotherm Seminar No. 30, Hamburg, Germany, 27 February–1 March 1992.
19. Hunold, D.; Ratzesberger, R.; Tamme, R. Heat transfer mechanism in latent-heat thermal energy storage medium temperature application. In Proceedings of the 6th International Symposium on Solar Thermal Concentrating Technologies, Mojacar, Spain, 28 September–2 October 1992.
20. Michels, H.; Hahne, E. Cascaded latent heat storage for solar thermal power stations. In Proceedings of the Eurosun, Freiburg, Germany, 16–19 September 1996.
21. Michels, H.; Pitz-Paal, R. Cascaded latent heat storage for parabolic trough solar power plants. *Sol. Energy* **2007**, *81*, 829–837. [CrossRef]
22. Vasiliev, L. Latent heat storage modules for preheating internal combustion engines: Application to a bus petrol engine. *Appl. Therm. Eng.* **2000**, *20*, 913–923. [CrossRef]
23. Cabeza, L.; Roca, J.; Nogu_es, M.; Zalba, B.; Mar, J. Transportation and conservation of temperature sensitive materials with phase change materials: State of the art. In Proceedings of the IEA ECES IA Annex 17 2nd Workshop, Ljubljana, Slovenia, 3–5 April 2002.

24. Mulligan, J.; Colvin, D.; Bryant, Y. Microencapsulated phase-change material suspensions for heat transfer in spacecraft thermal systems. *J. Spacecr. Rocket.* **1996**, *33*, 278–284. [CrossRef]

25. Guarino, F. PCM thermal storage design in buildings: Experimental studies and applications to solaria in cold climates. *Appl. Energy* **2017**, *185*, 95–106. [CrossRef]

26. Zalba, B. Free-cooling of buildings with phase change materials. *Int. J. Refrig.* **2004**, *27*, 839–849. [CrossRef]

27. Kenneth, I.P. *Solar Thermal Storage Using Phase Change Material for Space Heating of Residential Buildings*; Research Article on Net; University of Brighton, School of the Environment: Brighton, UK, 2002.

28. American Society of Heating, Refrigerating and Air-Conditioning Engineers. *ASHRAE Applications Handbook, I-P and SI edn*; American Society of Heating, Refrigerating, and Air-Conditioning Engineers, Inc.: Atlanta, GA, USA, 2003.

Article

Experimental and Numerical Study of the Flow and Heat Transfer in a Bubbly Turbulent Flow in a Pipe with Sudden Expansion

Pavel Lobanov [1], Maksim Pakhomov [2,*] and Viktor Terekhov [2]

[1] Laboratory of Problems of Heat and Mass Transfer, Kutateladze Institute of Thermophysics, Siberian Branch of Russian Academy of Sciences, Acad. Lavrent'ev Avenue 1, 630090 Novosibirsk, Russia

[2] Laboratory of Thermal and Gas Dynamics, Kutateladze Institute of Thermophysics, Siberian Branch of Russian Academy of Sciences, Acad. Lavrent'ev Avenue 1, 630090 Novosibirsk, Russia

* Correspondence: pakhomov@ngs.ru

Received: 3 July 2019; Accepted: 15 July 2019; Published: 17 July 2019

Abstract: The flow patterns and heat transfer of a downstream bubbly flow in a sudden pipe expansion are experimentally and numerically studied. Measurements of the bubble size were performed using shadow photography. Fluid phase velocities were measured using a PIV system. The numerical model was employed the Eulerian approach. The set of RANS equations was used for modelling two-phase bubbly flows. The turbulence of the carrier liquid phase was predicted using the Reynolds stress model. The peak of axial and radial fluctuations of the carrier fluid (liquid) velocity in the bubbly flow is observed in the shear layer. The addition of air bubbles resulted in a significant increase in the heat transfer rate (up to 300%). The main enhancement in heat transfer is observed after the point of flow reattachment.

Keywords: turbulent bubbly flow; sudden pipe expansion; measurements; modeling; wall friction; heat transfer modification

---

1. Introduction

Gas–liquid bubbly flows are frequently used in many engineering and practical applications. These flows are usually turbulent with a strong interfacial coupling between the fluid carrier and dispersed phases. A study of such flows is usually complicated by flow separation at sharp edges, and heat transfer. Flow recirculation largely determines the mean and turbulent characteristics and has a significant influence on the transport processes [1,2]. The modeling of bubble motion and their distributions over a duct or pipe cross-section is very important for safety and for the simulation of various emergency scenarios in steam and nuclear power equipment.

Experimental works [3,4] were probably among the first studies of bubbly flows downstream of a sudden expansion. They studied bubbly flows behind a sudden expansion in a horizontal duct. The pressure drop, void fraction, bubble size, and the mean and fluctuation velocities of the phases, were measured in these papers. Aloui and Souhar [3] and Rinne and Loth [4] experimentally observed a reduction in diameter of bubbles along the duct length due to the growth in the longitudinal pressure gradient and the bubbles break up in separation zone.

Later experimental [5–8] and numerical [9,10] studies of separated isothermal bubbly flows were continued. A theoretical and experimental study of upward bubbly flow in a pipe with sudden expansion [5] was carried out on. The mean and fluctuating bubble velocities, local void fraction, bubble distributions, pressure drop and friction on the wall were measured. The distributions of the bubble's velocity fluctuations were similar to that of the carrier fluid phase. Measurements were performed in the horizontal flow of the air–oil mixture [6], which was transient in the annular flow regime. Gas concentration increased immediately after the cross-section of the flow separation. Measurements by [7] were performed in an upward, vertical flow of a liquid (water) and CO_2 bubbles. The distribution of the mean carrier fluid velocity was strongly dependent on the bubbles' diameter. A review of the literature for two-phase flow in sudden expansions and contractions was performed in [8]. Wang et al. [8] have found that none of the existing eight correlations can accurately predict the experimental database. Most of these correlations greatly overpredicted the results for mini test sections. Some of the correlations also underpredicted the data for large test sections.

There are only a few papers that consider the numerical modeling of isothermal bubbly flows in a pipe with sudden expansion [9,10]. Krepper et al. [9] presented a mathematical model for polydisperse flow in a vertical pipe with sudden expansion. The numerical predictions for bubbly pipe flow with an obstacle demonstrated very difficult nature of such flows [9]. The Eulerian model was developed to simulate bubbly flows in a pipe with sudden expansion in [10]. Predictions were carried out using the CFD commercial package STAR-CCM+. The developed model was validated against measurements of Bel Fdhila [11]. The eddy-viscosity $k-\varepsilon$ model showed poor reproduction of the bubbly flow structure because it disregarded much of the information on fluid flow turbulence. The Second-Moment Closure and LES methods captured the main characteristic turbulent length-scales responsible for the diffusion of the bubble gas phase, and moreover, these methods predicted the flow patterns and bubble distributions across the pipe section [10].

Recently published numerical works [12,13] investigated heat transfer in bubbly flows without an abrupt pipe or channel expansion. These studies showed the significant influence of bubble diameter and gas volumetric flow rate ratios on the flow patterns and heat transfer. Heat transfer in bubbly flow increased up to three times that of one-phase fluid flow. The liquid turbulence was modeled using an isotropic $k-\varepsilon$ model [12]. The DNS of bubbly flow with heat transfer was performed [13]. The numerical [14,15] and experimental papers [15] concerned study of bubbly flows with a sudden pipe expansion. The numerical model [14] employed the Eulerian approach. The turbulence of the carrier fluid flow was predicted with the use of a Reynolds stress model. The effects of inlet gas concentrations, Reynolds numbers, and bubble diameters on the flow structure and heat transfer were investigated. The measurements of heat transfer in gas–liquid flow in the pipe with sudden expansion and validation analysis of model [14] was performed [15].

The aim of this work is to carry out a further validation of the numerical model [14] for predictions of the flow structure in pipe sudden expansion. Authors of [15] used for comparison analysis only their measured results of heat transfer in the gas–liquid flow downstream of the pipe with sudden expansion. In the paper we performed the validations with our experimental results of mean flow structure, measurements of [11] and LES simulations of [10]. The measured and predicted results for the mean axial carrier fluid (liquid) and gas (air bubbles) phases in a few stations along the pipe length are presented. The LES predictions of [10] and measurements of [11] were performed for much larger Reynolds number ($Re = 1.1 \times 10^5$) and another pipe geometry ($2R_1 = 50$ mm, $2R_2 = 100$ mm, $H = 25$ mm, and $ER = (R_2/R_1)^2 = 4$). It has expanded the possibility of using of the model. We have also presented the numerical results of the flow structure predictions (turbulent kinetic energy profiles and wall friction coefficient distributions for various β). All these abovementioned numerical results did not present in [15].

This study may be of interest to scientists and engineers dealing with the enhancement of heat transfer in power equipment. The paper does not have any direct practical application. The article shows the physical mechanisms of the control on flow structure and heat transfer enhancement in bubbly flow in a pipe sudden expansion.

2. Experimental Facility and Image Processing Methods

2.1. Experimental Facility

The experimental facility consists of a closed circuit in a liquid and an open circuit in the gas phase (see Figure 1). The test liquid, that is, distilled water, is pumped from the main tank, 1, to the test Section 2, through the water supply line, 3. The volume of liquid in the main tank is 40 L. To remove excess heat from the tank, a heat exchanger, through which cooling water was pumped, was installed. The liquid temperature in the tank was measured by a resistive temperature sensor installed in the tank and automatically controlled by management of the cooling water flow rate by a type TRM-2M programmable controller. This makes it possible to keep the liquid temperature at the beginning of the test section in the 24.9–25.1 °C range, according to an analysis of the signal from the temperature sensor mounted in the test section's inlet, which is a thin film platinum sensor. The nominal resistance of this sensor is 1000 Ohm at 0 °C. The response time of this type of sensor in water flow is about 0.2 s. The measurement uncertainty of temperature measurement in the test section inlet is about ±0.15 °C.

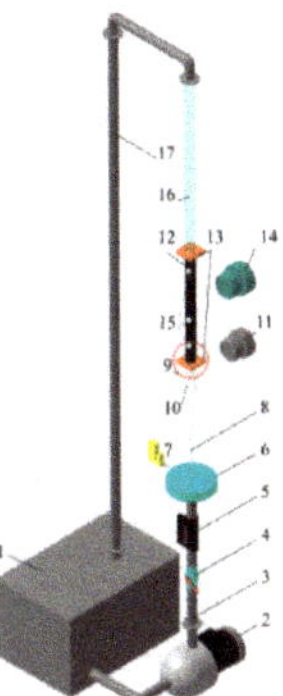

Figure 1. The scheme of the experimental facility. 1—main tank; 2—pump; 3—supply line; 4—control valve; 5—ultrasound flow meter; 6—gas–liquid mixer; 7—gas flow controller; 8—the pipe of small diameter (before expansion); 9—expansion area; 10—box filled by water; 11—video recorder; 12—test section; 13—copper conductors; 14—infrared (IR) camera; 15—resistive temperature detectors (RTDs); 16—the pipe of large diameter (after expansion); 17—downcomer.

A Grundfos CHI 4–40 ($Ql \leq 4.5$ m^3/h) is used to pump the liquid. The liquid flux is directed upstream, the liquid flow rate is controlled using a valve, 4. The flow rate is monitored using an ultrasonic flow meter, 5. In this paper the effect of intermediate sized bubbles ($d = 1 \div 3$ mm) on the heat transfer in the pipe with sudden expansion is studied. Bubbles are generated by feeding air into the liquid flow through 12 capillaries, where the length of each capillary is 50 mm and inner diameter is 0.7 mm. The capillaries are uniformly distributed across the pipe channel's cross-section. The gas flow rate is determined using a gas flow controller, 7, produced by Bronkhorst ($Qg \leq 1$ L/min). The measurement error of the gas and liquid flow rates is ±2% of the measured value. Then, a gas–liquid mixture is fed into the test section. It is a small-diameter pipe, 8 ($2R_1 = 15$ mm), which is inserted through the adapter and into a large diameter pipe. The inner diameter of the pipe, 16, is $2R_2 = 42$ mm, and the step height is $H = 13.5$ mm. The expansion ratio is defined as ER $= (R_2/R_1)^2 = 7.84$. The end of the small pipe is mounted flush with the plane of the adapter located in the flow separation zone. To ensure that a fully developed gas–liquid flow is obtained in the inlet of the test section, the length of the flow-stabilization area before the measurement region is $140R_1$. The experimental setup and the method for heat transfer measurements was presented in [15] in details.

and g_{ut} is the coefficient of involvement of dispersed phase into the large-eddy temperature fluctuation motion of fluid phase [12]:

$$g_{ut} = T_{LP,t}/\tau_\Theta - 1 + \exp(T_{LP,t}/\tau_\Theta), \tau_\Theta = C_{Pb}\rho_b d^2/(12 Y\lambda), Y = 1 + 0.3 Re_b^{1/2} Pr^{1/3} \tag{2}$$

where $T_{LP,t}$ is the time of interaction of bubbles with thermal turbulent eddies $T_{LP,t} = 0.6 k/\varepsilon$ [12].

The similar simplifications were adopted in many papers. The non-steady-state pressure on the gas "liquid–bubble" interface is affected by the turbulence of the fluid phase and a non-stationary interfacial slip velocity [12]. The influence of turbulence by now is rather difficult to calculate. A simpler problem in the mathematical plan is the calculation of the effect of the averaged interphase velocity [23–27]. When calculating the pressure on the interphase surface, one can use the Lamb' expression for the potential flow of a particle by a liquid flow [28]:

$$P_{in} = P_b = P - C\rho\alpha|\mathbf{U}_R|^2. \tag{3}$$

In this study, the value of the constant is assumed $C = 0.5$ [25,26]. The pressure value at the "liquid–bubble" interface is equal to that one inside the gas bubble for $P_b << P$, $P_{in} = P_b$ by [23,24]. $Pr^T = 0.85$ is the turbulent Prandtl number. The Boussinesq hypothesis is used for calculation of turbulent heat flux in the fluid phase

$$\langle u_j \theta \rangle = -\frac{\nu^T}{Pr^T}\frac{\partial T}{\partial x_j}.$$

3.3. The Turbulence Model of the Carrier Phase (Liquid)

In the present study, the low-Reynolds number elliptic blending second-moment closure from [29] was employed:

$$\begin{aligned}
\nabla(\alpha_1\rho\langle \mathbf{u}'\mathbf{u}'\rangle) &= \alpha_1\left[P + \phi - \varepsilon_d + \nabla\left(\nu\delta + \frac{C_\mu T^T}{\sigma_k}\langle \mathbf{u}'\mathbf{u}'\rangle\right)\nabla\langle \mathbf{u}'\mathbf{u}'\rangle\right] + S_k \\
\nabla(\alpha_1\rho\varepsilon) &= \alpha_1\left[\frac{1}{T^T}(C_{\varepsilon 1}P_2 - C_{\varepsilon 2}\varepsilon) + \nabla\left(\nu\delta + \frac{C_\mu T^T}{\sigma_\varepsilon}\langle \mathbf{u}'\mathbf{u}'\rangle\right)\nabla\varepsilon\right] + S_\varepsilon \\
\chi &- (L_T)^2\nabla^2\chi = 1.
\end{aligned} \tag{4}$$

Here, T^T and L_T are the time of turbulent macro-scale and turbulent macro-scale length, δ is the Kronecker symbol, χ is the blending coefficient. The value χ changes from zero at the wall to unity far from it.

$$T^T = \max\left(k/\varepsilon; 6\sqrt{\nu/\varepsilon}\right), L_T = 0.45\left[\frac{k^{3/2}}{\varepsilon}; 80\left(\frac{\nu}{\varepsilon}\right)^{3/4}\right] \tag{5}$$

The terms S_k and S_ε determine the turbulence generation in the carrier fluid (liquid) by the bubbles presence [23]:

$$S_k = C_4\frac{3}{4}C_D\alpha\frac{\rho|\mathbf{U}_R|^3}{d}, S_\varepsilon = C_3\frac{k}{\varepsilon}S_k, \tag{6}$$

where $C_3 = 1.92$, $C_4 = 0.1$, $C_{\varepsilon 4} = 1.44$, C_D, α and d are the drag coefficient, the void fraction and the gas bubble diameter. The turbulent kinetic energy of the carrier fluid of the two-phase bubbly flow k is calculated using the formula: $k = 0.5\langle u'u'\rangle$.

3.4. The System of Basic Equations for the Dispersed Phase (Gas Bubbles)

The set of mean governing equations for the dispersed phase has the form [14]:

$$
\begin{aligned}
&\frac{\partial(\alpha_b \rho_b \mathbf{U}_b)}{\partial t} + \nabla(\alpha_b \rho_b \mathbf{U}_b) = 0, \\
&\frac{D(\alpha_b \rho_b \mathbf{U}_b)}{Dt} = -\nabla(\alpha_b P_b) - \nabla(\rho_b E\langle \mathbf{u'u'}\rangle) + \mathbf{M}^b, \\
&\frac{D(\alpha_b \rho_b C_{Pb} \mathbf{U}_b T_b)}{Dt} = -h\alpha_b(T - T_b)\frac{\rho_b}{\tau_\Theta} + (1-\alpha_b)\rho C_P \frac{DT}{Dt} - \frac{D_b^\Theta}{\tau_\Theta}\nabla\alpha_b, \\
&\rho_b = P_b/\left(\overline{R}_b T_b\right),
\end{aligned}
\tag{7}
$$

Here, t is the time, $\frac{D\phi}{Dt} = \frac{\partial\phi}{\partial t} + \mathbf{U}\nabla\phi$ is the material derivative, E is the expression of [20]

$$
E = \frac{(1-A)(1-A\Omega)}{1+\Omega}, \quad A = 1.5\rho_0/(1+0.5\rho_0), \quad \rho_0 = \rho/\rho_b
$$

$\Omega = \tau/T_{LP}$ is the dimensionless timescale, D_b and D_b^Θ are the turbulent diffusion tensor and turbulent heat flux in the dispersed phase [20];

$$
\tau = \frac{4d(1 + C_{VM}\rho_0)}{3C_D|\mathbf{U}_R|\rho_0}, \quad \tau_\Theta = C_{Pb}\rho_b d^2/(6\lambda Y)
\tag{8}
$$

are dynamic and thermal relaxation times [20], the added mass coefficient is $C_{VM} = 0.5$ [30], and C_D is the drag coefficient [31].

3.5. Interface Forces

The interfacial force is usually consists of a few components: drag, virtual mass, gravity, lift, turbulent dispersion and wall lubrication:

$$
\begin{aligned}
&\mathbf{M}_l = -\mathbf{M}_b = \mathbf{F}_{Drag} + \mathbf{F}_{VM} + \mathbf{F}_{GA} + \mathbf{F}_L + \mathbf{F}_{TD} + \mathbf{F}_{WL}, \\
&\mathbf{F}_{Drag} = \frac{\alpha_b \mathbf{U}_R}{\tau}, \mathbf{F}_{VM} = \alpha_b A \mathbf{U}\nabla\cdot\mathbf{U}, \mathbf{F}_{GA} = \frac{(1-\rho_0)\alpha_b}{1+C_{VM}\rho_0}\mathbf{g}, \\
&\mathbf{F}_L = \frac{C_L \alpha_b \rho_0}{1+C_{VM}\rho_0}\mathbf{U}_R \times (\nabla\times\mathbf{U}), \mathbf{F}_{TD} = -\frac{C_{TD}\alpha_b}{Sc^T}\frac{\mu^T}{\tau}\left(\frac{\nabla\cdot\alpha_b}{\alpha_b} - \frac{\nabla\cdot\alpha_l}{\alpha_l}\right), \\
&\mathbf{F}_{WL} = C_W \frac{d\rho_b \alpha_b |\mathbf{U}_R|^2}{2}\left(\frac{1}{y^2} - \frac{1}{(2R-y)^2}\right).
\end{aligned}
\tag{9}
$$

The interphase interaction $\mathbf{M}_l$ is considered by taking into account the effect of following forces: drag $\mathbf{F}_{Drag}$, added mass $\mathbf{F}_{VM}$, gravity $\mathbf{F}_{GA}$, lift $\mathbf{F}_L$, turbulent diffusion (dispersion) $\mathbf{F}_{TD}$ and wall lubrication $\mathbf{F}_{WL}$. The interfacial interaction $\mathbf{M}_l$ in Equation (9) takes into account only the averaged fluid parameters in the paper.

The drag coefficient C_D of the bubbles is calculated by the formula [31]:

$$
C_D = \begin{cases} \frac{24}{Re_b}\left(1 + \frac{3}{16}Re_b^{0.687}\right), & Re_b \leq 10^3 \\ 0.44, & Re_b > 10^3 \end{cases}.
\tag{10}
$$

It was obtained the C_D in tap water is the same as for solid particles (see paper [14]).

The well-known lift force formulation for two-phase flows, which has a positive coefficient C_L, acts in the direction of decreasing liquid velocity. The expression for prediction of lift force has the form [32] and C_L is the lift coefficient [32]:

$$
C_L = \begin{cases} \min[0.288\tanh(0.121Re_b), f(Eo_b)], & Eo_b \leq 4 \\ f(Eo_b), & 4 \leq Eo_b \leq 10 \\ -0.27, & Eo_b > 10 \end{cases}
\tag{11}
$$

here $f(\mathrm{Eo}_b)$ is the correction function taking into account the bubble deformation, and d_H is the the maximum horizontal dimension of the bubble by [33]

$$f(\mathrm{Eo}_b) = 0.0011\mathrm{Eo}_b^3 - 0.0159\mathrm{Eo}_b^2 - 0.0204\mathrm{Eo}_b + 0.474, d_H = d\left(1 + 0.163\mathrm{Eo}^{0.757}\right)^{1/3} \qquad (12)$$

The coefficient C_L changes its sign at a bubble diameter of $d = 5.8$ mm for our conditions. It was previously shown that this expression of the lift force can be sufficiently used for predictions of bubbly flows in complex geometrical conditions, as example in fuel columns with grid spacers [34]. $C_{TD} = 0.1$ is the coefficient of turbulent diffusion [23]. In order to handle this behavior of bubbles near the wall an additional wall lubrication force is introduced [35]. The formula for F_{WL} has been modified for the flow in a pipe [36] and the coefficient C_W has the form [36]:

$$C_W = \begin{cases} \exp(-0.933\mathrm{Eo} + 0.179), & 1 \le \mathrm{Eo} \le 5 \\ 0.007\mathrm{Eo} + 0.04, & 5 \le \mathrm{Eo} \le 33 \end{cases} \cdot \qquad (13)$$

4. Numerical Procedures, Boundary Conditions

The mean transport equations for fluid (liquid) and dispersed (gas bubbles) phases and the turbulence model are solved using a control volumes method on a staggered grid. The QUICK scheme is used to approximate the convective transport. The central difference scheme of the second-order accuracy is performed for the diffusion terms. The SIMPLEC algorithm is employed for coupling velocity and pressure.

All velocity components, temperatures of the phases and turbulence levels are uniform at the inlet. The symmetry conditions are set on the pipe axis for gas and dispersed phases. No-slip conditions are set on the wall surface for the carrier phase. At the outlet edge, the computational domain condition $\partial\phi/\partial r = 0$ is set for all variables. The first cell was located at a distance $y_+ = yu^*/\upsilon = 0.3$–$0.5$ for all simulations, where u^* is the friction velocity obtained for the one-phase flow in the inlet. At least 10 control volumes were generated to ensure resolution of the mean velocity field and turbulence quantities in the viscosity-affected near-wall region ($y_+ < 10$).

The correctness of a numerical simulation is strongly dependent upon the quality of the grid. Grid sensitivity study was performed to determine the optimum grid resolution to give the mesh-independent solution. Grid dependence was verified for three different grid sizes: 128×50, 256×100, 450×200, and 550×200 control volumes (CVs) in the axial and radial directions (see Figure 4). The distributions of the Nusselt number along the longitudinal coordinate are presented in Figure 4a. The Nusselt number was calculated according to the following relationship for the case $q_W = \mathrm{const}$:

$$\mathrm{Nu} = Hq_W/[\lambda(T_{m2} - T_{m1})] \qquad (14)$$

here H is the step height, q_W is the heat flux density for the pipe wall, λ is the fluid's coefficient of heat conductivity, and T_{m2} and T_{m1} are the mean liquid temperatures at the pipe axis in the outlet and inlet sections respectively, while the mean liquid temperatures are calculated using the formula:

$$T_m = \frac{2}{U_1 R_1^2} \int_0^{R_2} T U r dr \qquad (15)$$

The results obtained with the grid of 128×50 cells deviate from those using the other two grids for all pipe lengths. The predicted values of the Nusselt number obtained for grids with 256×100, 450×200, and 550×200 control volumes (CVs) almost overlap each other at all pipe lengths. The basic grid with 256×100 CVs was chosen for all numerical investigations performed. The profiles of turbulent kinetic energy (TKE) in radial direction are shown in Figure 4b for a few grids. The predicted values of TKE obtained for grids with 256×100, 450×200, and 550×200 control volumes almost overlap each other over the pipe radius.

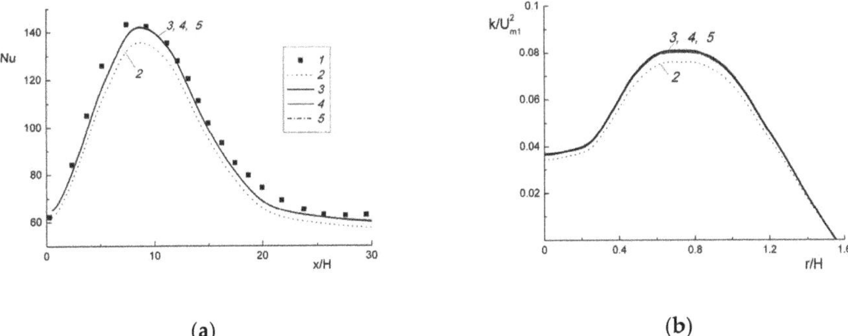

Figure 4. Grid independence tests for local heat transfer along the longitudinal coordinate (**a**) and radial profile of kinetic energy of turbulence at x/H = 5 (**b**). Points (1) are authors' measured data, curves (2–5) are authors' simulations results. $Re_H = 2.09 \times 10^4$, $2R_1 = 15$ mm, $H = 13.5$ mm, $2R_2 = 42$ mm, $ER = (R_2/R_1)^2 = 7.84$, x/H = 5, $q_W = const = 8.95$ kW/m^2, $T_1 = T_{b1} = 298$ K, d = 1.7 mm. 2—128×50, 3—256×100, 4—450×200, 5—550×200 CVs.

The solution methodology of the developed model is shown in the Figure 5.

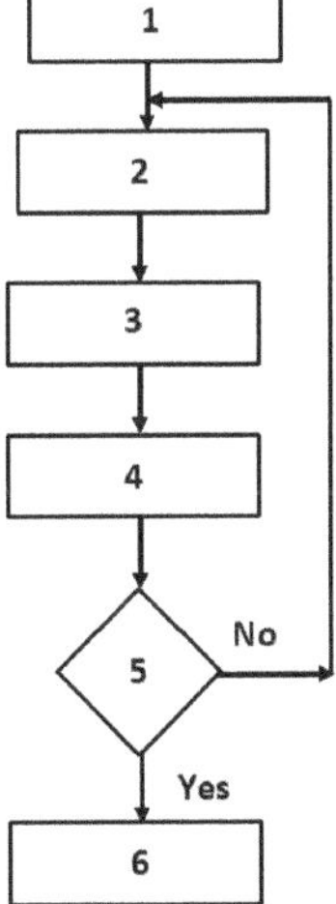

Figure 5. The solution methodology of the developed model. 1—Initial values input for all flow parameters, 2—Predictions of all flow parameters (two-fluid model + SMC), 3—Calculation of T and T_b, 4—Prediction of void fraction, 5—Convergence, 6 – Stop.

5. Results and Discussion

We estimated the effect of bubble break up and coalescence on heat transfer modification and we obtained that the influence is not significant (up to 10%) for the conditions of the study. Therefore, we have not been taking into account these effects and it allows us simplifying the mathematical model. The initial size of spherical bubbles is obtained from our measurements.

5.1. Flow Structure

The measured and predicted profiles for the mean velocities of the carrier fluid phase and gas bubbles are presented in Figure 6 at a few stations downstream of the pipe's abrupt expansion. The first two sections are situated in the flow recirculation region and two other stations are set in the flow relaxation zone. Open symbols are the one-phase fluid flow and solid symbols are the carrier fluid in the bubbly flow. Solid lines are the predicted data for the fluid phase and dashed curves are the predicted results for the air bubbles. The liquid velocity for bubbly flows (*2, 4*) is higher than those for the one-phase fluid flow (*1, 3*). The predicted mean axial velocities for air bubbles (*5*) are higher than those ones for the liquid velocity in the presence of air bubbles due to the upward direction of the two-phase flow. Thus, it is seen that the axial velocities of the liquid in bubbly flow have negative values at first two sections (x/H = 4 and 8). The carrier fluid in the bubbly flow forms a zone with negative magnitude of mean axial velocity in the near-wall area. These conclusions agree with those for the one-phase separated flow [1,2,37].

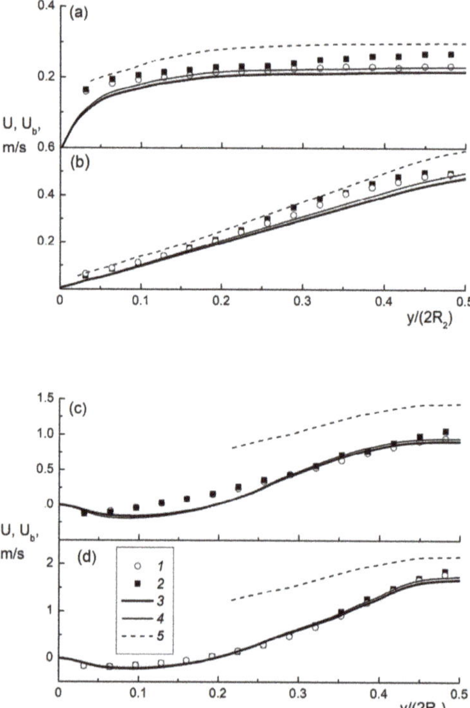

Figure 6. The distributions of mean axial fluid and gas bubbles velocities at x/H = 16 (**a**), 12 (**b**), 8 (**c**) and 4 (**d**). Re_H = 2.09 × 10^4, d = 1.7 mm, β = 3.5‰. Points and curves are authors' measurements and simulations respectively. *1,3*—β = 0 (single-phase fluid flow), *2,4*—carrier fluid phase (β = 3.5%), *5*—gas bubbles (β = 3.5%).

Figure 7a shows the normalized kinetic energy of turbulence for the fluid in the two-phase flow profiles. The unpredicted tangential normal stress w'w' was equal to the radial normal stress v'v':

$$2k = u'u' + v'v' + w'w' \approx u'u' + 2v'v'. \tag{16}$$

A significant increase in the fluid carrier phase turbulent kinetic energy (TKE) is observed in the case of bubbly flow. The increase in the intensity of the fluid phase turbulence level in the two-phase flow (up to 20–30%) is shown in comparison with the single-phase flow (*1*). The additional production of carrier fluid phase turbulence is explained by vortex formation upon streamlining of the gas bubbles by the carrier fluid flow. As is expected, the fluid's maximum turbulent kinetic energy observes in the shear mixing layer. The profiles of TKE at a small value of β agree qualitatively with those for the case of one-phase flow in a pipe with abrupt expansion.

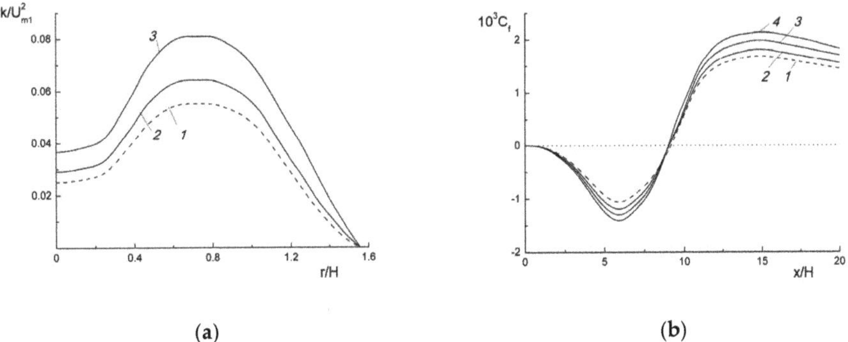

(a)   (b)

Figure 7. The profiles of TKE of the carrier fluid in radial direction at $x/H = 5$ (**a**) and wall friction coefficient (**b**). $\mathrm{Re}_H = 2.09 \times 10^4$, $2R_1 = 15$ mm, $H = 13.5$ mm, $2R_2 = 42$ mm, $d = 1.7$ mm. 1—$\beta = 0$ (one-phase water flow), 2—3.5%, 3—5.2%.

The distributions of the wall friction coefficient,

$$C_f = 2\tau_W / \left(\rho U_{m1}^2\right), \tag{17}$$

along the pipe length are shown in Figure 7b. The increase in the β leads to a significant increase in the absolute value of the wall friction coefficient. Note that the minimum value of friction on the wall, located in the recirculation area, and it shifts slightly toward the inlet cross-section with increasing concentration.

5.2. Heat Transfer

Figure 8 shows distributions of the parameter for heat transfer enhancement $\mathrm{Nu}/\mathrm{Nu}_{0,max}$ in bubbly flow in the pipe with sudden expansion with various β. The local Nusselt number is calculated using the Formula (14). Here $\mathrm{Nu}_{0,max}$ is the maximal value of the Nusselt number for the one-phase flow. Points and lines are the results of measurements and numerical simulations performed by the authors, and the dashed lines represent the calculation for a one-phase flow ($\beta = 0$) under otherwise identical conditions. For visual clarity, only every seventh data point in the experimental data is shown in the figure.

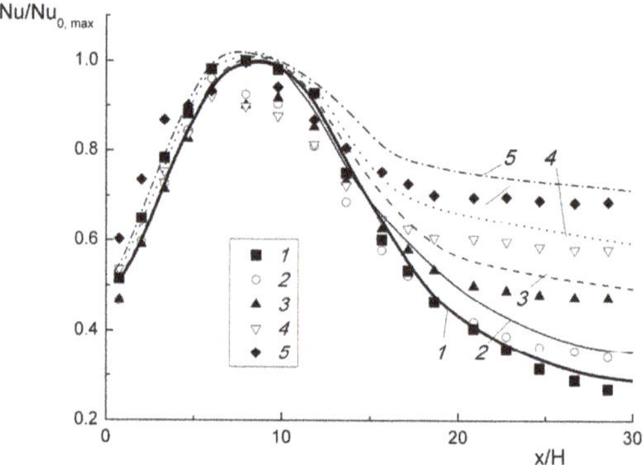

Figure 8. The distributions of heat transfer enhancement ratios along the axial coordinate at $Re_H = 1.02 \times 10^4$. Points and lines are authors' measurements and computations respectively. $d = 1.7$–2 mm, $2R_1 = 15$ mm, $H = 13.5$ mm, $2R_2 = 42$ mm, ER $= (R_2/R_1)^2 = 7.84$, $q_W = $ const $= 8.95$ kW/m², $T_1 = T_{b1} = 298$ K. 1—$\beta = 0$ (one-phase flow), 2—1.2%, 3—3.6, 4—7, 5—10.1.

The increase in heat transfer reaches almost three times for the case of $Re_H = 1.02 \times 10^4$ and $\beta = 10.1\%$ according to the measurements and simulations. The increased heat transfer coefficient is caused by significant deformation in the velocity profiles of the two-phase flow compared to the single-phase flow and by the increase in the liquid velocity gradient near the pipe wall (see Figure 7). A significant growth in wall friction is obtained in the bubbly flow, even when very small gas concentrations are present, as previously detected for turbulent [37] and laminar [38] flow regimes. The heat transfer enhancement is numerically predicted mainly in the flow relaxation zone. Actually, these conclusions are in agreement with the measured one. Only a few bubbles penetrate into the flow recirculation region and bubbles are available only in the core zone and shear layer region. We obtain a small effect of the gas bubbles on the measured and predicted heat transfer in the flow recirculation region. Authors of recent numerical work [14] showed that small bubbles ($d < 1.5$ mm) caused heat transfer intensification over the entire length of the recirculation zone, while the larger ones caused intensification mostly in the flow relaxation region. The bubbles migrate towards the wall after the reattachment point of the flow and accumulate there due to the action of the lift force; it leads to an increase in fluid phase turbulence and heat transfer in this region. The length of the zone of the heat transfer intensification is limited for the case of one-phase fluid flow. It is shown, that relatively small amount of the gas bubbles leads to the significant increase of the region where the intensification of heat transfer can be obtained.

5.3. The Comparison with Results of Other Papers

The modeling of bubble distributions across the pipe radius is crucial point for the predictions of interfacial term in turbulent bubbly flows. We did not measure the radial bubble distributions and therefore the developed model was additionally validated against measured and numerical results for isothermal, upward bubbly flows downstream from a sudden pipe expansion. Points are the measurements [11], solid lines represent the LES computations [10], and the dashed lines are authors' simulations. The bulk velocity was $U_{m1} = 1.78$ m/s, which corresponds to the Reynolds number $Re_H = 1.11 \times 10^5$. The pipe diameter, before the abrupt, was $2R_1 = 50$ mm, while after expansion, $2R_2 = 100$ mm, which corresponded to the step height of $H = 25$ mm, and ER $= (R_2/R_1)^2 = 4$ and the inlet bubble diameter was $d = 2$ mm. The measurements were carried out at five stations located downstream from the expansion edge $x = 70$ mm ($x/H = 2.8$), 130 mm (5.2), 250 mm (10) and 320 mm (12.8). The measurements [11] were performed without interphase heat transfer and realized for very different values of Reynolds number and flow geometry than that one of [15].

The development of the axial fluid phase velocity along different sections of the pipe is shown in Figure 9a. The maximal difference between measurements of [11], LES results of [10] and our RANS predictions are observed in the flow recirculation region (the maximal discrepancy is up to 15%). The predicted radial profiles for the liquid phase axial mean fluid phase velocity (a), liquid velocity fluctuations (b) and bubble volume fraction (c) are presented in Figure 9. The gas bubble volume fraction showed a strongly pronounced maximum in the shear layer for first two sections. It can be explained by turbophoresis (turbulent transport), turbulent diffusion forces [10,22] and bubbles accumulating in zones with a high value for fluid flow pulsations. Figure 9b shows the distributions of axial fluid (liquid) phase velocity pulsations at a few stations along the pipe length. After the bubbly flow reattachment at $x/H \approx 9$, along with its relaxation, there is a transition to the turbulent fully developed flow. The magnitude of axial fluid velocity fluctuations in this zone has the highest value too and the shear layer is a "trap" for them. The profiles for the void fractions in the recirculation zone are characterized by an almost zero value of the bubble concentration at first two stations ($x/H < 10$). It causes heat transfer enhancement predicted in our simulations, lesser than that one in the flow relaxation area. The radial profile for the bubble concentration became more uniform in the region after the flow reattachment ($x/H > 10$) compared to one in the first two stations (in the region of flow recirculation). The crest of the fluid flow pulsations disappeared and lead to the redistribution of bubbles across the pipe's cross section as they moved toward the wall (see Figure 9c). The values of bubble volume fractions have almost zero magnitude in pipe's near-wall region. The maximal difference between measured and numerical results is up to 20%. The authors results agree well with experiments and LES data.

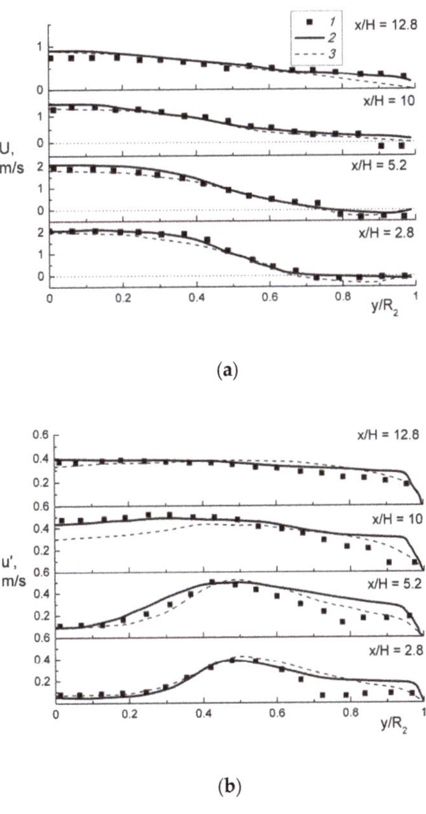

(a)

(b)

Figure 9. *Cont.*

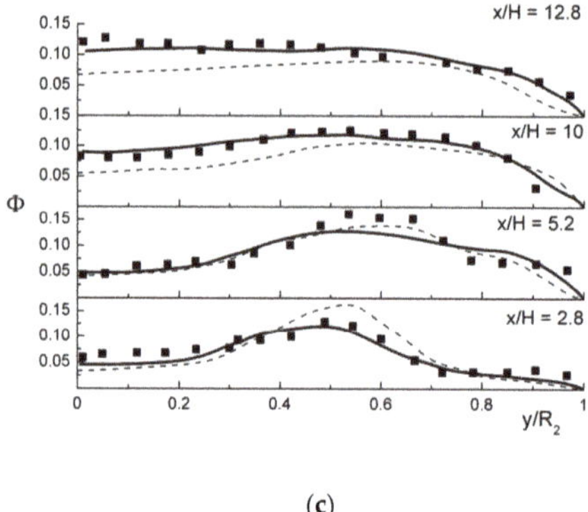

(c)

Figure 9. The profiles of mean axial fluid velocity (**a**), fluid phase axial velocity fluctuations (**b**) and gas volume fraction (**c**) across the radial coordinate in bubbly isothermal upward flow in the pipe with sudden expansion. *1*—measured results of [15], *2*—LES data [10], *2*—authors' predictions. $Re_H = 1.11 \times 10^5$, $U_{m1} = 1.78$ m/s, $d_1 = 2$ mm, $2R_1 = 50$ mm, $H = 25$ mm, $2R_2 = 100$ mm, $ER = (R_2/R_1)^2 = 4$, $T_1 = T_{b1} = 300$ K.

6. Conclusions

The turbulent flow patterns and the heat transfer for a bubbly turbulent flow in a pipe with sudden expansion were experimentally and numerically investigated. An experimental study of structure of bubbly flow was carried out using PIV–LIF approach and shadow photography. Experimental and numerical investigations were performed using a range of Reynolds numbers of $Re_H = (1–2) \times 10^4$ and $\beta = 0–10\%$. The numerical model was employed the Eulerian approach. The set of RANS equations was used for modelling two-phase bubbly flows. The turbulence of the carrier liquid phase was predicted using the Reynolds stress model.

A significant growth (up to 30%) in wall friction in bubbly flow downstream of pipe with abrupt expansion is obtained. The mean and fluctuating flow structures in bubbly flow with small values of $\beta \leq 10\%$ is similar to that of a one-phase fluid flow. The increase in the intensity of the fluid phase turbulence level in the two-phase flow (up to 20–30%) is shown. The peak of axial and radial fluctuations of the carrier fluid (liquid) velocity in the bubbly flow is observed in the shear layer. It is experimentally and numerically shown the heat transfer in bubbly flow is significantly enhances (up to three times). This effect is augmented with increasing gas volumetric flow rate ratios β. The main enhancement in heat transfer is observed after the point of flow reattachment. In general, the Nusselt number distributions in the bubbly flow have qualitatively similar character as that one for single-phase fluid turbulent separated flows.

Two regions can be found for the influence of bubbles on the flow. Similar behaviour of two-phase flow was previously found by the authors in the model of pressurized water reactor downstream a spacer grid [39]. In the recirculation zone the heat transfer is determined by large vortices structures. The effect of bubbles on the heat transfer in this region is minimal. The main increase in heat transfer is observed in the flow relaxation region after the point of flow reattachment. It is experimentally and numerically shown that the addition of air bubbles causes a significant increase in the heat transfer rate (up to three times), and these effects increase with increasing gas volumetric flow rate ratios. Intensification of the heat transfer in this region relates to the additional flow turbulisation by bubbles and the reorganisation of the liquid velocity profile which leads to the increase of the velocity gradient in the near wall region. Early a similar degree of the increasing of wall shear stress and heat transfer coefficient in turbulent two-phase bubbly flows was found in [40,41].

Author Contributions: The developing of the experimental circuit, the test section for optical measurements, and the purposed method of two-phase flow study by means of PIV/PLIF, experimental study, experimental data processing and description of experimental facility were performed by P.L. M.P. and V.T. developed the numerical model and performed the numerical part of the paper.

Funding: The experimental study of hydrodynamic characteristics of the flow was performed by financial support of the Russian Foundation for Basic Research (Project No. 18-58-45006), heat transfer measurements were performed under the state contract with IT SB RAS (AAAA-A18-118051690120-2) and all simulations of the paper were carried out under the state contract with IT SB RAS (AAAA-A17-117030310010-9).

Acknowledgments: P. Lobanov is grateful to O.N. Kashinsky for providing the heat transfer measurement unit.

Conflicts of Interest: The authors declare no conflict of interest.

Nomenclature

C_D	drag coefficient
C_f	friction coefficient
C_P, C_{Pb}	specific heat capacity of fluid phase and gas bubbles ($J\,kg^{-1}\,K^{-1}$)
D	diffusion coefficient, ($m^2\,s^{-1}$)
d	bubble diameter, (m)
Eo	$Eo = g(\rho - \rho_b)d^2/\sigma$ Eötvös number
Eo_b	$Eo_b = g(\rho - \rho_b)D_H^2/\sigma$ modified Eötvös number
H	step height, (m)
α	heat transfer coefficient, ($W\,K^{-1}\,m^{-2}$)
J, J_b	superficial velocity of the fluid and the gaseous phases ($m\,s^{-1}$)
k	$k = \langle u'u' \rangle/2$ kinetic energy of turbulence, ($m^2\,s^{-2}$)
Nu	$Nu = Hq_W/[\lambda(T_{m2} - T_{m1})]$ Nusselt number
P	pressure, ($N\,m^{-2}$)
Pr	$Pr = \mu C_P/\lambda$ Prandtl number
q_W	heat flux density, ($W\,m^{-2}$)
R	radius of a pipe, (m)
R_b	specific gas constant, ($J\,kg^{-1}\,K^{-1}$)
Re_H	HU_{m1}/υ Reynolds number
Re_b	$Re_b = |U_R|d/\upsilon$ Reynolds number, based on the slip velocity
R	radial coordinate, (m)
T	temperature, (K)
T	time, (s)
$|U_R|$	$U_R = |U - U_b|$ mean slip velocity, ($m\,s^{-1}$)
U, V	mean velocities in the axial and radial directions, ($m\,s^{-1}$)
U_{m1}	mean velocity at the inlet section, ($m\,s^{-1}$)
U_*	friction velocity, ($m\,s^{-1}$)
u', v'	velocity fluctuations in the axial and radial directions, ($m\,s^{-1}$)
We	$We = \rho|U_R|^2 d/\sigma$ Weber number
x	axial coordinate, (m)
y	distance normal from the wall, (m)
Greek	
Φ	volume fraction
α	gas void fraction
β	$\beta = J_b/(J_b + J)$ gas volumetric flow rate ratio
ε	dissipation of the turbulent kinetic energy, ($m^2 s^{-3}$)
λ	heat conductivity, ($W\,K^{-1}\,m^{-1}$)
μ	dynamic viscosity, ($kg\,m^{-1}\,s^{-1}$)
υ	kinematic viscosity, ($m^2\,s^{-1}$)
ρ	density, ($kg\,m^{-3}$)
σ	surface tension, ($N\,m^{-1}$)

τ	$\tau = \frac{4d(1+C_{VM}\rho_0)}{3C_D\|U_R\|\rho_0}$ dynamic relaxation time, (s)
τ_W	wall friction, (N m^{-2})
τ_Θ	$\tau_\Theta = C_{Pb}\rho_b d^2/(6\lambda\mathrm{Nu}_b)$ thermal relaxation time, (s)
Subscripts	
0	single-phase flow
1	initial condition
l	liquid
W	wall
b	gas bubble
m	mean-mass
Superscripts	
T	turbulent parameter
Acronym	
CFD	Computational Fluid Dynamics
LES	Large-Eddy Simulation
LIF	Laser Induced Fluorescence
QUICK	Quadratic Upstream Interpolation for Convective Kinematics
SIMPLEC	Semi-Implicit Pressure Linked Equation Consistent
TKE	Turbulent kinetic energy
PIV	Particle Image Velocimetry
RANS	Reynolds averaged Navier-Stokes
SMC	Second moment closure

References

1. Eaton, J.K.; Johnston, J.P. Review of research on subsonic turbulent flow reattachment. *AIAA J.* **1981**, *19*, 1093–1100. [CrossRef]
2. Ota, T.A. Survey of heat transfer in separated and reattached flows. *Appl. Mech. Rev.* **2000**, *53*, 219–235. [CrossRef]
3. Aloui, F.; Souhar, M. Experimental study of a two-phase bubbly flow in a flat duct symmetric sudden expansion. Part I: Visulization, pressure and void fraction. *Int. J. Multiph. Flow* **1996**, *22*, 651–665. [CrossRef]
4. Rinne, A.; Loth, R. Development of local two-phase flow parameters for vertical bubbly flow in a pipe with sudden expansion. *Exp. Therm. Fluid Sci.* **1996**, *13*, 152–166. [CrossRef]
5. Aloui, F.; Doubliez, L.; Legarand, J.; Souhar, M. Bubbly flow in an axisymmetric sudden expansion: Pressure drop, void fraction, wall shear stress, bubble velocities and sizes. *Exp. Therm. Fluid Sci.* **1999**, *19*, 118–130. [CrossRef]
6. Ahmed, W.H.; Ching, C.Y.; Shoukri, M. Development of two-phase flow downstream of a horizontal sudden expansion. *Int. J. Heat Fluid Flow* **2008**, *29*, 194–206. [CrossRef]
7. Voutsinas, A.; Shakouchi, T.; Tsujimoto, K.; Ando, T. Investigation of bubble size effect on vertical upward bubbly two-phase pipe flow consisted with an abrupt expansion. *J. Fluid Sci. Technol.* **2009**, *4*, 442–453. [CrossRef]
8. Wang, C.C.; Tseng, C.Y.; Chen, Y.I. A new correlation and the review of two-phase flow pressure change across sudden expansion in small channels. *Int. J. Heat Mass Transfer.* **2010**, *53*, 4287–4295. [CrossRef]
9. Krepper, E.; Beyer, M.; Frank, T.; Lucas, D.; Prasser, H.-M. CFD modelling of polydispersed bubbly two-phase flow around an obstacle. *Nucl. Eng. Des.* **2009**, *239*, 2372–2381. [CrossRef]
10. Papoulias, D.; Splawski, A.; Vikhanski, A.; Lo, S. Eulerian multiphase predictions of turbulent bubbly flow in a step-channel expansion. In Proceedings of the Nineth International Conference on Multiphase Flow ICMF-2016, Firenze, Italy, 22–27 May 2016.
11. Bel Fdhila, R. Analyse Experimental et Modelisation d'un Ecoulement Vertical a Bulles Dans un Elargissement Brusque. Ph.D. Thesis, Institut National Polytechnique de Toulouse, Toulouse, France, 1991.
12. Zaichik, L.I.; Alipchenkov, V.M. A statistical model for predicting the fluid displaced/added mass and displaced heat capacity effects on transport and heat transfer of arbitrary density particles in turbulent flows. *Int. J. Heat Mass Transfer.* **2011**, *54*, 4247–4265. [CrossRef]

13. Dabiri, S.; Tryggvason, G. Heat transfer in turbulent bubbly flow in vertical channels. *Chem. Eng. Sci.* **2015**, *122*, 106–113. [CrossRef]

14. Pakhomov, M.A.; Terekhov, V.I. Modeling of the flow patterns and heat transfer in a turbulent bubbly polydispersed flow downstream of a sudden pipe expansion. *Int. J. Heat Mass Transfer.* **2016**, *101*, 1251–1262. [CrossRef]

15. Lobanov, P.D.; Pakhomov, M.A. Experimental and numerical study of heat transfer enhancement in a turbulent bubbly flow in a pipe sudden expansion. *J. Eng. Thermophys.* **2017**, *26*, 377–390. [CrossRef]

16. Cerqueira, R.F.L.; Paladino, E.E.; Ynumaru, B.K.; Maliska, C.R. Image processing techniques for the measurement of two-phase bubbly pipe flows using particle image and tracking velocimetry (PIV/PTV). *Chem. Eng. Sci.* **2018**, *189*, 1–23. [CrossRef]

17. Drew, D.A.; Lahey, R.T., Jr. Application of general constitutive principles to the derivation of multidimensional two-phase flow equations. *Int. J. Multiph. Flow* **1979**, *5*, 243–264. [CrossRef]

18. Nigmatulin, R.I. Spatial averaging in the mechanism of heterogeneous and dispersed systems. *Int. J. Multiph. Flow* **1979**, *5*, 353–385. [CrossRef]

19. Drew, D.A. Mathematical modeling of two-phase flow. *Ann. Rev. Fluid Mech.* **1983**, *15*, 261–291. [CrossRef]

20. Zaichik, L.I.; Skibin, A.P.; Solov'ev, S.L. Simulation of the distribution of bubbles in a turbulent liquid using a diffusion-inertia model. *High Temp.* **2004**, *42*, 111–118. [CrossRef]

21. Kashinsky, O.N.; Lobanov, P.D.; Pakhomov, M.A.; Randin, V.V.; Terekhov, V.I. Experimental and numerical study of downward bubbly flow in a pipe. *Int. J. Heat Mass Transfer.* **2006**, *49*, 3717–3727. [CrossRef]

22. Mukin, R.V. Modeling of bubble coalescence and break-up in turbulent bubbly flow. *Int. J. Multiph. Flow* **2014**, *62*, 52–66. [CrossRef]

23. Lopez de Bertodano, M.; Lee, S.J.; Lahey, R.T., Jr.; Drew, D.A. The prediction of two-phase turbulence and phase distribution using a Reynolds stress model. *ASME J. Fluids Eng.* **1990**, *112*, 107–113. [CrossRef]

24. Nigmatulin, R.I. *Dynamics of Multiphase Media*; Hemisphere: New York, NY, USA, 1991; Volume 1.

25. Lopez de Bertodano, M.; Lahey, R.T., Jr.; Jones, O.C. Phase distribution in bubbly two-phase flow in vertical ducts. *Int. J. Multiph. Flow* **1994**, *20*, 805–818. [CrossRef]

26. Politano, M.; Carrica, P.; Converti, J. A model for turbulent polydisperse two-phase flow in vertical channel. *Int. J. Multiph. Flow* **2003**, *29*, 1153–1182. [CrossRef]

27. Chahed, J.; Roig, V.; Masbernat, L. Eulerian–Eulerian two-fluid model for turbulent gas–liquid bubbly flows. *Int. J. Multiph. Flow* **2003**, *29*, 23–49. [CrossRef]

28. Lamb, H. *Hydrodynamics*; Cambridge University Press: Cambridge, UK, 1932.

29. Fadai-Ghotbi, A.; Manceau, R.; Boree, J. Revisiting URANS computations of the backward-facing step flow using second moment closures. Influence of the numerics. *Flow Turbul. Combust.* **2008**, *81*, 395–410. [CrossRef]

30. Drew, D.A.; Lahey, R.T., Jr. The virtual mass and lift force on a sphere in rotating and straining in viscid flow. *Int. J. Multiph. Flow* **1987**, *13*, 113–121. [CrossRef]

31. Loth, E. Quasi-steady shape and drag of deformable bubbles and drops. *Int. J. Multiph. Flow* **2008**, *34*, 523–546. [CrossRef]

32. Tomiyama, A.; Tamai, H.; Zun, I.; Hosokawa, S. Transverse migration of single bubbles in simple shear flows. *Chem. Eng. Sci.* **2002**, *57*, 1849–1858. [CrossRef]

33. Wellek, R.M.; Agrawal, A.K.; Skelland, A.H.P. Shapes of liquid drops moving in liquid media. *AIChE J.* **1966**, *12*, 854–860. [CrossRef]

34. Cong, T.; Zhang, X. Numerical study of bubble coalescence and breakup in the reactor fuel channel with a vaned grid. *Energies* **2018**, *11*, 256. [CrossRef]

35. Antal, S.P.; Lahey, R.T., Jr.; Flaherty, J.F. Analysis of phase distribution in fully developed laminar bubbly of two-phase flow. *Int. J. Multiph. Flow* **1991**, *17*, 635–652. [CrossRef]

36. Tomiyama, A. Struggle with computational bubble dynamics. In Proceedings of the Third International Conference on Multiphase Flow ICMF'98, Lyon, France, 8–12 June 1998.

37. Kashinskii, O.N.; Timkin, L.S.; Gorelik, R.S.; Lobanov, P.D. Experimental study of the friction stress and true gas content in upward bubbly flow in a vertical tube. *J. Eng. Phys. Thermophys.* **2006**, *79*, 1117–1129. [CrossRef]

38. Rivière, N.; Cartellier, A.; Timkin, L.; Kashinsky, O. Wall shear stress and void fraction in Poiseuille bubbly flows: Part II: Experiments and validity of analytical predictions. *Eur. J. Mech. B/Fluids* **1999**, *18*, 847–867. [CrossRef]

39. Kashinsky, O.N.; Lobanov, P.D.; Kurdyumov, A.S.; Pribaturin, N.A.; Volkov, S.E. Experimental Modeling of Light Phase Effect on Heat Transfer in Rod Bundle. In Proceedings of the 2013 21st International Conference on Nuclear Engineering (American Society of Mechanical Engineers), Chengdu, China, 29 July–2 August 2013; p. V004T09A081.

40. Lobanov, P.D. Wall Shear Stress and Heat Transfer of Downward Bubbly Flow at Low Flow Rates of Liquid and Gas. *J. Eng. Thermophys.* **2018**, *27*, 232–244. [CrossRef]

41. Kashinsky, O.N.; Randin, V.V.; Chinak, A.V. The effect of channel orientation on heat transfer and wall shear stress in the bubbly flow. *Thermophys. Aeromech.* **2013**, *20*, 391–398. [CrossRef]

Article

Numerical Study of Heated Tube Arrays in the Laminar Free Convection Heat Transfer

Zuzana Brodnianská * and Stanislav Kotšmíd

Faculty of Technology, Technical University in Zvolen, Študentská 26, 960 01 Zvolen, Slovakia; stanislav.kotsmid@gmail.com
* Correspondence: zuzana.brodnianska@tuzvo.sk; Tel.: +421-45-5206678

Received: 16 January 2020; Accepted: 19 February 2020; Published: 21 February 2020

Abstract: Laminar free convection heat transfer from a heated cylinder and tube arrays is studied numerically to obtain the local and average Nusselt numbers. To verify the numerical simulations, the Nusselt numbers for a single cylinder were compared to other authors for the Rayleigh numbers of 10^3 and 10^4. Furthermore, the vertically arranged heated tube arrays 4×1 and 4×2 with the tube ratio spacing $S_V/D = 2$ were considered, and obtained average Nusselt numbers were compared to the existing correlating equations. A good agreement of the average Nusselt numbers for the single cylinder and the bottom tube of the 4×1 tube array is proved. On the other hand, the bottom tubes of the 4×2 tube array affect each other, and the Nusselt numbers have a different course compared to the single cylinder. The temperature fields for the tube array 4×4 in basic, concave, and convex configurations are studied, and new correlating equations were determined. The simulations were done for the Rayleigh numbers in the range of 1.3×10^4 to 3.7×10^4 with a tube ratio spacing S/D of 2, 2.5, and 3. On the basis of the results, the average Nusselt numbers increase with the Rayleigh numbers and tube spacing increasing. The average Nusselt number and total heat flux density for the convex configuration increase compared to the base one; on the other hand, the average Nusselt number decreases for the concave one. The results are applicable to the tube heaters constructional design in order to heat the ambient air effectively.

Keywords: heat transfer; free convection; cylinder; tube array; numerical investigation

---

1. Introduction

Laminar free convection heat transfer from heated tubes to air has been used in various technical applications such as heat exchangers, electronic components, heat storage equipment, and waste heat recovery systems. The advantage of free convection heat transfer is a creation of thermal comfort without a draft and dust fragments swirling in the space. This energy-saving way of heat transfer is reliable and the operation without fans is cost-effective and noise reducing. On the other hand, heat transfer at free convection is lower; therefore, a sufficiently large heat transfer surface and its suitable shape and arrangement is needed to fulfil effective heat transfer. Owing to this, the research of a chimney effect over the heat transfer surfaces is justified and required.

Several authors have dealt with the numerical calculations of heat transfer from a heated cylinder or tubes array at free convection. Churchill and Chu [1] developed an empirical expression for the average Nusselt number Nu_{av} of a horizontal cylinder at presented Rayleigh numbers Ra and Prandtl numbers Pr. The temperature fields and airflows around the horizontal cylinder at free convection were researched by Morgan [2]. Kuehn and Goldstein [3] studied laminar free convection heat transfer from the horizontal cylinder by solving the Navier–Stokes and energy equations using an elliptical numerical procedure for $10^0 \leq Ra \leq 10^7$.

The effect of vertical spacing on free convection heat transfer for a pair of heated horizontal tubes arranged one above the other was studied by Sparrow and Niethammer [4]. They investigated how the heat transfer characteristics of a top cylinder are affected by a bottom one, while changing Ra in the range of 2×10^4–2×10^5 and the tube spacing from 2 to 9 cylinder diameters. It was found that the Nusselt numbers of the top cylinder are strongly affected by the tube spacing. A decrease of the Nusselt numbers occurs at small spacing, an enhancement prevails at higher one. Regarding the temperature differences, their effect on the Nusselt numbers is higher at small tube spacing, unlike the higher one. Saitoh et al. [5] presented bench-mark solutions with accuracy to at least three decimal places for $Ra = 10^3$ and 10^4, where Nu_ϕ, Nu_{av}, isotherms, streamlines, and vorticities were obtained. The laminar free convection around an array of two isothermal tubes arranged above each other for Ra in the range of 10^2 to 10^4 was studied numerically by Chouikh et al. [6]. Decreased Nu at close spacing and enhanced Nu at a large one occurred at the upper tube. Within the same tube spacing, heat transfer at the upper tube increases with Ra. Herraez and Belda [7] used the holographic interferometry method for the temperature field visualization around the tubes of different diameters when changing the surface temperature. Furthermore, the functions of an exponential form were defined, and Nu numbers were calculated for the range of $Gr{\cdot}Pr = 1.2 \times 10^3$–$1.6 \times 10^5$.

Corcione [8] numerically studied a steady laminar free convection from horizontal isothermal tubes set in a vertical array. In this research, a computer code based on the SIMPLE-C algorithm was developed and simulations for the arrays of 2–6 tubes with the center-to-center distance from 2 to more than 50 tube diameters were performed for Ra in the range between 5×10^2 and 5×10^5. Moreover, the heat transfer correlating equations for individual tubes in the array and for the whole tube array were proposed. Furthermore, a steady laminar free convection from a pair of vertical tube arrays for Ra in the range of 10^2 and 10^4 was numerical studied by Corcione [9]. The pairs of tube arrays consisted of 1–4 tubes with the center-to-center horizontal and vertical spacing from 1.4 to 24 and from 2 to 12 tube diameters, respectively.

Ashjaee and Yousefi [10] experimentally investigated the free convection heat transfer from the horizontal tubes arranged above each other and the tubes shifted in a horizontal direction. The tube spacing varied from 2 to 5 tube diameter in a vertical direction and from 0 to 2 tube diameter in a horizontal direction for the inclined array between $Ra = 10^3$ and 3×10^3. The Mach–Zehnder interferometer was used to visualize the temperature fields. It was found out that the location of each tube affects the flow and heat transfer on the other tubes in the array. Heo and Chung [11] numerically investigated the natural convection heat transfer of two staggered cylinders for laminar flows. They used the Ansys Fluent software to examine the effect of varying the Prandtl number and the vertical and horizontal pitch-to-diameter ratios for Ra of 1.5×10^8. The heat transfer rates of the upper cylinder were affected by thermal plumes from the lower cylinders, and lower cylinders were not affected by the upper cylinders. When the vertical pitch is very small, the upper cylinders are affected.

Cernecky et al. [12] visualized the temperature fields around the two heated tubes arranged above each other at the surface temperature of 323 K. The local heat transfer coefficients were determined from the holographic interferogram images of the temperature fields and the results were compared with numerical simulations. New criterion equations for calculating Nu of electrically heated tubes at free convection heat transfer were presented by Malcho et al. [13]. The vertical center-to-center distance of the tubes was gradually changed from 20 to 100 mm with a step of 20 mm. The surface temperatures of the tubes were 30, 45, 60, 75, 90, and 105 °C at a fluid temperature of 20 °C. Lu et al. [14] numerically investigated heat transfer from the vertical tube arrays (2–10 horizontal tubes) in molten salts at $Ra = 2 \times 10^3$–5×10^5. It was found out that the tube spacing affects the average heat transfer rate around the whole tube array. The research resulted in a determination of the heat transfer dimensionless correlating equations for any individual tube in two vertically aligned horizontal tubes.

Kitamura et al. [15] experimentally investigated the free convection heat transfer around a vertical row of 10 heated tubes with the diameters of 8.4, 14.4, and 20.4 mm at modified Ra in the range of 5×10^2 to 10^5. The thermal plumes arising from the upstream tubes remained laminar throughout the rows when the gaps between the tubes were smaller than 20.6 mm. When the gaps were higher than 30.6 mm, the thermal plumes began to sway and undergo the turbulent transition on the halfway of the rows. Cernecky et al. [16] visualized the temperature fields in a set of four tubes arranged above each other at horizontal shift of 1/4 and 1/2 of the tube diameter by the holographic interferometry. Subsequently, the local heat transfer coefficients around the tubes were evaluated. The spacing between the tubes in vertical direction has a significant effect on the heat transfer parameters compared with the horizontal spacing of the tube centers. The heat transfer parameters on the other tubes in the direction of free convection flow varied depending on the tube position and geometry of the array. Ma et al. [17] studied free convection for a single cylinder by the Computational Fluid Dynamics with laminar and different turbulent models. Moreover, a thermal chimney was studied, and the effect of horizontal spacing arrangement of tubes on temperature and velocity fields was clarified.

The mentioned results were compared with those obtained from our simulations and are presented in this paper. First, a numerical simulation for a single cylinder at $Ra = 10^3$ and 10^4 was performed, and the local Nusselt numbers Nu_ϕ were compared with other authors. Subsequently, the results of Nu_{av} for the tube arrays 4×1 and 4×2 are presented according to arrangement in Figure 1. The main part of the presented contribution are the results of Nu_ϕ and Nu_{av} for the tube array 4×4 at basic, concave, and convex configurations. The simulations were performed for Ra in the range of 1.3×10^4 to 3.7×10^4 and for vertical and horizontal spacing between the tubes $S/D = 2, 2.5$ and 3. New correlating equations for the Nusselt numbers and the temperature fields for the tube arrays were created. The goal of this paper is to compare the base configuration of the tube arrays 4×4 (Figure 2a) with the concave (Figure 2b) and convex one (Figure 2c) and to increase the Nu_{array} and q_{array}, respectively, by means of a suitable configuration. Furthermore, the goal is to determine new correlating equations for the 4×4 tube array of basic, concave, and convex configurations with subsequent evaluation of the tubes' arrangement and spacing effects on Nu_{array}. The results are applicable to the design of heating equipment where free convection is used such as the tube heaters with a hot water circuit and electric heaters which are suitable for bathroom and toilet heating.

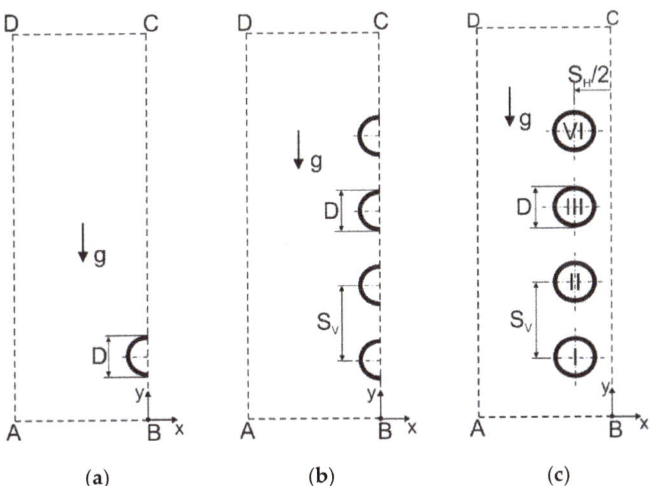

Figure 1. Arrangement of the single cylinder and tube arrays 4×1 and 4×2: (**a**) single cylinder; (**b**) 4×1 tube array; (**c**) 4×2 tube array.

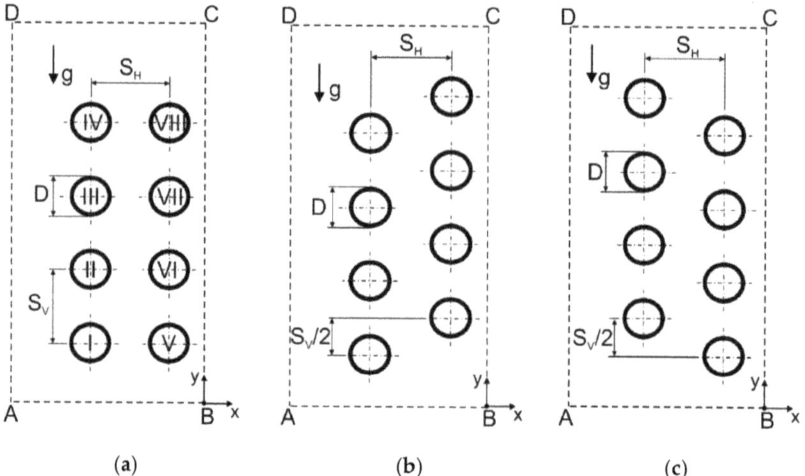

Figure 2. Arrangement of the tube arrays 4 × 4: (**a**) base configuration; (**b**) concave configuration; (**c**) convex configuration.

2. Mathematical Formulation

Free convection heat transfer is conducted between the heated cylinder surface and ambient environment (air) and between the heated tube array and ambient air, respectively. In the mathematical formulation, the uniform and constant temperature of the cylinder surface (tube array surfaces) T_s is considered, and the ambient air is taken with the temperature T_f without the interfering airflow. The air, flown at free convection, is considered as steady, two-dimensional (x, y), laminar, and incompressible. The properties of air are shown in Table 1.

Table 1. The properties of air.

Property and Unit	Values
Density ρ (kg/m^3)	1.196
Specific heat c (J/kgK)	piecewise polynomial
Thermal conductivity λ (W/mK)	$-2.48 \times 10^{-8} \cdot T^2 + 8.92 \times 10^{-5} \cdot T + 1.12 \times 10^{-3}$
Dynamic viscosity μ (kg/ms)	$-3.76 \times 10^{-11} \cdot T^2 + 6.95 \times 10^{-8} \cdot T + 1.12 \times 10^{-6}$

The mass (continuity equation), momentum equations, and the energy equation for incompressible and compressible flows are the following:

$$\frac{\partial(\rho u)}{\partial x} + \frac{\partial(\rho v)}{\partial y} = 0, \tag{1}$$

Momentum equation in x direction:

$$\rho\left(u\frac{\partial u}{\partial x} + v\frac{\partial u}{\partial y}\right) = \mu\left(\frac{\partial^2 u}{\partial x^2} + \frac{\partial^2 u}{\partial y^2}\right) - \frac{\partial p}{\partial x}, \tag{2}$$

Momentum equation in y direction:

$$\rho\left(u\frac{\partial v}{\partial x} + v\frac{\partial v}{\partial y}\right) = \mu\left(\frac{\partial^2 v}{\partial x^2} + \frac{\partial^2 v}{\partial y^2}\right) - \frac{\partial p}{\partial y} + \rho g\beta(T_s - T_f), \tag{3}$$

Energy equation:

$$\left(u\frac{\partial T}{\partial x} + v\frac{\partial T}{\partial y}\right) = \alpha\left(\frac{\partial^2 T}{\partial x^2} + \frac{\partial^2 T}{\partial y^2}\right).\tag{4}$$

The local Nusselt numbers Nu_ϕ of any cylinder are calculated as [8]:

$$Nu_\phi = \frac{qD}{\lambda_f\left(T_s - T_f\right)} = -\left.\frac{\partial T}{\partial r}\right|_{r=0.5}.\tag{5}$$

The average Nusselt numbers Nu_{av} of any cylinder are calculated as [8]:

$$Nu_{av} = \frac{Q}{\lambda_f\pi\left(T_s - T_f\right)} = -\frac{1}{\pi}\int_0^\pi \left.\frac{\partial T}{\partial r}\right|_{r=0.5} d\phi.\tag{6}$$

The average Nusselt number of the whole array Nu_{array} is obtained as an arithmetic average value of Nu_{av} from the individual tubes in the array [8]:

$$Nu_{array} = \frac{1}{N}\sum_{i=1}^{N} Nu_{av}.\tag{7}$$

3. Arrangement of Investigated Tube Arrays and Boundary Conditions

The numerical simulations of temperature fields and calculation of the local and average Nusselt numbers were performed in the Ansys Fluent software. The following boundary conditions are applied:

(a) at the right symmetry line B-C (symmetry):

$$\frac{\partial V}{\partial X} = 0;\ U = 0;\ \frac{\partial T}{\partial X} = 0,\tag{8}$$

(b) on the cylinder surfaces (wall):

$$U = 0;\ V = 0;\ T = 1,\tag{9}$$

(c) at the bottom boundary line A-B (velocity inlet):

$$\frac{\partial V}{\partial Y} = 0;\ U = 0;\ T = 0\quad if\quad V \geq 0,\tag{10}$$

(d) at the left boundary line A-D (pressure outlet):

$$V = 0;\ \frac{\partial U}{\partial X} = 0;\ \frac{\partial T}{\partial X} = 0\quad if\quad U \geq 0,\tag{11}$$

(e) at the top boundary line C-D (pressure outlet):

$$\frac{\partial V}{\partial Y} = 0;\ U = 0;\ \frac{\partial T}{\partial Y} = 0\quad if\quad V \geq 0.\tag{12}$$

Inlet boundary: velocity inlet and constant temperature of 295 K. Outlet boundary, left wall: pressure outlet = 0 gauge pressure and constant temperature of 295 K. Right wall: symmetry. Tube walls: constant temperature from 313 to 373 K (Figures 1 and 2).

A preliminary study of the mesh grid size was carried out where quadrilateral elements were used. As shown in Table 2, five different numbers of elements were created around the half tube where the average Nusselt numbers of the single cylinder arrangement were being evaluated. On the basis of this study, 50 elements around the half tube are sufficient from the accuracy point of view. Owing to having more information about the local Nusselt numbers, 90 elements around the half tube (2°

angle increment) were used despite the computational time increase. Furthermore, the element length, perpendicular to the tube, was studied. Table 2 shows the numbers of elements on the length of 8 mm perpendicular to the tube with 90 elements around the half tube. It is clear that 15 is a sufficient number of elements. Considering the elements quality (aspect ratio, angles, etc.), 20 elements were chosen. The average element length size close to the tube is 0.3 mm.

Table 2. Mesh independence test for single cylinder.

No. of Elements Around the Half Tube	Average *Nu*	No. of Elements Perpendicular to the Tube	Average *Nu*
36	4.749	10	4.747
50	4.759	15	4.721
60	4.721	20	4.710
80	4.719	30	4.700
90	4.710	45	4.695

The governing Equations (1)–(4) considering the boundary conditions (8)–(12) were solved by the finite volume method using the Ansys Fluent software. Both transient and steady analyses were compared with each other from an accuracy point of view. The comparison was performed on the single cylinder simulation. While the transient analysis was done with variable air properties, the steady analysis was done with both variable and constant air properties. Regarding the variable properties, thermal conductivity and viscosity were considered as polynomial functions of temperature while specific heat was considered as a piecewise-polynomial function of temperature.

On the basis of negligible differences shown in Figure 3, the steady pressure-based analyses with the variable air properties were performed for the other arrangements. The Boussinesq approximation was used as a computational model and the pressure–velocity coupling was handled by the SIMPLE-C algorithm. For a momentum and energy gradient, the Quick and Second order upwind schemes were used. The solution was considered to be fully converged when the residuals of continuity, x-velocity, y-velocity, and energy met the convergence criterion 10^{-6}. Finally, Nu_ϕ were exported and subsequently evaluated.

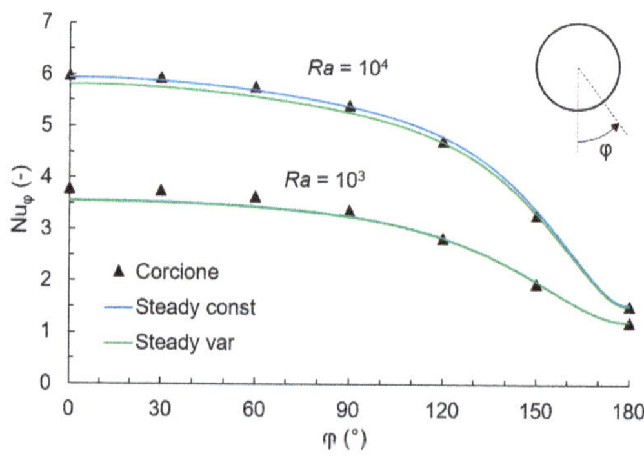

Figure 3. The comparison of the local Nusselt numbers Nu_ϕ of the single cylinder for various computational models at $Ra = 10^3$ and $Ra = 10^4$.

4. Results and Discussion

4.1. Results of the Single Cylinder

Laminar free convection heat transfer from the single isothermal circular cylinder with a diameter of 20 mm was studied (Figure 1a). The simulations were performed for $Ra = 10^3$ and $Ra = 10^4$. For a comparison, numerical simulations for the steady analysis with constant and variable material properties were performed (Figure 3).

The results of Corcione [8] were compared with ours. The discrepancies of Nu_ϕ values for $Ra = 10^3$ are in the range of 0.37–5.90% for the steady state at constant air properties, and 0.14% to 6.11% for the steady state at variable air properties. The discrepancies of Nu_ϕ values for $Ra = 10^4$ are in the range of 0.23–3.14% for the steady state at constant air properties and 0.33–3.07% for the steady state at variable air properties.

The present numerical simulations quantitatively match the data of the numerical simulations by Corcione [8], Saitoh et al. [5], and Wang et al. [18] for the single cylinder shown in Figure 4. The present numerical simulations were performed at the steady state analysis and variable air properties, described in Section 3.

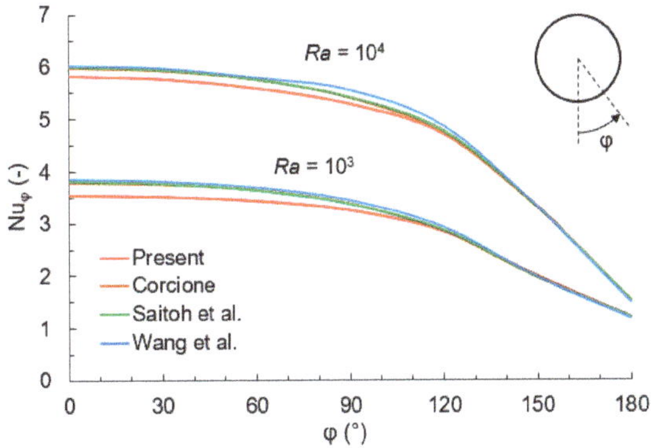

Figure 4. The comparison of the local Nusselt numbers Nu_ϕ of the single cylinder with other authors at $Ra = 10^3$ and $Ra = 10^4$.

In comparison with Saitoh et al. [5], the discrepancies of Nu_ϕ values are in the range of 0.02–6.70% for $Ra = 10^3$, and 0.36–3.00% for $Ra = 10^4$, respectively. In comparison with Wang et al. [18], the discrepancies of Nu_ϕ values are in the range of 1.05–7.84% for $Ra = 10^3$ and 0–5.13% for $Ra = 10^4$, respectively.

The numerical simulations for the single cylinder were performed to create a suitable computational model which was verified with other authors. On the basis of this model, the other simulations for the tube arrays 4 × 1, 4 × 2, and 4 × 4 were performed.

4.2. Results of the Tube Array 4 × 1a

We have studied the laminar free convection heat transfer from the vertical array of four isothermal circular tubes with a diameter of 20 mm. The simulations were performed for the tubes at a center-to-center dimensionless vertical distance of $S_V/D = 2$ (Figure 1b) at $Ra = 10^3$ and $Ra = 10^4$. The courses of the local Nusselt numbers Nu_ϕ for the individual tubes in the array are shown in Figure 5a,b for the angles of $\phi = 0°$ to 180°. Owing to the model symmetry, the courses of Nu_ϕ are identical for the right side as well.

As it is shown in Figure 5, the highest values of Nu_ϕ are achieved on the tube I circumference. For $Ra = 10^4$, Nu_ϕ increases for all tubes I–IV in comparison with $Ra = 10^3$. A significant increase of Nu_ϕ on the downstream area of the tube II ($\phi = 0°–70°$) is shown in comparison with $Ra = 10^3$ (Figure 5b). Due to the higher temperature difference between the tube surface and ambient air at $Ra = 10^4$, heat transfer to the tube II is affected by the heat flux from the tube I, and therefore Nu_ϕ increases. Simultaneously, the tubes III and IV are affected by the heat flux of previous tubes. When comparing the same positions of the tubes at $Ra = 10^3$ and 10^4 and the angles of $\phi = 0°–180°$, Nu_ϕ increases for $Ra = 10^4$ in the range of 35–61% for the tube I, 30–171% for the tube II, 29–158% for the tube III, and 29–121% for the tube IV.

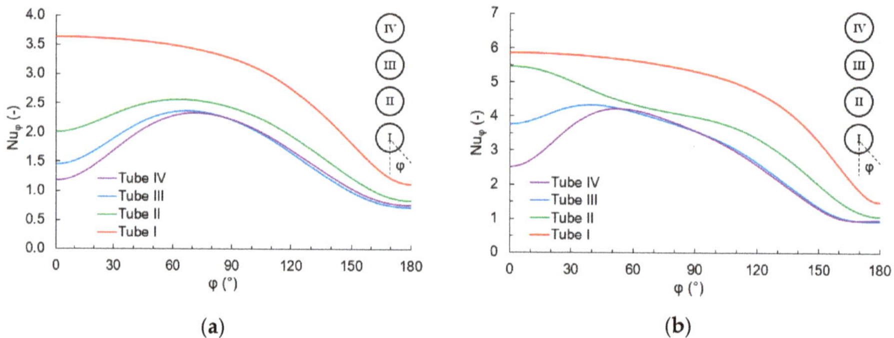

Figure 5. The course of the local Nusselt numbers Nu_ϕ for the individual tubes in the tube array 4×1: (**a**) $Ra = 10^3$; (**b**) $Ra = 10^4$.

The ratio of the average Nusselt number for the ith cylinder and single cylinder Nu_{iav}/Nu_{0av} for a vertical column of four horizontal cylinders at $Ra = 10^4$ is shown in Figure 6a. The presented values were compared with Razzaghpanah and Sarunac [19] on the basis of Table 3 and Figure 9 [19] and Corcione [8] on the basis of Equation (17) [8]. The average Nu for the first cylinder in a column is the same as for the single cylinder. The relative error of the other presented tubes for the ratio Nu_{iav}/Nu_{0av}, compared with Razzaghpanah and Sarunac [19], is up to 5.2%. Figure 6b compares the presented average Nusselt numbers of the ith cylinder with Equation (22) of the authors Razzaghpanah and Sarunac [19] and Equations (11)–(13) of the authors Kitamura et al. [15].

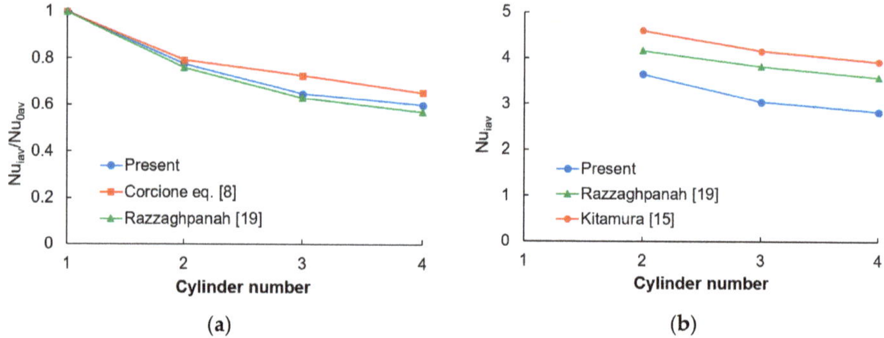

Figure 6. Variation of the (**a**) ratio Nu_{iav}/Nu_{0av} vs. cylinder number; (**b**) average Nusselt number of the ith cylinder.

Corcione [8] obtained numerical results for the average Nusselt number of the tube array Nu_{array} which may be correlated to the Rayleigh number Ra, the tube ratio spacing S_V/D, and the number of the tubes in the array N:

$$Nu_{array} = Ra^{0.235}\left\{0.292ln\left[\left(\tfrac{S_V}{D}\right)^{0.4} \times N^{-0.2}\right] + 0.447\right\},$$
$$2 \leq N \leq 6; \quad 5\times 10^2 \leq Ra \leq 5\times 10^5, \tag{13}$$
$$S_V/D \leq 10 - \log(Ra),$$

with the percent standard deviation of error $E_{SD} = 2.25\%$ and error E in the range of -4.79% to +5.27%. The values Nu_{array} obtained by our simulations lie out of the error range according to Corcione [8] by 4.9% for $Ra = 10^3$ and 3.7% for $Ra = 10^4$, respectively.

4.3. Results of the Tube Array 4 × 2

We have studied the laminar free convection heat transfer from the two vertical arrays of four isothermal circular tubes with a diameter of 20 mm. The simulations were performed for the center-to-center dimensionless vertical distance $S_V/D = 2$ and horizontal distance $S_H/D = 2$ (Figure 1c) at $Ra = 10^3$ and $Ra = 10^4$.

The courses of the local Nusselt numbers Nu_ϕ for the individual tubes in the array 4 × 2 are shown in Figure 7. The courses of Nu_ϕ are valid for the left tube array; the right tube array has mirrored Nu_ϕ courses. As shown in Figure 7a, the highest values of Nu_ϕ for $Ra = 10^3$ are achieved on the tube I in the range of 1.167–3.615. For $Ra = 10^4$, the local Nusselt numbers on the tube I are in the range of 1.518–6.118, which is an increase of 30–69% in comparison with $Ra = 10^3$. When comparing the same positions of the tubes at $Ra = 10^3$ and 10^4 and the angles of $\phi = 0°–360°$, Nu_ϕ increase for $Ra = 10^4$ in the range of 30–70% for the tube I, 48–183% for the tube II, 55–245% for the tube III, and 48–265% for the tube IV.

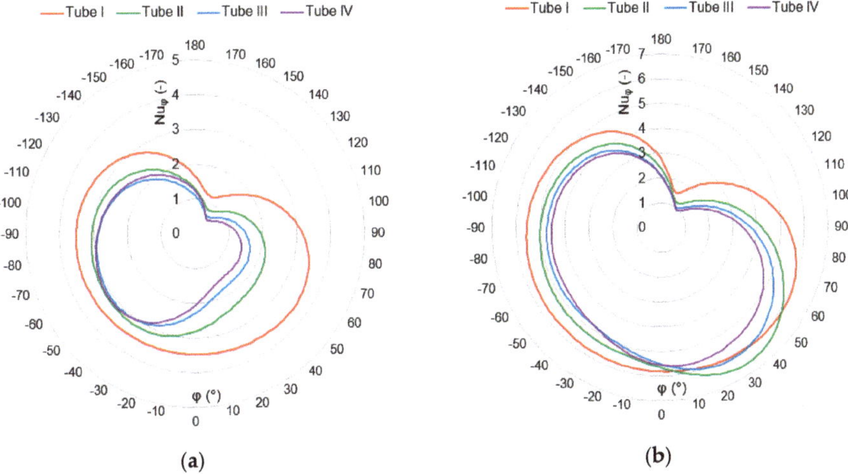

Figure 7. The course of the local Nusselt numbers Nu_ϕ for the individual tubes in tube array 4 × 2: (**a**) $Ra = 10^3$; (**b**) $Ra = 10^4$.

Corcione [9] obtained numerical results for the average Nusselt number of the double tube array Nu_{array} which may be correlated to the Rayleigh number Ra, the tube ratio spacing S_V/D and S_H/D, and the number of the tubes in each vertical tube array of the pair N:

$$Nu_{array} = 0.43 Ra^{0.235}(S_H/D)^{0.14}(S_V/D)^{0.2} \times N^{-0.1},$$
$$2 \leq N \leq 4;\ 2.4 - 0.2\log(Ra)\ \leq S_H/D < 5, \tag{14}$$
$$2 \leq S_V/D < 5;\ 10^2 \leq Ra \leq 10^4,$$

with the percent standard deviation of error $E_{SD} = 2.12\%$ and error E in the range of -5.32% to +5.67%. The values Nu_{array} obtained by our simulations lie out of the error range according to Corcione [9] by 1.2% for $Ra = 10^3$. For $Ra = 10^4$, our values lie in the error range according to Corcione [9].

The values of average Nusselt numbers Nu_{av} for the single cylinder and tube arrays 4 × 1 and 4 × 2 are given in Table 3 for $Ra = 10^3$ and $Ra = 10^4$. The values of Nu_{av} increase with Ra, specifically from 62% to 108% at $Ra = 10^4$ against $Ra = 10^3$ for individual tubes within the same arrangement. For the tube array 4 × 1, the values of Nu_{av} decrease from the tube I to the tube IV. For the tube array 4 × 2, the columns affect each other, where the values of Nu_{av} are higher from 2% to 35% on each tube compared with the tube array 4 × 1.

Table 3. Average Nusselt numbers Nu_{av} for the single cylinder and tube arrays 4×1, 4×2 at $Ra = 10^3$ and $Ra = 10^4$.

	$Ra = 10^3$				$Ra = 10^4$			
Single Cylinder	**2.906**				**4.710**			
Tube Array 4×1	I	II	III	IV	I	II	III	IV
	2.875	1.974	1.692	1.645	4.722	3.654	3.044	2.827
Tube Array 4×2	I	II	III	IV	I	II	III	IV
(Left Column)	2.946	2.248	1.928	1.861	4.816	4.344	4.015	3.824

The temperature fields around the single cylinder and created thermal chimney at $Ra = 10^3$ and 10^4 are shown in Figure 8a,b. The temperature fields of the tube arrays 4×1 and 4×2 together with created thermal chimneys at $Ra = 10^3$ and 10^4 are shown in Figure 8c–f.

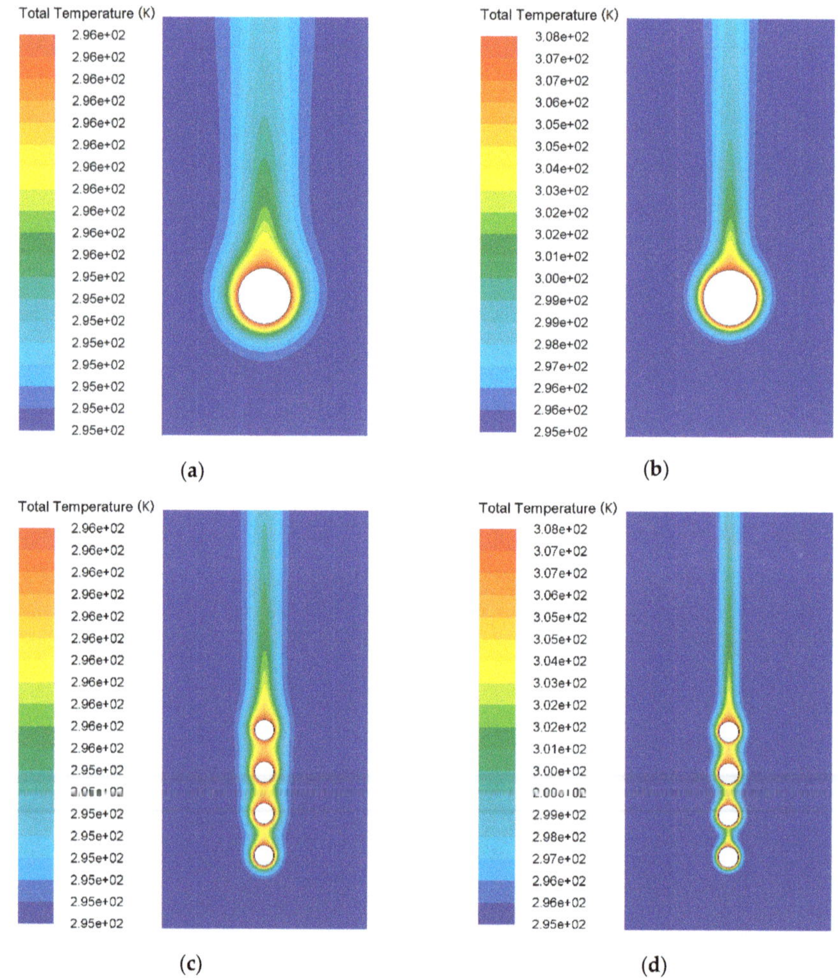

Figure 8. *Cont.*

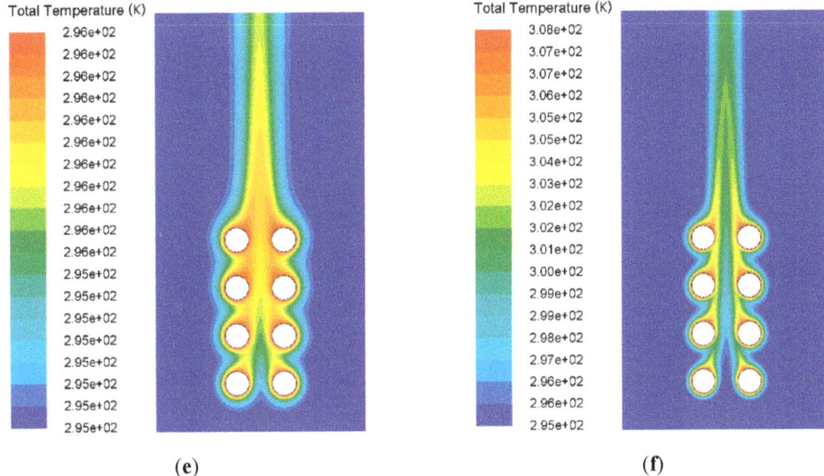

Figure 8. The temperature fields of the cylinder and tube arrays 4×1, 4×2: (**a**) single cylinder at $Ra = 10^3$; (**b**) single cylinder at $Ra = 10^4$; (**c**) tube array 4×1 at $Ra = 10^3$; (**d**) tube array 4×1 at $Ra = 10^4$; (**e**) tube array 4×2 at $Ra = 10^3$; (**f**) tube array 4×2 at $Ra = 10^4$.

As the thermal chimney develops upwards, its temperature decreases due to heat transfer to the cold ambient air. A density difference between the thermal chimney and ambient air induces a buoyancy force; near the heated cylinder, the buoyancy force is higher.

In Figure 8c,d, it can be seen a gradual increase of the thermal boundary layer thickness in a vertical direction of the tube array 4×1. The widening of the thermal boundary layer causes the diminishing of the temperature gradient on the surfaces and the reduction of the local and average Nusselt numbers Nu_ϕ and Nu_{av}, respectively. In Figure 8e–f, a mutual effect of two tube columns in the tube array 4×2 can be seen, where an in-draft heat occurs between the first and second columns and a common thermal chimney is achieved over the tube array. For $Ra = 10^4$, a thinner thermal boundary layer around the tube circumferences can be observed. Owing to this, the values of Nu_ϕ and Nu_{av} increase compared with $Ra = 10^3$.

The comparison of the local Nusselt numbers Nu_ϕ of the single cylinder with the bottom tube of the tube array 4×1 and with the bottom left tube of the tube array 4×2 is shown in Figure 9 for $Ra = 10^3$ and 10^4. The courses of Nu_ϕ for the tube array 4×2 are valid for the left tube column; the right tube column has mirrored Nu_ϕ courses. As shown, the values Nu_ϕ for the bottom tube of the 4×1 array are similar to the values Nu_ϕ for the single cylinder at $Ra = 10^3$ and 10^4. When comparing the bottom tube of the 4×2 array and the single cylinder, there are different courses due to the right tube column effect.

4.4. Results of the Tube Array 4×4 (Base, Concave, Convex)

The laminar free convection heat transfer from the four vertical arrays of four isothermal circular tubes with the diameter of 20 mm has been studied. Moreover, the effect of the tube ratio spacing on the Nusselt numbers had been investigated for different Ra before. The tube spacing in the vertical direction S_V and horizontal one S_H increased in the same way; therefore, general tube ratio spacing S/D is defined. The simulations were performed for the S/D ratios of 2, 2.5, and 3 (Figure 2) for the tube array 4×4 at three different configurations (Base—Figure 2a, Concave—Figure 2b, Convex—Figure 2c). The courses of the Nu_ϕ values in the tube array 4×4 for base, concave, and convex configurations at $Ra = 1.3 \times 10^4$ and 3.7×10^4 with a S/D ratio of 2 are shown in Figure 10. The courses for the same parameters and the S/D ratio of 2.5 are shown in Figure 11 and for the S/D ratio of 3 in Figure 12. When increasing the Rayleigh number, the convection increases in strength. Furthermore, the Nu_{av} and Nu_{array} values increase with the S/D ratio and Ra increasing.

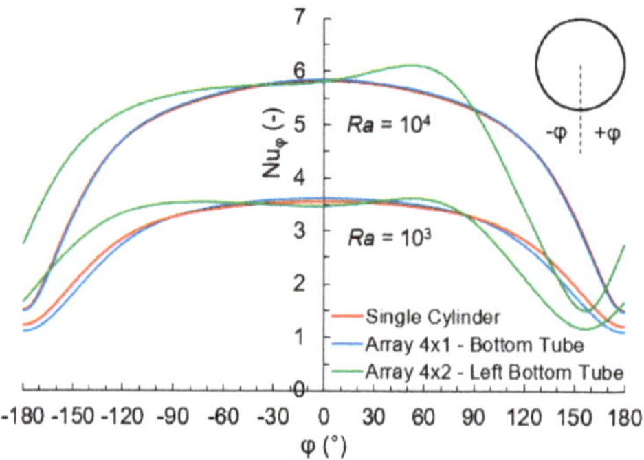

Figure 9. The comparison of the local Nusselt numbers Nu_ϕ of the single cylinder with the bottom tube of the tube arrays at $Ra = 10^3$ and $Ra = 10^4$.

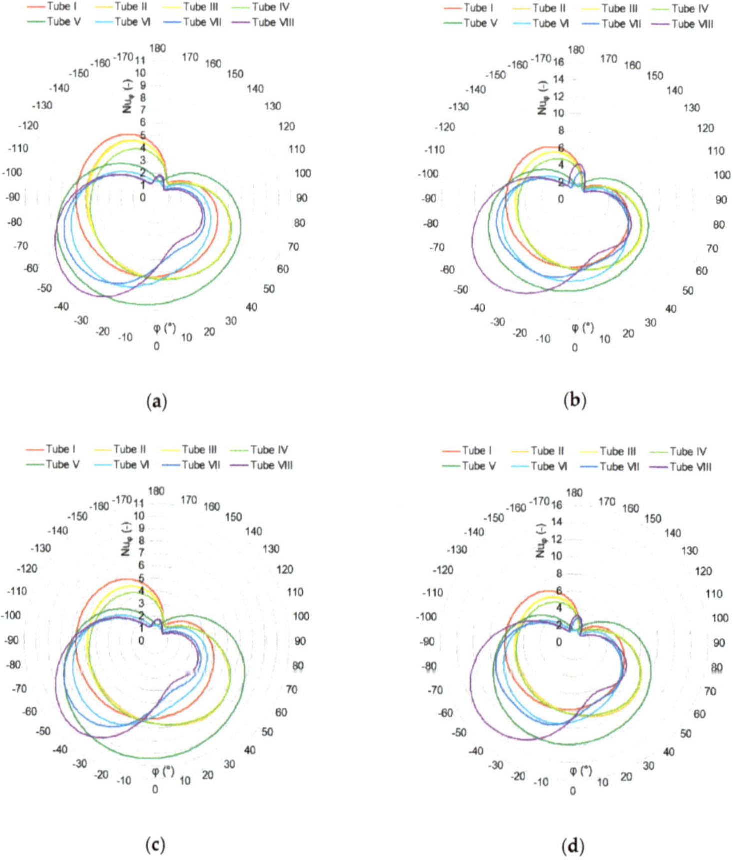

Figure 10. *Cont.*

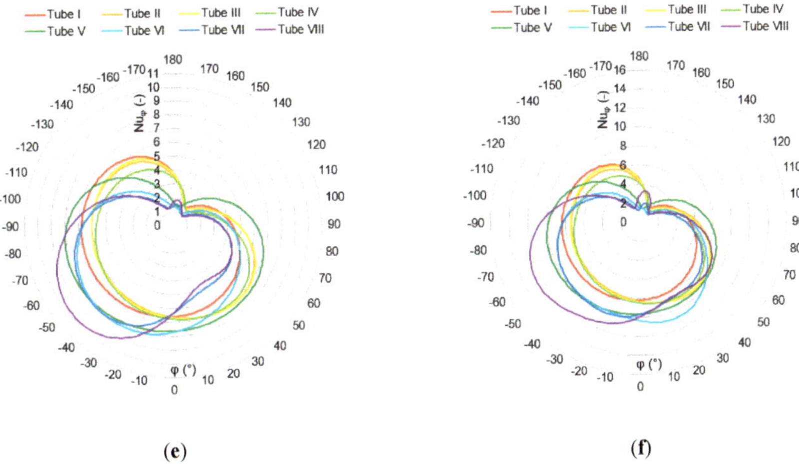

(e)   (f)

Figure 10. The local Nusselt numbers Nu_ϕ for different configurations of the tube array 4×4 with the tube ratio spacing S/D of 2: (**a**) base configuration at $Ra = 1.3 \times 10^4$; (**b**) base configuration at $Ra = 3.7 \times 10^4$; (**c**) concave configuration at $Ra = 1.3 \times 10^4$; (**d**) concave configuration at $Ra = 3.7 \times 10^4$; (**e**) convex configuration at $Ra = 1.3 \times 10^4$; (**f**) convex configuration at $Ra = 3.7 \times 10^4$.

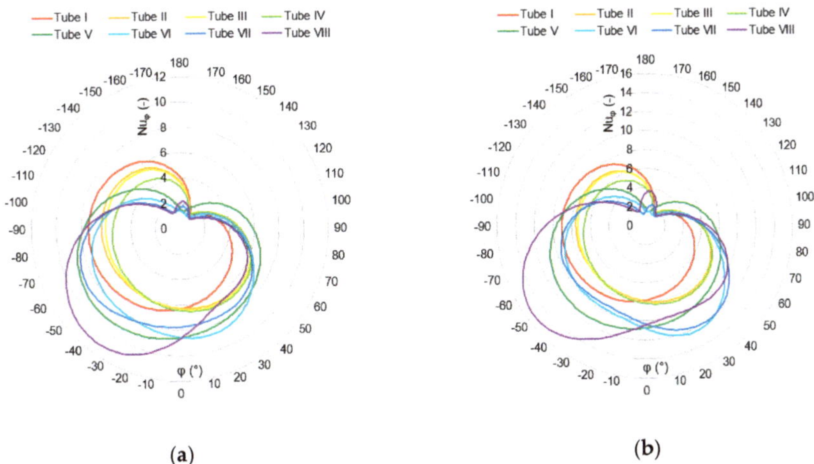

(a)   (b)

Figure 11. *Cont.*

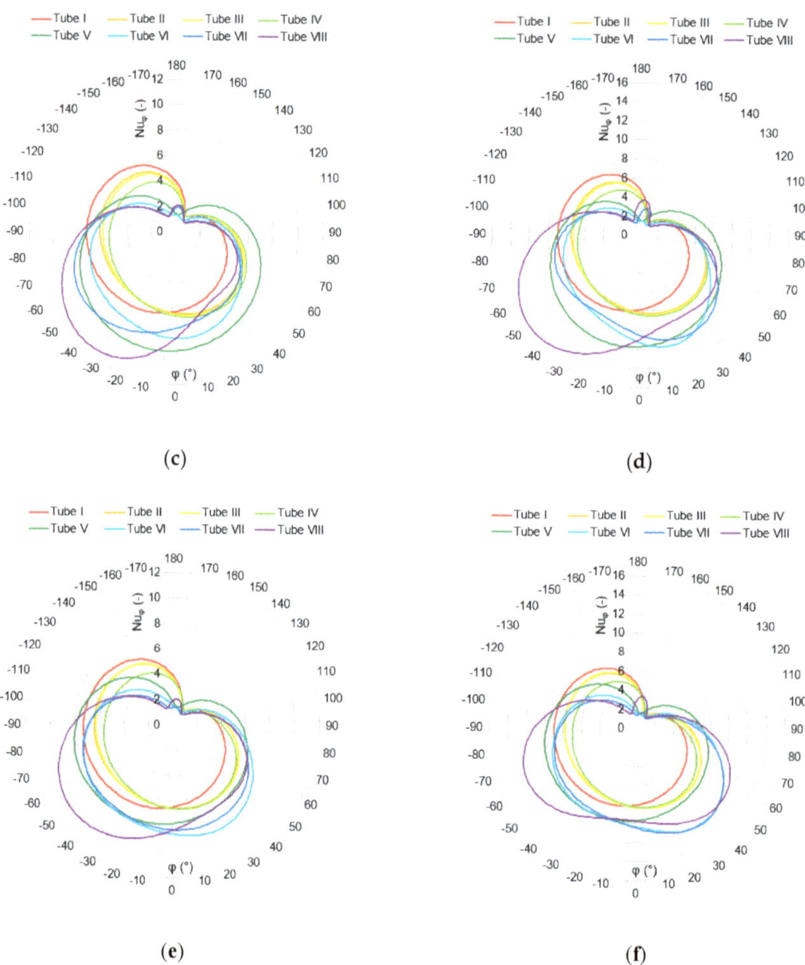

(c)

(d)

(e)

(f)

Figure 11. The local Nusselt numbers Nu_ϕ for different configurations of the tube array 4×4 with the tube ratio spacing S/D of 2.5: (**a**) base configuration at $Ra = 1.3 \times 10^4$; (**b**) base configuration at $Ra = 3.7 \times 10^4$; (**c**) concave configuration at $Ra = 1.3 \times 10^4$; (**d**) concave configuration at $Ra = 3.7 \times 10^4$; (**e**) convex configuration at $Ra = 1.3 \times 10^4$; (**f**) convex configuration at $Ra = 3.7 \times 10^4$.

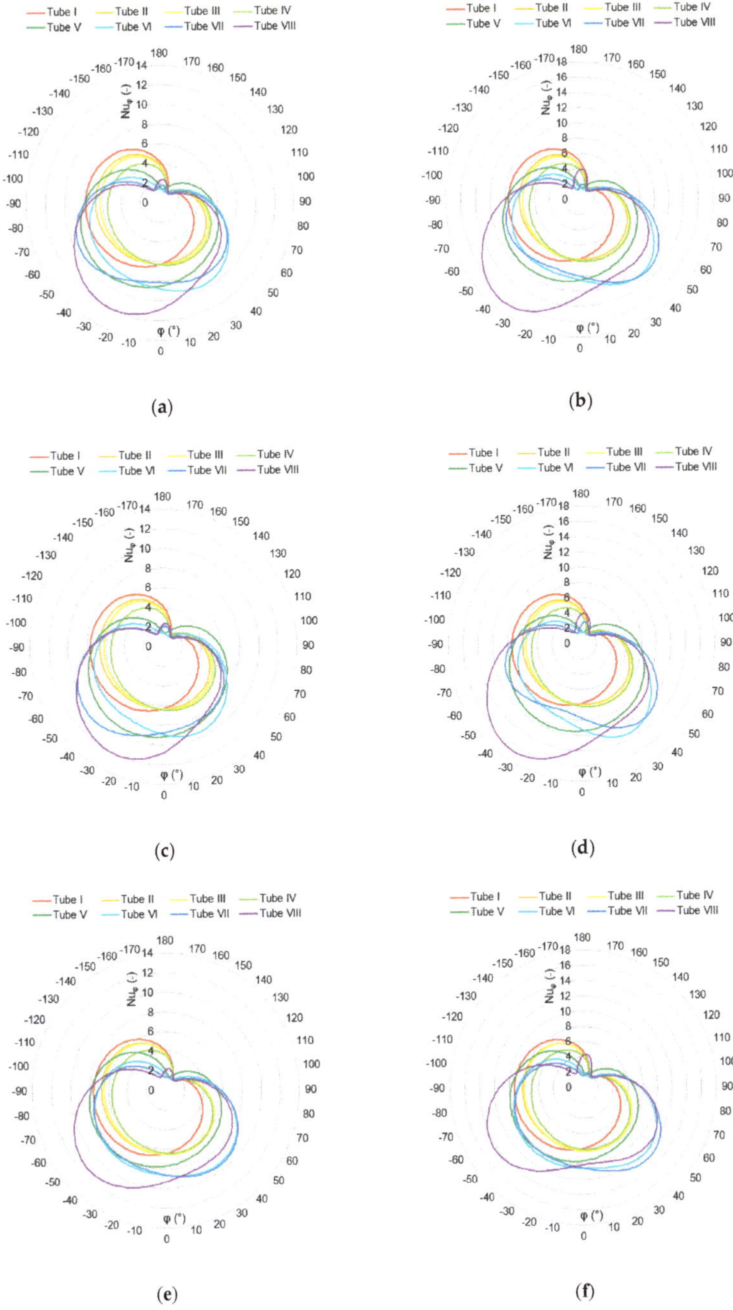

Figure 12. The local Nusselt numbers Nu_ϕ for different configurations of the tube array 4×4 with the tube ratio spacing S/D of 3: (**a**) base configuration at $Ra = 1.3 \times 10^4$; (**b**) base configuration at $Ra = 3.7 \times 10^4$; (**c**) concave configuration at $Ra = 1.3 \times 10^4$; (**d**) concave configuration at $Ra = 3.7 \times 10^4$; (**e**) convex configuration at $Ra = 1.3 \times 10^4$; (**f**) convex configuration at $Ra = 3.7 \times 10^4$.

The ratio of presented Nu for the in-line bundle of horizontal cylinders and the single cylinder for $Ra = 10^4$ was compared with Razzaghpanah and Sarunac [20] on the basis of Figure 10 [20]. For the ratio S_L/D and $S_T/D = 2$, the values $Nu_{array} = 4.726$ and $Nu_0 = 4.710$ were obtained. The mentioned ratios $Nu_{array}/Nu_0 = 1.003$ and $Nu_0/Nu_{array} = 0.997$ are close to those obtained by Razzaghpanah and Sarunac [20] (1.073 and 0.900, respectively).

Subsequently, the average Nusselt numbers of the whole array Nu_{array} for three different configurations (base, concave, convex) with the tube ratio spacing S/D of 2, 2.5, and 3 were compared for Ra in the range of 1.3×10^4 to 3.7×10^4 (Figure 13). On the basis of the results, the Nu_{array} value increases with the increase of Ra and tube spacing. The highest values Nu_{array} are achieved at the convex configuration. The discrepancies of Nu_{array} between the concave and convex configurations are up to 3.7% depending on Ra. Owing to negligible error ranges for the correlating equations, these are interchangeable with similar discrepancies. The Nu_{array} values for the base configuration lie between the values for the concave and convex ones for all considered Ra.

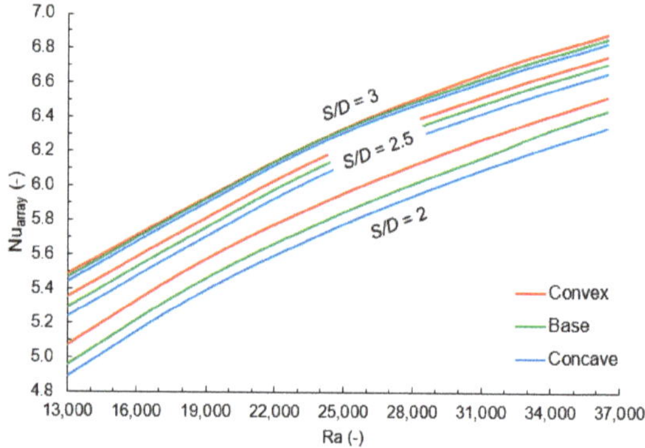

Figure 13. The average Nusselt numbers Nu_{array} for the base, concave, and convex configurations of the tube array 4×4 with the different tube ratio spacing S/D.

The temperature fields for the tube arrays 4×4 with the tube ratio spacing S/D of 2 obtained by the numerical simulations are shown in Figure 14. The temperature fields for the tube array 4×4 with the tube ratio spacing S/D of 2.5 and 3 are shown in Figures 15 and 16, respectively. For the middle columns of the tubes (V to VIII–Figure 2), it is clear that the thermal plume from the lower tubes affects the local and average Nusselt numbers of the upper tubes. The upper tubes create a quasi-forced convection which tends to a heat transfer increase from the upper tubes. It causes a lower temperature difference between the tube surface and the incoming air, which tends to a heat transfer decrease at the tubes downstream that is of a major importance at close spacing.

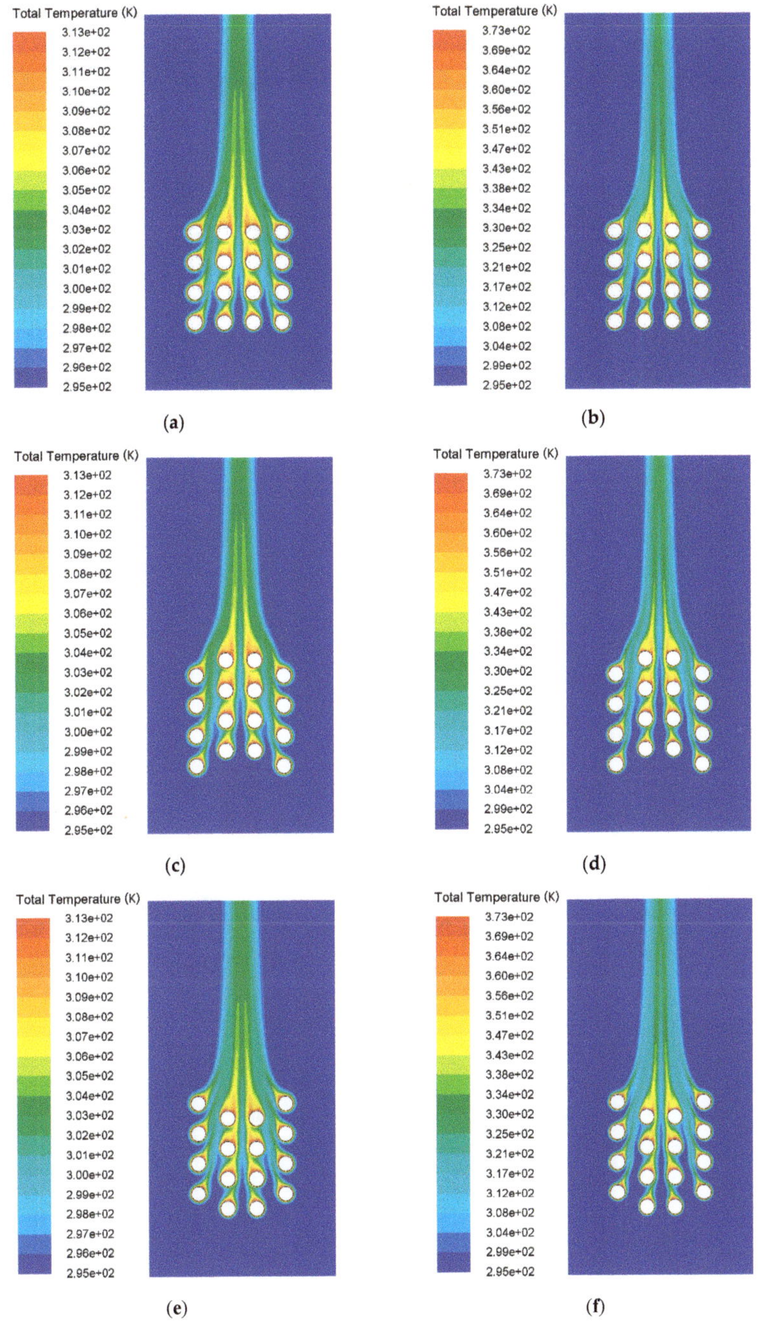

Figure 14. The temperature fields of the tube arrays 4×4 with the tube ratio spacing $S/D = 2$: (**a**) base configuration at $Ra = 1.3 \times 10^4$; (**b**) base configuration at $Ra = 3.7 \times 10^4$; (**c**) concave configuration at $Ra = 1.3 \times 10^4$; (**d**) concave configuration at $Ra = 3.7 \times 10^4$; (**e**) convex configuration at $Ra = 1.3 \times 10^4$; (**f**) convex configuration at $Ra = 3.7 \times 10^4$.

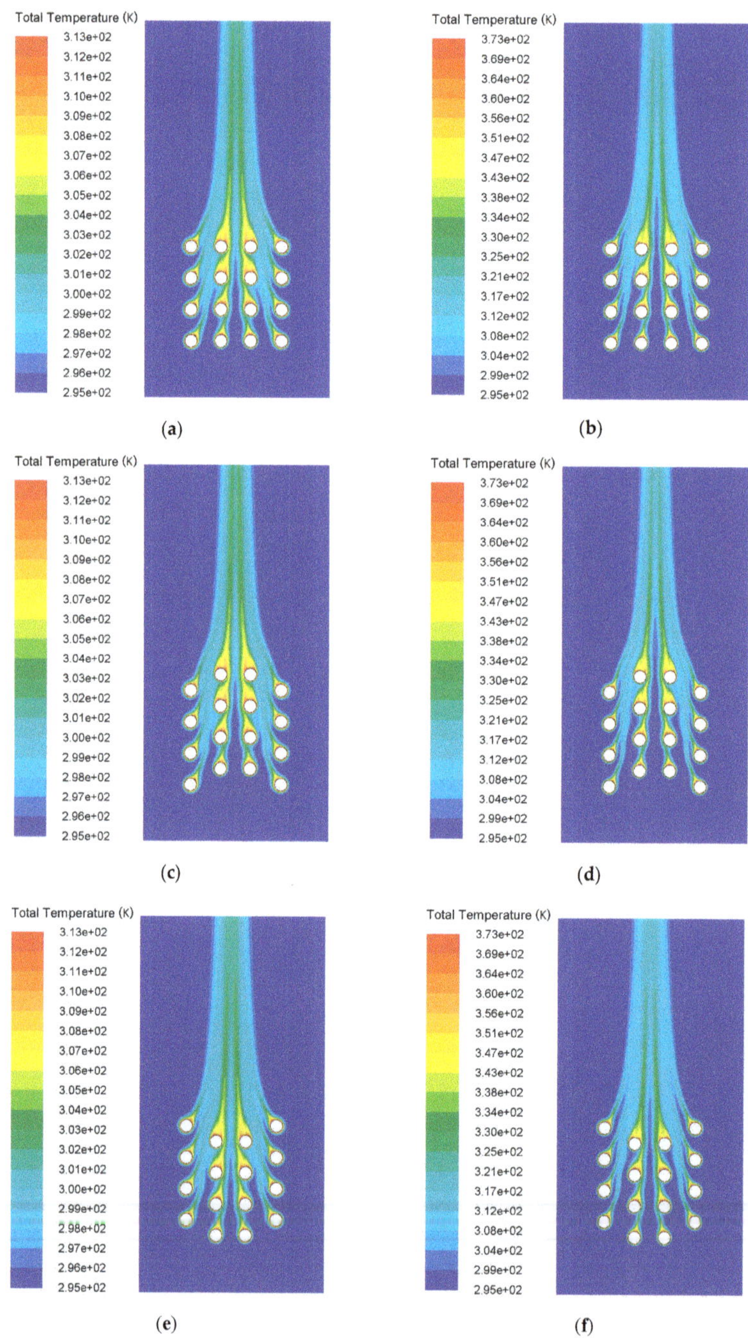

Figure 15. The temperature fields of the tube arrays 4×4 with the tube ratio spacing $S/D = 2.5$: (**a**) base configuration at $Ra = 1.3 \times 10^4$; (**b**) base configuration at $Ra = 3.7 \times 10^4$; (**c**) concave configuration at $Ra = 1.3 \times 10^4$; (**d**) concave configuration at $Ra = 3.7 \times 10^4$; (**e**) convex configuration at $Ra = 1.3 \times 10^4$; (**f**) convex configuration at $Ra = 3.7 \times 10^4$.

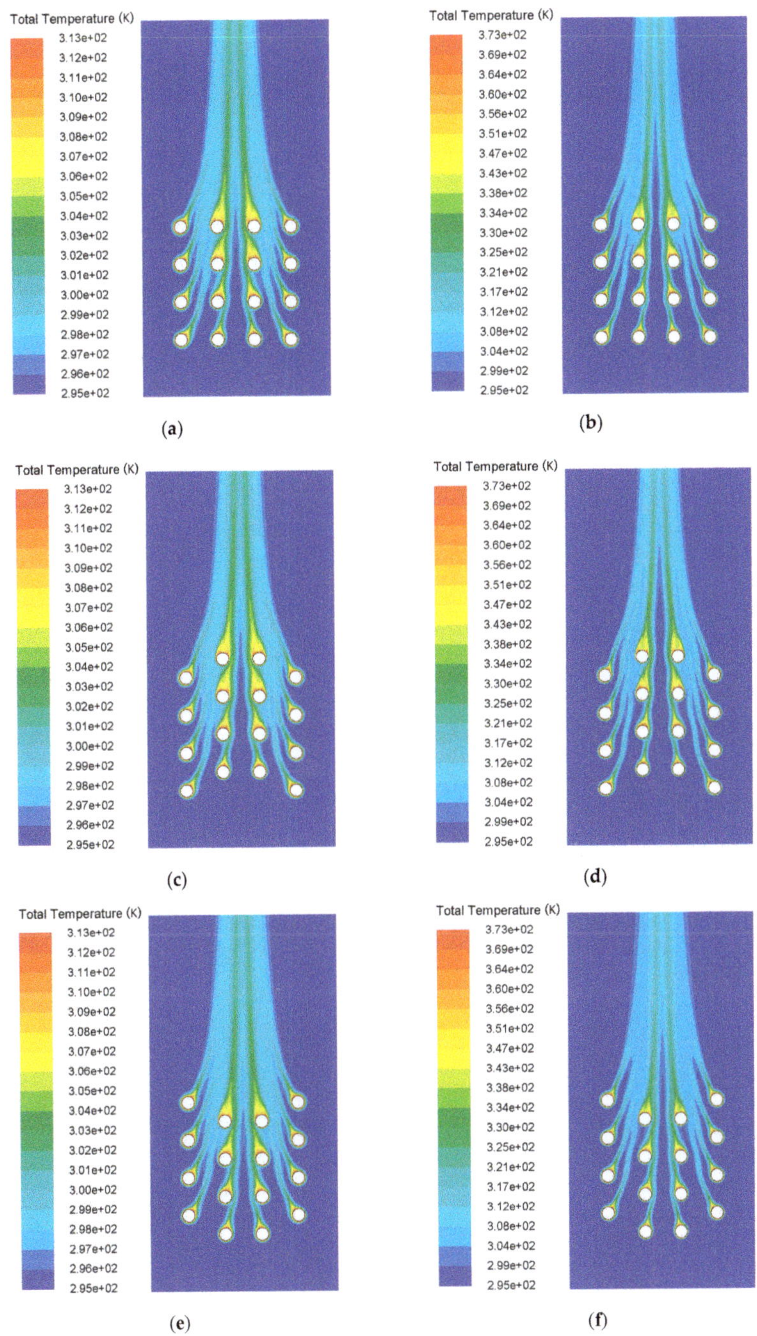

Figure 16. The temperature fields of the tube arrays 4×4 with the tube ratio spacing $S/D = 3$: (**a**) base configuration at $Ra = 1.3 \times 10^4$; (**b**) base configuration at $Ra = 3.7 \times 10^4$; (**c**) concave configuration at $Ra = 1.3 \times 10^4$; (**d**) concave configuration at $Ra = 3.7 \times 10^4$; (**e**) convex configuration at $Ra = 1.3 \times 10^4$; (**f**) convex configuration at $Ra = 3.7 \times 10^4$.

The outer column's air flow (I to IV–Figure 2) is affected by the inner tube columns and the heat is directed to the central part of the tube array with a subsequent thermal plume creation. It may be seen that the Rayleigh number increases the "chimney effect". As the temperature fields around the tubes do not affect each other significantly, the boundary layer thickness decreases and Nu_{av} increases with the higher ratio S/D. The effect of the tube spacing on the temperature field shapes is obvious in Figures 14–16. On the basis of the results, we can state that the higher tube spacing, the more effective heat transfer.

The effect of the S/D value and the tube array configuration for $Ra = 1.3 \times 10^4$ and 3.7×10^4 is obvious in Tables 4 and 5. The Nu_{av} value for the tubes I–VIII, the Nu_{array} value for whole tube arrays, and the q_{array} value were evaluated for all configurations and S/D ratios. When increasing the S/D ratio, also Nu_{av}, Nu_{array}, and q_{array} increase, where the tube array configuration has a crucial effect.

Table 4. Average Nusselt numbers for different ratios S/D and configurations of 4×4 tube arrays at $Ra = 1.3 \times 10^4$.

S/D	Config.	Average Nusselt Numbers, Nu_{av} (-)								Nu_{array} (-)	q_{array} (W/m^2)
		I	II	III	IV	V	VI	VII	VIII		
	Base	5.273	5.169	5.163	4.996	5.970	4.519	4.212	4.339	4.955	118.371
2	Concave	5.198	5.118	5.136	5.003	6.081	4.294	4.114	4.194	4.892	116.869
	Convex	5.302	5.225	5.278	4.968	5.819	4.901	4.398	4.672	5.070	121.117
	Base	5.279	5.162	5.129	4.909	6.018	5.255	5.108	5.471	5.291	126.381
2.5	Concave	5.238	5.145	5.114	4.945	6.111	5.018	5.023	5.342	5.242	125.201
	Convex	5.287	5.219	5.219	4.869	5.837	5.556	5.216	5.629	5.354	127.875
	Base	5.256	5.161	5.095	4.820	5.984	5.690	5.672	6.057	5.467	130.550
3	Concave	5.244	5.152	5.082	4.859	6.077	5.501	5.668	5.971	5.444	130.014
	Convex	5.264	5.208	5.168	4.793	5.795	5.863	5.686	6.122	5.487	131.049

Table 5. Average Nusselt numbers for different ratios S/D and configurations of 4×4 tube arrays at $Ra = 3.7 \times 10^4$.

S/D	Config.	Average Nusselt Numbers, Nu_{av} (-)								Nu_{array} (-)	q_{array} (W/m^2)
		I	II	III	IV	V	VI	VII	VIII		
	Base	6.485	6.412	6.432	6.206	7.444	5.998	5.869	6.695	6.443	768.405
2	Concave	6.392	6.376	6.396	6.237	7.556	5.664	5.690	6.462	6.346	756.880
	Convex	6.522	6.477	6.544	6.098	7.229	6.436	5.979	6.895	6.523	777.818
	Base	6.466	6.343	6.350	6.056	7.452	6.792	6.633	7.626	6.715	800.298
2.5	Concave	6.429	6.330	6.335	6.122	7.572	6.481	6.510	7.526	6.663	794.168
	Convex	6.496	6.438	6.459	5.984	7.226	7.093	6.815	7.549	6.757	805.471
	Base	6.427	6.332	6.304	5.957	7.393	7.195	7.127	8.141	6.860	817.236
3	Concave	6.422	6.321	6.296	6.019	7.512	6.956	7.049	8.108	6.835	814.315
	Convex	6.484	6.441	6.419	5.949	7.177	7.376	7.252	7.994	6.887	820.744

As it is shown in Table 4, Nu_{array} and q_{array} values decrease by 1.27% at the concave configuration compared with the base one for $S/D = 2$ and $Ra = 1.3 \times 10^4$. On the contrary, these representative values increase by 2.32% at the convex configuration compared with the base one for the same parameters. Additionally, they increase by 3.64% at the convex configuration compared with the concave one. For $S/D = 2.5$, the same comparison is carried out, where Nu_{array} and q_{array} decrease by 0.93% at the concave configuration and increase by 1.19% at the convex configuration compared with the base one. When comparing the convex configuration with the concave one, Nu_{array} and q_{array} increase by 2.14%. For

$S/D = 3$, Nu_{array} and q_{array} decrease by 0.42% at the concave configuration and increase by 0.37% at the convex configuration compared with the base one. When comparing the convex configuration with the concave one, Nu_{array} and q_{array} increase by 0.79%.

As shown in Table 5, Nu_{array} and q_{array} values decrease by 1.51% at the concave configuration compared with the base one for $S/D = 2$ and $Ra = 3.7 \times 10^4$. These representative values increase by 1.24% at the convex configuration compared with the base one. Additionally, they increase by 2.79% at the convex configuration compared with the concave one. For $S/D = 2.5$, Nu_{array} and q_{array} decrease by 0.77% at the concave configuration and increase by 0.63% at the convex configuration compared with the base one. When comparing the convex configuration with the concave one, Nu_{array} and q_{array} increase by 1.41%. For $S/D = 3$, Nu_{array} and q_{array} decrease by 0.36% at the concave configuration and increase by 0.39% at the convex configuration compared with the base one. When comparing the convex configuration with the concave one, Nu_{array} and q_{array} increase by 0.76%. It is obvious that the increase of the S/D ratio and Ra values causes a decrease of Nu_{array} and q_{array} percentage discrepancies between the individual configurations.

New correlating equations for Nu_{array} were created for all Ra values in the range of 1.3×10^4 to 3.7×10^4 as follows:

1. The base configuration:

$$Nu_{array} = 0.475\,Ra^{0.2351}(S/D)^{0.1907}, \tag{15}$$

with the percent standard deviation of error $E_{SD} = 0.70\%$, and error E ranged from -0.94% to +1.37%,

2. The concave configuration:

$$Nu_{array} = 0.463\,Ra^{0.2348}(S/D)^{0.2132}, \tag{16}$$

with the percent standard deviation of error $E_{SD} = 0.65\%$, and error E ranged from -0.90% to +1.37%,

3. The convex configuration:

$$Nu_{array} = 0.520\,Ra^{0.2303}(S/D)^{0.1555}, \tag{17}$$

with the percent standard deviation of error $E_{SD} = 0.54\%$, and error E ranged from -0.84% to +1.14%.

The correctness of the correlating equations is shown in Figure 17, where the values of Nu_{array} from the correlating equations and numerical simulations were compared.

Considering the correlating equations for the individual tubes in the array, there is an effort to create the equation containing the tube position in numerical order from bottom to top. For the Nusselt numbers of individual tubes situated in the outer columns, it is possible to assume an approximation by a linear function with the maximum discrepancy of 3% for specific tube position, spacing, configuration, and Ra. However, the monotony is not uniform throughout the tube's interval. The correlating equation indicates a decreasing of Nu_{av} with increasing the tube position. On the contrary, in general, Nu_{av} for the tube 3 is higher than the value for the tube 2. This statement is valid for all configurations, tube spacing and Ra. The Nusselt numbers of the tubes for the inner columns do not have the same course character when changing the parameters. On the basis of the previous facts, the correlating equations for Nu_{av} are not created.

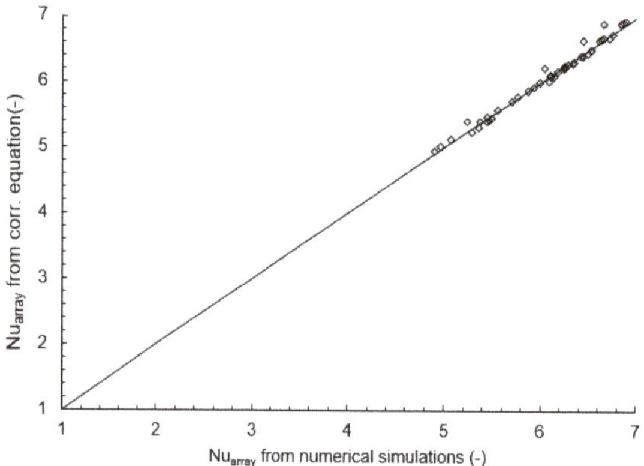

Figure 17. Comparison between presented correlating equations and numerical results.

5. Conclusions

The paper has reported on the results of the numerical investigation of laminar free convection heat transfer from various tube configurations. The single cylinder and the tube arrays of 4×1 and 4×2 arrangements at $Ra = 10^3$ and 10^4 were used for computational model verification. The main part of the research focused on the tube arrays in the configuration of 4×4 (base, concave, convex) at Ra in the range of 1.3×10^4 to 3.7×10^4. From the obtained local and average Nusselt numbers and temperature fields, the following conclusions can be drawn:

1. The local Nusselt numbers Nu_ϕ for the single circular cylinder at $Ra = 10^3$ and 10^4 were confirmed.
2. The average Nusselt numbers Nu_{av} for the tube array 4×1 lie out of the error range, according to the Corcione correlating Equation [8], by 4.9% for $Ra = 10^3$ and 3.7% for $Ra = 10^4$, respectively.
3. The average Nusselt numbers Nu_{av} for the tube array 4×2 lie out of the error range, according to the Corcione correlating Equation [9], by 1.2% for $Ra = 10^3$ and lie in the error range for $Ra = 10^4$.
4. There is a good agreement between the Nu_ϕ of the single cylinder and bottom tube of the tube array 4×1 at $Ra = 10^3$ and 10^4.
5. The left bottom tube for the tube array 4×2 is affected by the right tube column; therefore, the Nu_ϕ values have different courses for $Ra = 10^3$ and 10^4.
6. The Nu_{av} values increase with the S/D ratio and Ra increasing for the tube array 4×4 at the base, concave, and convex configurations.
7. The highest values of Nu_{av} for the tube array 4×4 are noticed at the convex configuration which has discrepancies up to 3.7% in comparison with the concave one depending on Ra in the range of 1.3×10^4 to 3.7×10^4. The value of Nu_{av} for the base configuration lies between the convex and concave ones for all Ra.
8. The concave configuration achieves a decrease of Nu_{array} and q_{array} compared with the base one. The convex configuration achieves an increase of these representative values compared with the base one.
9. The convex configuration achieves an increase of Nu_{array} and q_{array} compared with the concave one for all ratios S/D and Ra. When increasing the ratio S/D and Ra, a decrease of Nu_{array} and q_{array} percentage discrepancies between the individual configurations is reached.
10. The new correlating equations for Nu_{array} calculation concerning the tube array 4×4 were defined. The equations have negligible error ranges and are interchangeable with each other.

Author Contributions: Writing—original draft preparation; Conceptualization; Methodology and Validation, Z.B., S.K.; Software; Data Curation and Validation, S.K.; Formal analysis; Resources and Supervision, Z.B. All authors have read and agreed to the published version of the manuscript.

Funding: The paper was created within the project VEGA 1/0086/18 The Research of Temperature Fields in the System of Shaped Heat Transfer Surfaces, funded by the Ministry of Education, Science, Research, and Sport of the Slovak Republic.

Acknowledgments: The authors thanks to the very competent reviewers for the valuable comments and suggestions.

Conflicts of Interest: The authors declare no conflict of interest.

Abbreviations

Roman Symbols:

c	specific heat (J/kgK)
D	external diameter of the cylinder (m)
E	percentage error
g	gravitational acceleration (m/s^2)
h	heat transfer coefficient (W/m^{2}K)
p	pressure (Pa)
Q	heat transfer rate (W)
q	heat flux density (W/m^2)
N	number of cylinders in the array
$Pr = v/\alpha$	dimensionless Prandtl number
r	dimensionless radial coordinate
Nu	dimensionless Nusselt number
$Ra = \left[g\beta(t_s - t_f)D^3\right]/v\alpha$	dimensionless Rayleigh number
S	center-to-center separation distance (m)
T	temperature (K)
u, v	velocity components in x- and y- directions (m/s)

Greek Symbols:

α	thermal diffusivity (m^2/s)
Δ	difference of values
$\beta = 1/K$	volume-expansion coefficient (K^{-1})
ϕ	local section angle (°)
λ	thermal conductivity (W/mK)
μ	dynamic viscosity (kg/ms)
v	kinematic viscosity (m^2/s)
ρ	density (kg/m^3)
τ	viscous stress tensor (kg/s^{2}m)

Subscripts:

av	average value
$array$	total average value
s	surface
SD	standard deviation
f	fluid
H	horizontal
V	vertical

References

1. Churchill, S.W.; Chu, H.H.S. Correlating equations for laminar and turbulent free convection from a horizontal cylinder. *Int. J. Heat Mass Transf.* **1975**, *18*, 1049–1053. [CrossRef]
2. Morgan, V.T. The overall convective heat transfer from smooth circular cylinders. *Adv. Heat Transf.* **1975**, *11*, 199–264.
3. Kuehn, T.H.; Goldstein, R.J. Numerical solution to the navier-stokes equations for laminar natural convection about a horizontal isothermal circular cylinder. *Int. J. Heat Mass Transf.* **1980**, *23*, 971–979. [CrossRef]

4. Sparrow, E.M.; Niethammer, J.E. Effect of vertical separation distance and cylinder-to-cylinder temperature Imbalance on natural convection for a pair of horizontal cylinders. *J. Heat Transf.* **1981**, *103*, 638–644. [CrossRef]

5. Saitoh, T.; Sajiki, T.; Maruhara, K. Bench mark solutions to natural convection heat transfer problem around a horizontal circular cylinder. *Int. J. Heat Mass Transf.* **1993**, *36*, 1251–1259. [CrossRef]

6. Chouikh, R.; Guizani, A.; Maalej, M. Numerical study of the laminar natural convection flow around an array of two horizontal isothermal cylinders. *Int. Commun. Heat Mass* **1999**, *26*, 329–338. [CrossRef]

7. Herraez, J.V.; Belda, R. A study of free convection in air around horizontal cylinders of different diameters based on holographic interferometry. Temperature field equations and heat transfer coefficients. *Int. J. Therm. Sci.* **2002**, *41*, 261–267. [CrossRef]

8. Corcione, M. Correlating equations for free convection heat transfer from horizontal isothermal cylinders set in a vertical array. *Int. J. Heat Mass Transf.* **2005**, *48*, 3660–3673. [CrossRef]

9. Corcione, M. Interactive free convection from a pair of vertical tube-arrays at moderate rayleigh numbers. *Int. J. Heat Mass Transf.* **2007**, *50*, 1061–1074. [CrossRef]

10. Ashjaee, M.; Yousefi, T. Experimental study of free convection heat transfer from horizontal isothermal cylinders arranged in vertical and inclined arrays. *Heat Transf. Eng.* **2007**, *28*, 460–471. [CrossRef]

11. Heo, J.H.; Chung, B.J. Natural convection of two staggered cylinders for various prandtl numbers and vertical and horizontal pitches. *Heat Mass Transf.* **2014**, *50*, 769–777. [CrossRef]

12. Cernecky, J.; Jandacka, J.; Malcho, M.; Koniar, J. A study of free convection around a system of horizontal tubes arranged above each other. *JP J. Heat Mass Transf.* **2015**, *11*, 43–56. [CrossRef]

13. Malcho, M.; Jandacka, J.; Ochodek, T.; Kolonicny, J. Correlations for heat transport by natural convection of cylindrical surfaces situated one above the other. *Communications* **2015**, *17*, 4–11.

14. Lu, Y.; Yu, Q.; Du, W.; Wu, Y. Natural convection heat transfer of molten salts around a vertically aligned horizontal cylinder set. *Int. Commun. Heat Mass* **2016**, *76*, 147–155. [CrossRef]

15. Kitamura, K.; Mitsuishi, A.; Suzuki, T.; Kimura, F. Fluid flow and heat transfer of natural convection induced around a vertical row of heated horizontal cylinders. *Int. J. Heat Mass Transf.* **2016**, *92*, 414–429. [CrossRef]

16. Cernecky, J.; Brodnianska, Z.; Blasiak, P.; Koniar, J. The research of temperature fields in the proximity of a bundle of heated pipes arranged above each other. *J. Heat Transf.* **2017**, *139*, 082001. [CrossRef]

17. Ma, H.; He, L.; Rane, S. Heat transfer-fluid Flow interaction in natural convection around heated cylinder and its thermal chimney effect. In Proceedings of the International Conference on Innovative Applied Energy, St Cross College, University of Oxford, Oxford, UK, 14–15 March 2019.

18. Wang, P.; Kahawita, R.; Nguyen, T.H. Numerical computation of the natural convection flow about a horizontal cylinder using splines. *Numer. Heat Transf.* **1990**, *17*, 191–215. [CrossRef]

19. Razzaghpanah, Z.; Sarunac, N. Natural convection heat transfer from a vertical column of finite number of heated circular cylinders immersed in molten solar salt. *Int. J. Heat Mass Transf.* **2019**, *134*, 694–706. [CrossRef]

20. Razzaghpanah, Z.; Sarunac, N. Natural convection heat transfer from a bundle of in-line heated circular cylinders immersed in molten solar salt. *Int. J. Heat Mass Transf.* **2020**, *148*, 119032. [CrossRef]

Article

Semi-Analytical Model for Heat and Mass Transfer Evaluation of Vapor Bubbling

Giuseppe Starace, Lorenzo Carrieri and Gianpiero Colangelo *

Department of Engineering for Innovation, University of Salento, 73100 Lecce, Italy;
giuseppe.starace@unisalento.it (G.S.); carrierilorenzo@gmail.com (L.C.)
* Correspondence: gianpiero.colangelo@unisalento.it; Tel.: +39-0832-299440

Received: 30 January 2020; Accepted: 25 February 2020; Published: 2 March 2020

Abstract: Multi-stage refrigeration systems cover a wide range of possibilities and are diffusing more and more. The idea that inspired this work derived from the need to have a tool to model the energy behavior of the intercooler inside a multi-stage refrigeration system. In this work, a semi-analytical model of a single bubble, injected into the liquid of an intercooler of a multi-stage system, has been developed. The developed model is a set of equations derived from the Fourier equation for heat conduction in defined conditions and includes the effects of sensible and latent heat. The vapor bubble is supposed to be injected in the saturated liquid contained in a tank at a defined depth, at an intermediate pressure. The model has been implemented in Matlab and the results show the influence of the liquid surface tension, the injection depth and the thermal diffusivity of the vapor. The model developed here is a useful low-cost tool for evaluating heat transfer optimization of a separator/intercooler of a multi-stage refrigeration system.

Keywords: Semi-analytical model; vapor; liquid; bubble; two-phase heat transfer

---

1. Introduction

The growing demand for energy for industrial applications and the need to protect the environment for future generations requires effort to make the production processes more efficient, from the early design stages, through to the adoption of innovative techniques and systems both with low environmental impact and economic sustainability. In industrial refrigeration applications, the reduction of energy requirements is made, at a thermodynamic cycle level, through multi-stage intercooled compression techniques, which limit the increase of refrigerant temperature during the compression phase, approximating an almost isothermal process (ideally with a succession of small intercooled compressions). In a two-stage scheme, the low-temperature heat sink is obtained internally at an intermediate pressure and is formed by the expansion of saturated refrigerant after exiting the condenser [1]. In each desuperheating stage, the temperature of the vapor tends to near the saturation value, through a heat transfer process, carried out either by injecting liquid into the discharge line of the low-pressure compressor or by bubbling the vapor into the liquid, as shown in Figure 1.

In the first scheme (Figure 1a) the desuperheating process is due to the liquid vaporization injected by the pump, supplied by the receiver. The second scheme exploits the bubbling of the superheated vapor in a low-temperature saturated liquid bath, through an open tube, at a certain depth. Here, the intense convective action, induced by the bubbling, makes the heat transfer effective, allowing the vapor to reach the saturation condition. This solution makes it possible to avoid an injection pump, with a considerable system simplification and reduction of maintenance costs. Due to its characteristics, the bubbling technique is particularly attractive, as it combines economic and energy benefits in a simple technical solution.

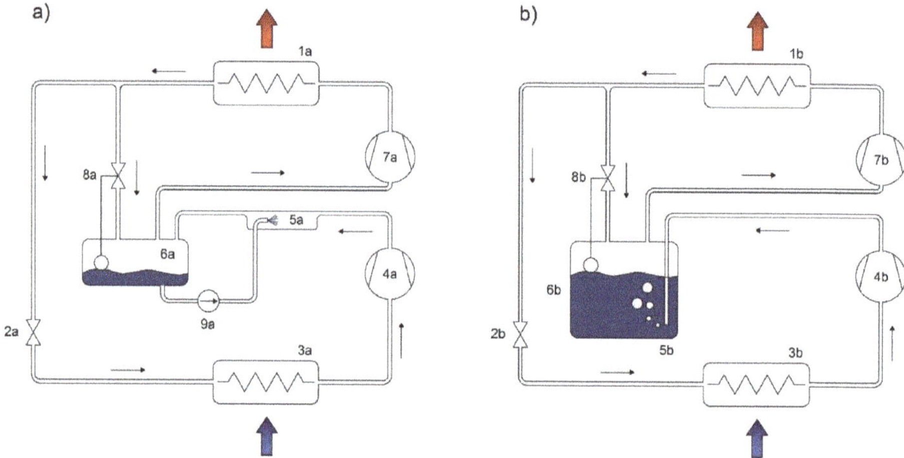

Figure 1. Desuperheating techniques: (**a**) liquid injection method and (**b**) the bubbling method. Legend: 1a and 1b—condenser; 2a and 2b—expansion valves; 3a and 3b—evaporator; 4a and 4b—low-pressure compressor; 5a—liquid vaporization section; 5b—superheated vapor injection section; 6a and 6b—receiver at intermediate pressure; 7a and 7b—high-pressure compressor; 8a and 8b—expansion valve driven by refrigerant level; 9a—liquid injection pump.

The main design parameter is the injection depth, it must guarantee the minimum residence time of the vapor, to allow complete desuperheating with the minimum backpressure at the exit of the low-pressure stage. The optimization of the device requires the formulation of a thermal model that considers the complexity of the convective and two-phase heat transfer phenomena occurring.

The evolution of a vapor bubble within liquid has been the topic of several studies, limited to the class of pool-type processes.

The Rayleigh model, reported by Farajisarir [2], proposes a mechanism for developing a vapor bubble that expands in a superheated, infinitely extended, inviscid and incompressible liquid bath, starting from an initial radius, neglecting the effects of surface tension and considers the growth phenomenon due to differences in pressure across the liquid-vapor interface (inertial model).

Van Stralen [3] described the Bosnjakovic model that analyzes the phenomenon by considering the thermal and mass exchanges with the liquid on the liquid—vapor frontier (thermal model) as significant for the development of the bubble and this model has been of interest for this work.

One more model is that proposed by Plesset et al. [4] that considers both the inertial and the thermal mechanisms. Solutions to the Plesset model, valid for the transient of the bubble collapse starting from the assigned initial dimension, are provided by Akiyama [5] in the hypothesis of inertial collapse and by Florschuetz and Chao, reported in [6] and [7], for the case of thermal collapse.

The experimental campaign on this topic is hard to carry out as shown by Bohdal [8], who investigated heat transfer and pressure drop during bubbly boiling in a refrigeration fluid, elaborating an analytical one-D model of heat transfer and pressure drop using experimental data, assuring accuracy up to ±20%.

Sujatha et al. [9], instead, carried out an experimental campaign on a bubble absorber to obtain heat and mass transfer and pressure drop data, then compared the results with the numerical correlation relating Sherwood number, Reynolds number, Schmidt number and length to diameter ratio developed by the same authors. Their work is a useful tool to compare experimental data to numerical correlations.

Recently, Wu et al. [10] experimentally investigated the effects of vibration amplitude and frequency on the bubble absorption characteristics of R134a-DMAC in a vertical tube. Their work was strictly linked to the application of refrigeration systems.

Merrill et al. [11] analytically studied the bubble behavior, from inception to collapse, in a subcooled binary liquid solution using a finite difference method to solve the governing equations, describing bubble diameter and mass over time.

Surtaev et al. [12] experimentally investigated the influence of sub-atmospheric pressure on multiscale heat transfer during liquid pool boiling. Experiments were carried out in the pressure range of 8.8–103 kPa at saturated water boiling, obtaining simultaneously, an extensive data set of the effect of reduced pressure on the main characteristics of boiling, including heat transfer coefficients, nucleation site density, growth rate and departure diameter of vapor bubbles. They demonstrated that the growth rate of dry spots is constant in time and has a non-monotonic dependence on pressure.

Mimouni et al. [13] developed a model and a numerical simulation of upward bubbly flow (in the same motion conditions of the present work), evaluating interfacial momentum transfer, polydispersion and turbulence.

Yan et al. [14] investigated the effect of liquid properties on fluid dynamic parameters in bubble columns. In particular, they used the correction of surface tension and viscosity to highlight the influence of different liquid properties on the hydrodynamic parameters of bubble columns. Computational fluid dynamics coupled with the population balance model were used to simulate the effects of liquid viscosity and liquid surface tension on the hydrodynamic parameters of the bubble column.

Lobanov et al. [15] studied the flow patterns and heat transfer of downstream bubbly flow in a sudden pipe expansion. They used the Eulerian approach for the numerical model. The set of RANS equations was used for modeling two-phase bubbly flows. The turbulence of the liquid phase was predicted using the Reynolds stress model.

Saha and Sandilya [16] developed a simplified model for quick evaluation of the storage performance of an injection cooling system for liquid subcooling, without involving the complex transport phenomena-based conservation equations. Simulations were carried out and the model was validated, considering the storage of liquid oxygen, and obtained a satisfactory match between the experimental data and the model predictions. The present paper is a step forward, considering the limitations of the cited work.

Xu et al. [17] proposed a bubble diameter model for the boiling flows. The model considered heat transfer growth, the breakup or coalescence of collision and the liquid impacting separation as factors affecting bubble diameter. They developed three bubble diameter sub-models to calculate overall bubble diameter, based on a departure diameter.

Kumar et al. [18] carried out an experimental and theoretical campaign, obtaining measurements of the local void fraction, rise velocity and bubble diameter for cocurrent, wall-heated, upward bubbly flows in a refrigerant. Bubble size was correlated in terms of liquid subcooling and bulk bubble size in terms of void fraction. The results were used to develop bubbly flow models, applicable to heated two-phase flows at high-pressure.

Zarate et al. [19] studied the modeling and numerical simulation of the turbulent subcooled boiling flow of a refrigerant, through the two-fluid model conservation equations. The turbulent viscosity was considered to be comprised of shear-induced and bubble-induced components. The results were compared with experimental measurements.

However, none of the models of the cited papers can be applied directly to the bubbling analysis since the vapor is injected in superheated conditions and is, therefore, far from saturation. A new model is needed that takes into account these new conditions, as well as the vertical motions not to be neglected as they modify the saturation conditions while convection occurs.

In this paper, a heat transfer model, based on the Fourier law, that includes the effects of sensible and latent heat contributions, the effects of the liquid surface tension and the vertical motions on the convection and on the saturation conditions of the vapor at real operating conditions, is proposed.

The model has been applied to water, in order to have a reliable benchmark, due to the well-known thermophysical properties of this commonly used heat transfer fluid. On the other hand, the model could be used to simulate all heat transfer fluids and in particular refrigerants, which is the application this study has been conceived for. Thus, the model developed here can be seen as a reliable tool for evaluating the heat transfer optimization of a separator/intercooler in a multi-stage refrigeration system. Further, the proposed model has other potential applicability in the study of two-phase heat and mass transfer occurring in many heat exchangers.

2. Mathematical Model and Hypotheses

The proposed mathematical model has been developed for a single vapor bubble. The heat transfer equations have been written for discrete time intervals $\Delta\tau$, taking into account a semi-analytical approach, considering the sensible and latent heat in the operating conditions of the desuperheater.

The hypotheses that have been made for the formulation of the model are the following:

- Single spherical shape bubble of radius r_b, variable in time and space, starting from an assigned generation value r_{b0};
- Liquid phase at uniform temperature T_l, equal to the saturation value at the intermediate pressure;
- Pressure in the liquid linearly variable with the depth (Stevino's law);
- Non-negligible liquid surface tension effects (Laplace equation);
- Bubble in vertical motion under gravity force and hydrodynamic resistance;
- Flow regime with independent bubbles;
- Pure thermal conduction in vapor domain (Ra < 1000).

Mechanical and thermal models of the bubble inside the liquid are segregated. Their mutual interaction is set up according to the scheme proposed in Figure 2.

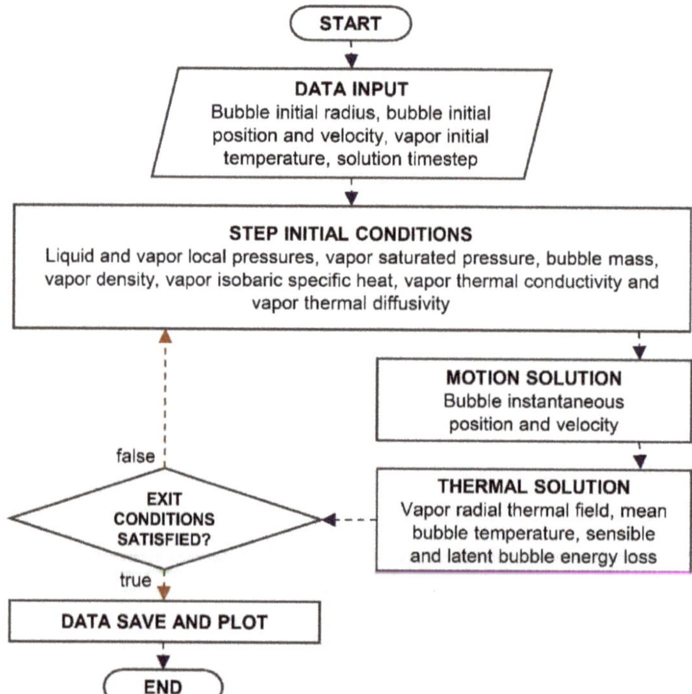

Figure 2. Concept flow-chart of the model.

Bearing in mind the scheme reported in Figure 3, the modeling of the motion of the analyzed bubble was described by considering only the vertical component as significant. With the *z* axis according to the direction of the gravity force and positive sign upwards, as in the scheme of Figure 3, the fundamental equation of motion has been expressed by Equation (1).

$$F_g - F_p - F_r = m_{bub}\ddot{z}_{bub} \tag{1}$$

where F_g is the buoyancy force, F_p is the gravity force and F_r is the drag force.

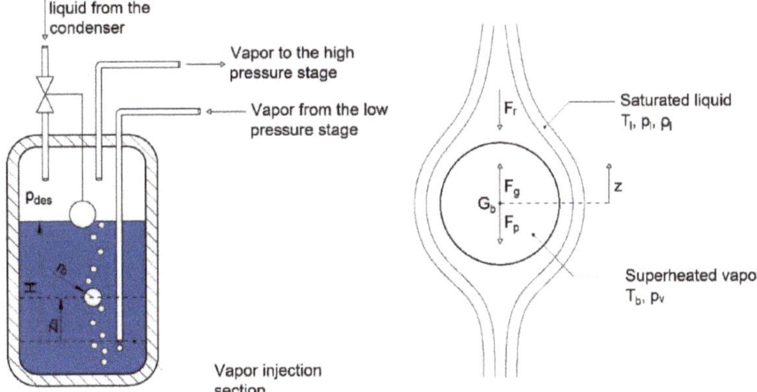

Figure 3. Bubble equilibrium of forces in the z direction.

Substituting each force with the relative analytical expression, Equation (2) is obtained:

$$\frac{4}{3}\rho_l \pi g r_{bub}^3 - \frac{4}{3}\overline{\rho}_{bub}\pi g r_{bub}^3 - \text{sgn}(\dot{z})\frac{1}{2}C_D A_{f,bub}\rho_l \dot{z}_{bub}^2 = \frac{4}{3}\overline{\rho}_{bub}\pi r_{bub}^3 \ddot{z}_{bub} \tag{2}$$

Equation (3) is obtained simplifying and rearranging the terms of Equation (2):

$$\overline{\rho}_b(t)\ddot{z}_b(t) + \frac{3\text{sgn}\left[\dot{z}_b(t)\right]}{8r_b(t)}\rho_l C_D(\dot{z}_b)\dot{z}_b^2(t) - g\left[\rho_l - \overline{\rho}_b(t)\right] = 0 \tag{3}$$

Equation (3) is an ordinary non-linear differential equation with non-constant coefficients, as the thermodynamic properties of the bubble are generally dependent on time (by temperature and pressure) and the hydrodynamic drag coefficient C_D is function of the bubble velocity. Searching for exact solutions in time intervals, small enough to neglect the temporal variation of the density $\overline{\rho}_b(t)$ and of the coefficient of resistance $C_D(\dot{z}_b)$, Equation (3), written for the generical time interval j, becomes:

$$\overline{\rho}_{b,j}\ddot{z}_{b,j}(t) + \frac{3\text{sgn}\left[\dot{z}_{b,j}(t)\right]}{8r_{b,j}(t)}C_{D,j}\rho_l \dot{z}_{b,j}^2(t) - g\left[\rho_l - \overline{\rho}_{b,j}\right] = 0 \tag{4}$$

Recognizing in Equation (4) a Riccati equation, the solutions are:

$$\begin{cases} z_{b,j}(t) = -\frac{1}{G_j}\ln\left\{\frac{1}{2}\left[e^{\sqrt{|F_j G_j|}t - G_j z_{0,j}}\left(1 - \frac{\dot{z}_{0,j}G_j}{\sqrt{|F_j G_j|}}\right) + e^{-\sqrt{|F_j G_j|}t - G_j z_{0,j}}\left(1 + \frac{\dot{z}_{0,j}G_j}{\sqrt{|F_j G_j|}}\right)\right]\right\} \\[2ex] \dot{z}_{b,j}(t) = -\frac{\sqrt{|F_j G_j|}e^{-\sqrt{|F_j G_j|}t - G_j[z_{0,j}+z_{b,j}(t)]}}{2G_j}\left[e^{2\sqrt{|F_j G_j|}t}\left(1 - \frac{\dot{z}_{0,j}G_j}{\sqrt{|F_j G_j|}}\right) - \left(1 + \frac{\dot{z}_{0,j}G_j}{\sqrt{|F_j G_j|}}\right)\right] \end{cases} \tag{5}$$

For the case of an ascending bubble, and:

$$\begin{cases} z_{b,j}(t) = z_{0,j} - \frac{1}{G_j}\ln\left[\cos\left(\sqrt{|F_jG_j|}\,t\right) - \frac{G_j\dot{z}_{0,j}}{\sqrt{|F_jG_j|}}\sin\left(\sqrt{|F_jG_j|}\,t\right)\right] \\[4mm] \dot{z}_{b,j}(t) = -\frac{1}{G_j}\left[\dfrac{-G_j\dot{z}_{0,j}\cos\left(\sqrt{|F_jG_j|}\,t\right) - \sqrt{|F_jG_j|}\sin\left(\sqrt{|F_jG_j|}\,t\right)}{\cos\left(\sqrt{|F_jG_j|}\,t\right) - \frac{G_j\dot{z}_{0,j}}{\sqrt{|F_jG_j|}}\sin\left(\sqrt{|F_jG_j|}\,t\right)}\right] \end{cases} \tag{6}$$

for the case of a descending bubble, with:

$$\begin{cases} G_j = -\dfrac{3C_{D,j}\rho_l}{8r_{b,j}\bar{\rho}_{b,j}} \\[4mm] F_j = -\dfrac{3C_{D,j}\rho_l g\left(\rho_l - \bar{\rho}_{b,j}\right)}{8r_{b,j}\bar{\rho}_{b,j}^2} \end{cases} \tag{7}$$

The heat transfer was modeled through a series of isobaric heat transfer equations of duration $\Delta\tau$ (time interval), valid for bubbles considered to be of constant radius in that time interval. Considering a uniform thermal field around the bubble and a constant Nusselt number at the liquid-vapor interface, a distributed 1-D unsteady heat transfer model has been adopted. The Fourier equation in spherical coordinates in the vapor domain (Equation (8)) has been used [6,7,20–22]:

$$\frac{1}{r^2}\frac{\partial}{\partial r}\left(r^2\frac{\partial T}{\partial r}\right) = \frac{\rho_v(r,t)c_{pv}(r,t)}{\lambda_v(r,t)}\frac{\partial T}{\partial t} = \frac{1}{\alpha_v(r,t)}\frac{\partial T}{\partial t} \tag{8}$$

Equation (8) is a non-linear differential equation with second order partial derivatives in space (r) and first order in time (t). The spatial and temporal non-linearities, introduced by the functions ρ_v, c_{pv} and α_v, dependent on the variable T (and thus on r and t) at a given pressure, can be relaxed by making a discretization in time and space. Assimilating the spherical domain to a system of M concentric shells of thickness Δr_k, each one characterized by average physical and thermodynamic properties, based on the volume V_k and relevant to the average thermal level in the integration interval considered, Equation (8) can be rewritten as a set of M linear differential equations to the partial derivatives:

$$\begin{cases} \frac{1}{r^2}\frac{\partial}{\partial r}\left(r^2\frac{\partial T_{1,j}}{\partial r}\right) = \frac{\rho_{v1,j}c_{pv1,j}}{\lambda_{v1,j}}\frac{\partial T_{1,j}}{\partial t} = \frac{1}{\alpha_{v1,j}}\frac{\partial T_{1,j}}{\partial t} & 0 \le r \le r_{1,j} \\[3mm] \frac{1}{r^2}\frac{\partial}{\partial r}\left(r^2\frac{\partial T_{2,j}}{\partial r}\right) = \frac{\rho_{v2,j}c_{pv2,j}}{\lambda_{v2,j}}\frac{\partial T_{2,j}}{\partial t} = \frac{1}{\alpha_{v2,j}}\frac{\partial T_{2,j}}{\partial t} & r_{1,j} \le r \le r_{2,j} \\[2mm] \quad\vdots \\[2mm] \frac{1}{r^2}\frac{\partial}{\partial r}\left(r^2\frac{\partial T_{M,j}}{\partial r}\right) = \frac{\rho_{vM,j}c_{pvM,j}}{\lambda_{vM,j}}\frac{\partial T_{M,j}}{\partial t} = \frac{1}{\alpha_{vM,j}}\frac{\partial T_{M,j}}{\partial t} & r_{(M-1),j} \le r \le r_{M,j} \end{cases} \tag{9}$$

The $2M$ boundary conditions and M initial conditions (valid $\forall r \in [r_{k-1}, r_k]$ with $k = 1,\dots,M$) guarantee a unique solution. The $2M$ boundary conditions are related to the characteristics of the phenomenon. Two cases can be distinguished: sensible only, and both sensible and latent heat transfer.

Sensible heat only case:

$$\begin{cases} \left.\dfrac{\partial T_{1,j}}{\partial r}\right|_{r=0} = 0 \\[3mm] \left.\lambda_{vk,j}\dfrac{\partial T_{k,j}}{\partial r}\right|_{r=r_{k,j}} = \left.\lambda_{v(k+1),j}\dfrac{\partial T_{(k+1),j}}{\partial r}\right|_{r=r_{k,j}} & \forall k = 1,\dots M-1 \\[3mm] T_{k,j}\left(r_{k,j},t\right) = T_{(k+1),j}\left(r_{k,j},t\right) & \forall k = 1,\dots M-1 \\[3mm] \left.\lambda_{vk,j}\dfrac{\partial T_{k,j}}{\partial r}\right|_{r=r_{M,j}} = -h_{c,j}\left[T_{k,j}(r,t) - T_l\right]_{r=r_{M,j}} \end{cases} \tag{10}$$

Sensible and latent heat case (associated with the phase change):

$$\begin{cases} \left.\dfrac{\partial T_{1,j}}{\partial r}\right|_{r=0} = 0 \\[2mm] \left.\lambda_{vk,j}\dfrac{\partial T_{k,j}}{\partial r}\right|_{r=r_{k,j}} = \left.\lambda_{v(k+1),j}\dfrac{\partial T_{(k+1),j}}{\partial r}\right|_{r=r_{k,j}} & \forall k = 1,\ldots M-1 \\[2mm] T_{k,j}\!\left(r_{k,j},t\right) = T_{(k+1),j}\!\left(r_{k,j},t\right) & \forall k = 1,\ldots M-1 \\[2mm] \left.T_{M,j}(r,t)\right|_{r=r_{M,j}} = T_{vsat,j} \end{cases} \tag{11}$$

In both cases:

$$T_{k,j}(r,0) = T_{0k,j}(r) \text{ con } r_{ik,j} \le r \le r_{ek,j} \forall k = 1,\ldots,M \tag{12}$$

where $T_{0k,j}(r)$ is the initial temperature field in the k-th shell in the time interval. Equation (10) and Equation (11) describe a non-homogeneous problem at the boundaries. By changing the type of variable, it can be changed to a homogeneous problem (as shown in Equation (13)), where Ω_j is equal to T_l for the case of sensible-only heat transfer and equal to $T_{vsat,j}$ in the case of a sensible-latent heat transfer one.

$$U_{k,j}(r,t) = r\left[T_{k,j}(r,t) - \Omega_j\right] \tag{13}$$

Separating the variables, the equations can be expressed by the unknown function $U_{k,j}$, that is a Sturm—Liouville problem with general solutions of Equation (14) [23]:

$$\begin{cases} T_j(r,t) = U_{k=1}^{M} T_{k,j}(r,t) \\[2mm] T_{k,j}(r,t) = \Omega_j + \dfrac{1}{r}\displaystyle\sum_{n=1}^{+\infty} c_{n,j}\Psi_{kn,j}(r)e^{-\beta_{n,j}^2 t} \\[2mm] c_{n,j} = \dfrac{\sum_{k=1}^{M}\int_{r_{(k-1),j}}^{r_{k,j}}\rho_{vk,j}c_{pvk,j}r\left[T_{0k,j}(r)-T_1\right]\Psi_{kn,j}(r)\,\mathrm{d}r}{\sum_{k=1}^{M}\int_{r_{(k-1),j}}^{r_{k,j}}\rho_{vk,j}c_{pvk,j}\Psi_{kn,j}^2(r)\,\mathrm{d}r} \\[2mm] \Psi_{kn,j}(r) = A_{kn,j}\cos\!\left(\sqrt{\dfrac{\beta_{n,j}^2}{\alpha_{vk,j}}}r\right) + B_{kn,j}\sin\!\left(\sqrt{\dfrac{\beta_{n,j}^2}{\alpha_{vk,j}}}r\right) \end{cases} \tag{14}$$

Equation (14), considering average thermodynamic properties in the domain in the case of sensible-only heat transfer, can be simplified as in Equation (15):

$$T_j(r,t) = T_1 + \frac{1}{r}\sum_{n=1}^{\infty} \frac{4\beta_{n,j}\int_0^{r_{b,j}} r\left[T_{0,j}(r) - T_1\right]\sin(\beta_{n,j}r)\,\mathrm{d}r}{2\beta_{n,j}r_{b,j} - \sin(2\beta_{n,j}r_{b,j})}\sin(\beta_{n,j}r)e^{-\alpha_{v,j}\beta_{n,j}^2 t} \tag{15}$$

and in the case of a sensible-latent one as in Equation (16):

$$T_j(r,t) = T_{vsat,j} + \frac{1}{r}\sum_{n=1}^{\infty} \frac{2}{r_{b,j}}\sin(\beta_{n,j}r)e^{-\alpha_{v,j}\beta_{n,j}^2 t}\int_0^{r_{b,j}} r\left[T_{0,j}(r) - T_{vsat,j}\right]\sin(\beta_{n,j}r)\,\mathrm{d}r \tag{16}$$

The mass flow rate $\dot{m}_{l,j}$, released during the process of vapor condensation, can be estimated by the energy balance on the bubble shell under saturation conditions, under the hypothesis of $\Delta r_{k,sat} \ll \Delta r_k$. In particular, with reference to Figure 4, the following equation can be written:

$$\dot{m}_{l,j} = \frac{4\pi r_{b,j}^2}{h_{vl,j}}\left[\lambda_{vM}\left.\frac{\mathrm{d}T_{j^-}}{\mathrm{d}r}\right|_{r=r_{b,j}} + h_{c,j}\left(T_{vsat,j} - T_1\right)\right] \tag{17}$$

where the first term in brackets is the portion of energy transferred by conduction from the inner layers of the bubble to the saturating shell, while the second term represents the portion transferred to the liquid by convection.

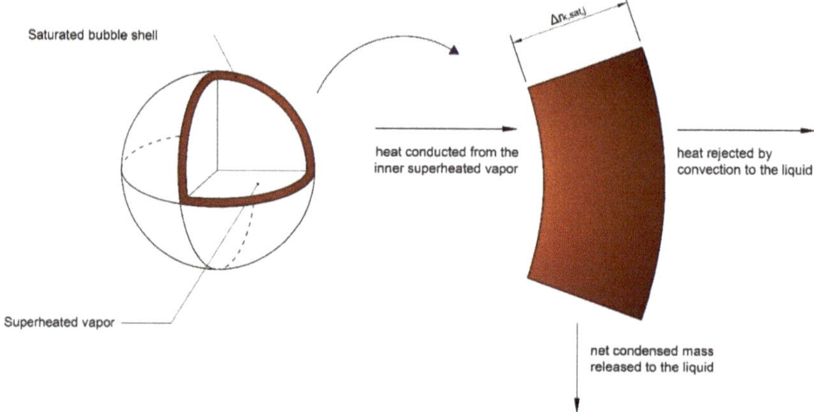

Figure 4. Energy flows in the shell under saturation conditions.

3. Results and Discussion

The proposed model has been implemented in a Matlab® environment, provided with an interactive graphic interface for the definition of the inputs and the analysis of the outputs. The simulation has been initialized by providing some initial values as: velocity, position, radius and temperature of the bubble, injection depth and the intermediate pressure as shown in Table 1. The simulated system was a two-stage refrigeration system, with water as refrigerant, operating between the evaporation and condensation temperatures (respectively of 273.15 and 318.15 K), varying the injection depth H, the vapor injection temperature T_{b0} and the bubble generation dimension r_{b0}, according to the values in Table 1.

Table 1. Initialization values of the simulation.

Variable	Value	Comments
Desuperheating Pressure, p_{des} [MPa]	2.42×10^{-3}	Geometric mean of the characteristic cycle pressures (corresponding to a liquid temperature of 293.72 K)
Initial Velocity, $\dot{z}_0$ [m/s]	0.0	With nozzle facing upwards
Initial Position, z_0 [m]	0.0	With reference to the injection section
Initial Radius, r_{b0} [mm]	{1.0; 5.0; 10.0}	
Injection Temperature, T_{b0} [K]	{383.2; 423.2}	
Injection Depth, H [m]	{0.1; 0.2; 0.3; 0.4; 0.5; 0.6; 0.7}	

The thermodynamic and transport properties of water were calculated through the correlations of the IAPWS (International Association for the Properties of Water and Steam) [24–27]. The average convective heat transfer coefficient at the liquid-vapor frontier surface was evaluated by the Whitaker correlation [28], while the interpolation proposed by Morrison [29] was used for the drag coefficient of the flow.

Variations of the thermodynamic injection conditions and the initial radius yielded similar trends as the results obtained by the simulations, as shown in Figure 5, where the comparison between velocity and position of the bubble is plotted as a function of the initial radius r_b0. During the cooling phase, velocity increases up to a maximum value and then decreases to zero. The buoyant force, generated by the difference in densities between liquid and vapor, that is the engine for the motion upward, tends to zero as the vapor cools down to reach the liquid condition.

The average temperature curves, shown in Figure 6, are of particular interest. Compared to the classic exponentially decreasing cooling trend, three different regimes can be observed: an initial phase, an intermediate plateau and a thermal decay up to the bubble collapse. The behavior can be explained by observing the radial thermal profile curves of the vapor in Figure 7. At the beginning of curve 1, during transient cooling, the high thermal gradient at the liquid-vapor interface yields the quick set-up of a thermal gradient (portion of the volume of the bubble which shows a temperature variation of at least 1% higher than the injection temperature), where the vapor is affected by the presence of the liquid, by changing its temperature with respect to the initial value. In this phase the average bubble temperature decreases quickly, until the saturation conditions of the liquid-vapor frontier are reached.

Due to the phase change, the temperature difference between the vapor surface and the liquid remains constant, increasing the heat transfer and determining, in relation to the reduced thermal diffusivity of the vapor, the "rigid" translation of the thermal profile towards the center of the bubble (curves 2 and 3). This explains the nearly constant average temperature over time during the intermediate flat phase. When the amplitude of the transition region equals the instantaneous radius of the bubble (curve 4), even the core of the vapor, that behaves as a thermal reservoir up to that moment, is affected by the heat flow and a sudden decrease in average temperature (curve 5) up to the point when bubble collapse occurs (final thermal decay phase).

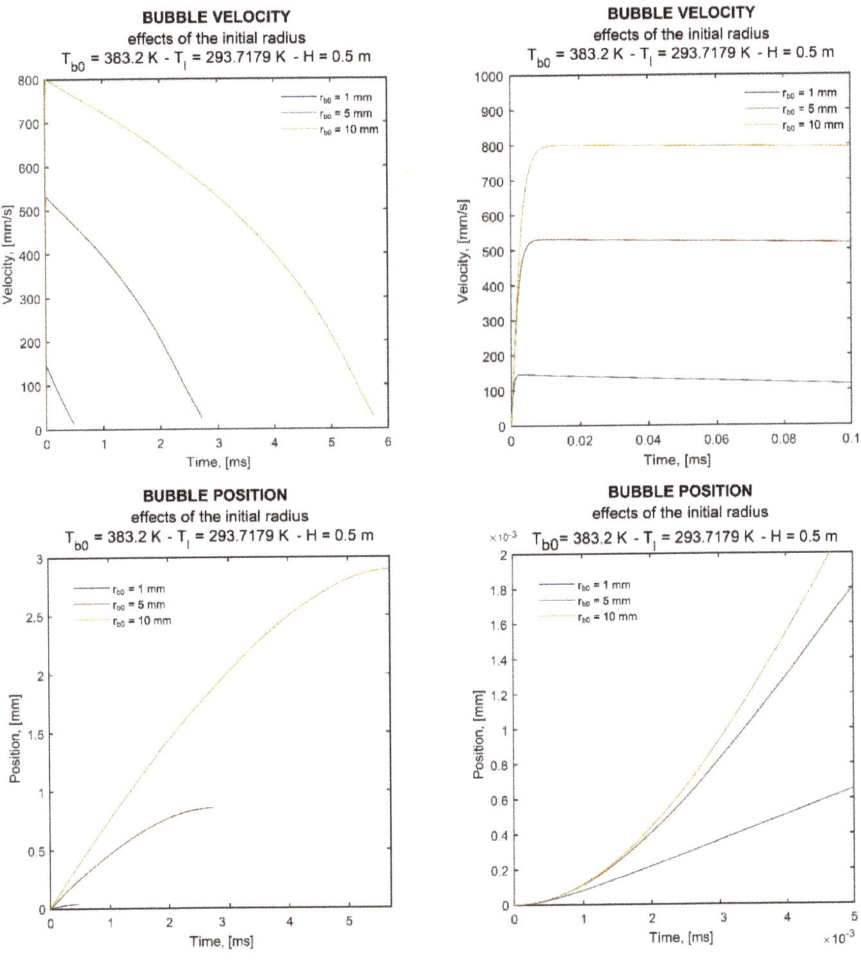

Figure 5. Velocity and position of the bubble.

thermal front to the innermost layers of the bubble and the effect is an increase (in module) in the thermal gradient *dT/dr* calculated on the border surface of the bubble, Equation (17). The increase in the condensed water mass flow rate (and therefore of the heat transferred to the liquid), expected with the increase in convection, is diminished by the effects of thermal diffusivity, which acts by limiting the increase in the energy release rate.

Once the injection conditions, T_{b0} and r_{b0}, have been fixed, the increase in depth H, on one hand governs the increase in heat transferred by the bubble, on the other hand it yields the increase in vapor density and thus, at a fixed volume, the increase in the total heat to be removed, is reliant to the saturation conditions. The increase both in heat transferred to the liquid and in the bubble energy as a consequence of the increase in its mass are competing and thus an optimal condition has to be found. The graphs in Figure 9 show the curves of the normalized residual energy (quantity of energy still kept by the vapor expressed as a fraction of the internal energy difference of the bubble between the injection conditions and the conditions of saturated vapor at the desuperheating temperature and pressure), for a bubble with r_{b0} = 10 mm. When injection depth increases, the thermal power transferred, with respect to the total transferable one, increases (curves with increasing slope) up to the maximum value, reached at an optimum depth H_{opt} = 0.4 m and then decreases again for $H > H_{opt}$ (with progressively less-inclined curves).

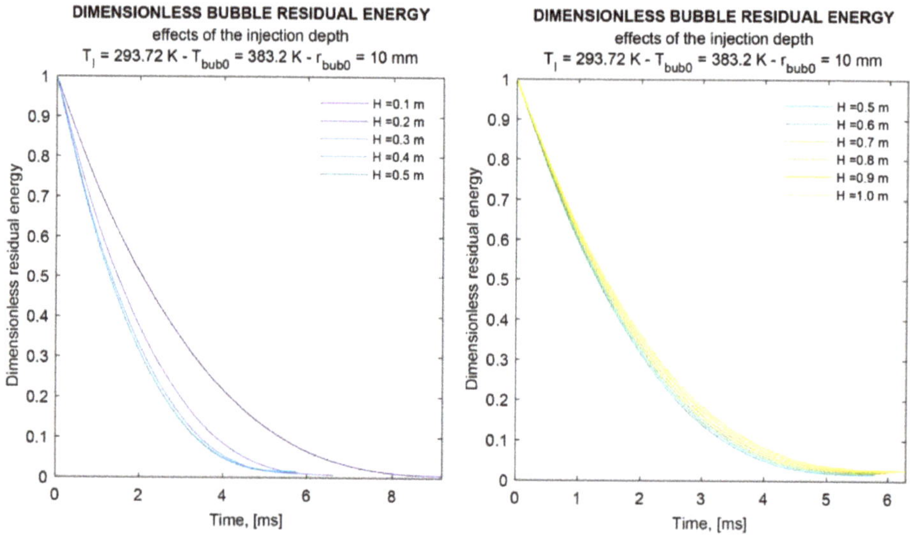

Figure 9. Bubble residual energy vs. time (T_{bub0} = 383.2 K).

When the injection conditions change, a similar behavior can be observed, even if with a different optimal depth. The bubble cooling rate depends on the balance between the energy to be removed (proportional to the mass) and the thermal power transferred to the liquid, both increasing with the injection depth H. Increasing the injection depth, starting from fixed T_{b0} and r_{b0}, the curves of the normalized bubble residual energy move towards the origin, showing an increase in the bubble cooling rate. In this condition, the increase in heat power transferred to the liquid is higher, with respect to the increase in the energy to be yielded due to the complete saturation of the vapor. The result is a reduction in cooling time. It can be observed that the same increases in depth always correspond to ever smaller displacements of the curves, which tend to overlap as $H \rightarrow H_{opt}$. By increasing the depth $H > H_{opt}$, the curves tend to move away from the origin, thus demonstrating a reduction in the bubble cooling rate.

In this condition, the energy that the vapor must release to reach saturation increases with depth in a preponderant manner with respect to the increase of the thermal power transferred to the liquid, causing an increase in cooling time.

At assigned vapor injection conditions and bubble initial size, an optimal H_{opt} can be found that maximizes the energy release rate towards the bath or minimizes the vapor desuperheating time. The vapor injection at depth H above the optimal value is not convenient since the only effect would be the increase in discharge conditions of the low-pressure compressor, which results in a decrease in cycle efficiency.

The saturation of the energy release rate, analyzing the radial thermal profile curves in the bubble, has been attributed to the thermal diffusivity of the vapor which acts by limiting the propagation of the thermal perturbation to the inner layers of the bubble. In these hypotheses it is evident that to speed up the cooling process, and therefore to exploit the increase in liquid ΔT vapor associated with the increase in injection depth, it is necessary to act in such a way as to increase the thermal diffusivity of the vapor. Once the temperature and pressure conditions of the desuperheater have been set, the only possibility is to increase the injection temperature. Further investigations were carried out by varying the vapor injection temperature on two new temperatures, respectively at 343.20 and 423.20 K as shown in Figures 10 and 11.

The direct comparison of the three curves of bubble residual energy at the same injection depth (the optimal H = 0.4 m) for the three analyzed temperatures (343.2, 383.2 and 423.2 K) is shown in Figure 12.

The analysis of Figures 9–11 shows the independence of the optimal injection depth from the inlet temperature in the liquid bath even if this affects the cooling speed as shown in Figure 12, where it is evident that by increasing the temperature of the vapor, the slope of the cooling curve increases.

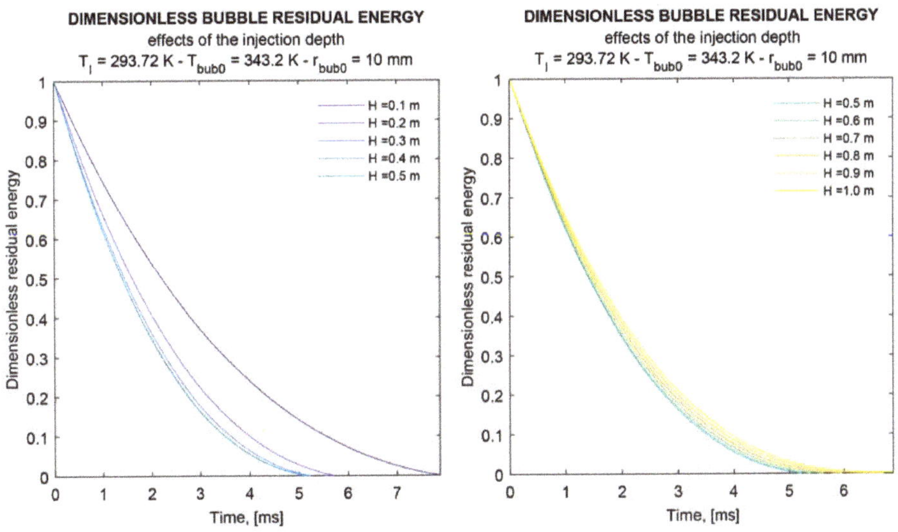

Figure 10. Bubble residual energy vs. time for low injection temperature (T_{bub0} = 343.2 K).

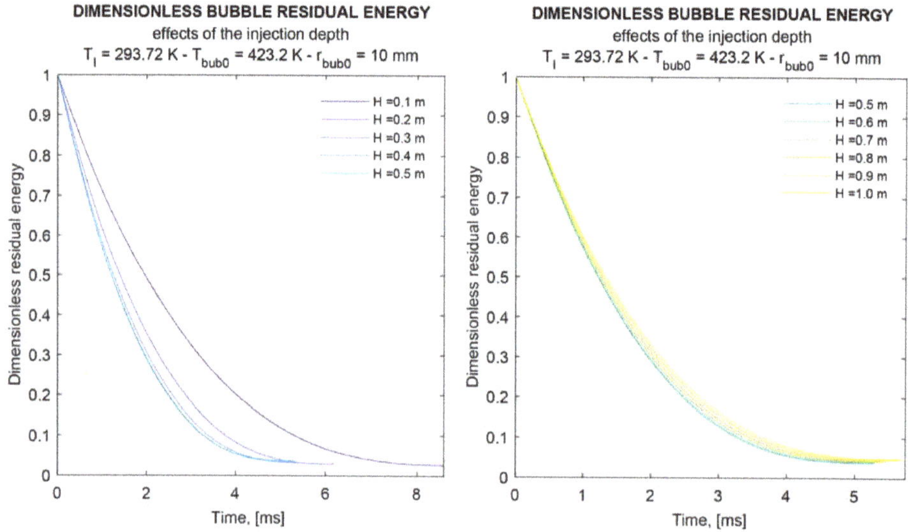

Figure 11. Bubble residual energy vs. time for high injection temperature (T_{bub0} = 423.2 K).

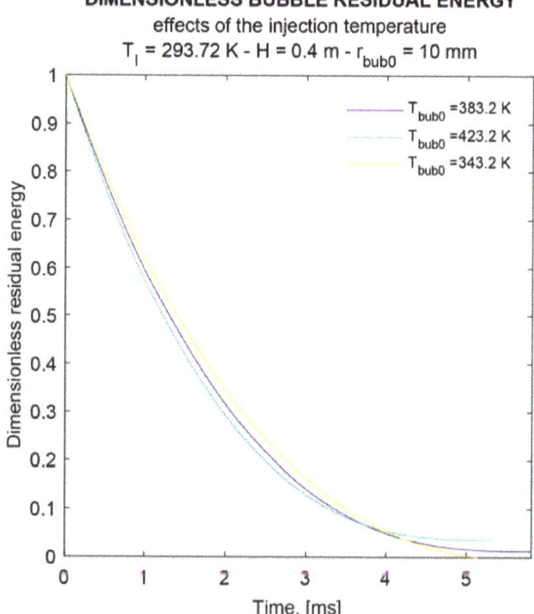

Figure 12. Bubble residual energy vs. time, temperature dependence.

As a natural extension of the results discussed so far, the effects of the variation of the desuperheating pressure (temperature of the liquid bath) on the optimal depth (H) were investigated. A simulation campaign with variable injection depth was then carried out by setting the initial temperature and radius of the bubble to 383.20 K and 10 mm, respectively. The desuperheating pressure was finally set to 0.0073849 MPa, corresponding to a saturation temperature of the liquid

313.15 K. The expected behavior was a reduction in the available temperature difference between vapor and liquid against a decrease in the cooling speed. It is therefore reasonable to suppose that the optimal injection depth (H) increases compared to the previous case to favor an increase in the interfaced liquid-vapor ΔT. The results of these simulations, shown in Figure 13, confirmed this behavior. It is observable that in the simulated conditions the cooling speed continues to increase as the depth of injection of the vapor increased and the process exhibits saturation at a new optimal depth around 1.4 m.

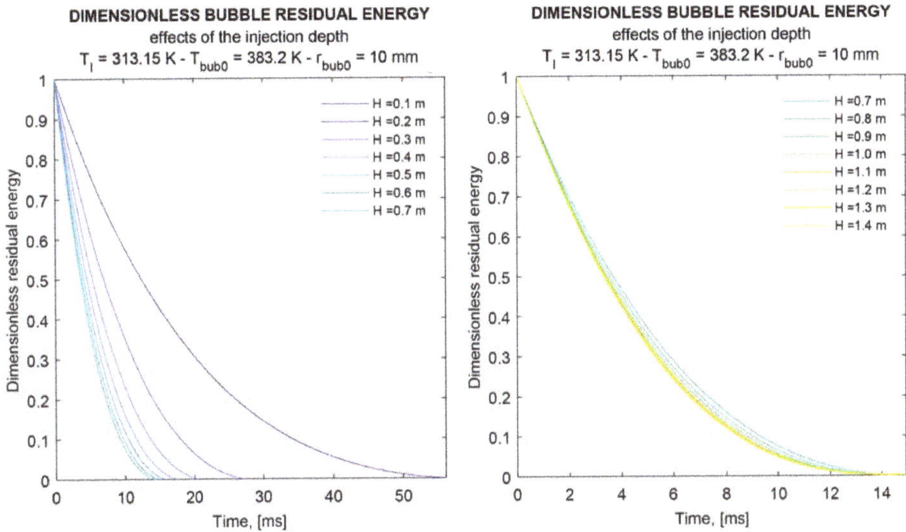

Figure 13. Bubble residual energy vs. time, desuperheating pressure dependence.

4. Conclusions

A semi-analytical model of heat transfer between liquid and vapor has been proposed, using a single bubble approach, with the adaptation of the Fourier thermal equation to real cooling conditions and includes the effects of sensible and latent heat. The model appears to be physically consistent and interpretable trends have been obtained. The liquid surface tension, the injection depth and the thermal diffusivity of the vapor have been investigated, finding antagonistic behavior in the cooling process. Their effect results in the existence of an optimal injection depth that optimizes the heat transfer at given thermodynamic conditions of the vapor entering the desuperheater and at given bubble size.

The model can be proposed as a valid tool to help the design of multi-stage systems. Further theoretical and experimental investigations have been planned and will be performed in the future for a full validation of the model and for extension to the general area of multiple bubbles flow.

Author Contributions: Conceptualization, G.S.; data curation, L.C.; formal analysis, G.C.; investigation, G.S. and G.C.; methodology, G.S., L.C.; writing—original draft, G.C., L.C.; writing—review and editing, G.S. and G.C. All authors have read and agreed to the published version of the manuscript.

Funding: This research received no external funding.

Conflicts of Interest: The authors declare no conflict of interest.

Nomenclature

r_b	bubble radius (mm)	r_{b0}	bubble initial radius (mm)
r	bubble radial coordinate (mm)	$r_{k-1,j}$	inner radius of the k-th bubble shell during the j-th time interval (mm)
$r_{k,j}$	outer radius of the k-th bubble shell during the j-th time interval (mm)		
T_l	desuperheater saturated liquid temperature (K)	$T_{vsat,j}$	saturated vapor temperature during the j-th time interval (K)
T	vapor radial thermal field (K)	T_j	vapor radial thermal field during the j-th time interval (K)
$T_{0k,j}$	initial temperature field across the k-th bubble shell during the j-th time interval (K)	$T_{k,j}$	vapor radial thermal field across the k-th bubble shell during the j-th time interval (K)
T_{b0}	vapor injection temperature (K)	ΔT	liquid–vapor temperature difference at the bubble boundary surface (K)
$U_{k,j}$	boundary value problem auxiliary variable (K × m)		
$\overline{\rho}_b$	mean vapor density (kg × m^{-3})	$\overline{\rho}_{b,j}$	mean vapor density during the j-th time interval (kg × m^{-3})
$\ddot{z}_b$	bubble vertical acceleration (m × s^{-2})	$\ddot{z}_{b,j}$	bubble vertical acceleration during the j-th time interval (m × s^{-2})
$\dot{z}_b$	bubble vertical velocity (m × s^{-1})	$\dot{z}_{b,j}$	bubble vertical velocity during the j-th time interval (m × s^{-1})
ρ_l	liquid density (kg × m^{-3})		
C_D	flow drag coefficient (dimensionless)	$C_{D,j}$	flow drag coefficient during the j-th time interval (dimensionless)
g	gravity acceleration (m × s^{-2})	t	time (s)
$z_{0,j}$	initial bubble position during the j-th time interval (m)		
j	time interval index	k	bubble shell index
ρ_v	vapor local density (kg × m^{-3})	$\rho_{v1,j}$	vapor density of the k-th bubble shell during the j-th time interval (kg × m^{-3})
c_{pv}	vapor isobaric specific heat (J × kg^{-1} × K^{-1})	$c_{pvk,j}$	vapor isobaric specific heat of the k-th bubble shell during the j-th time interval (J × kg^{-1} × K^{-1})
λ_v	vapor thermal conductivity (W × m^{-1} × K^{-1})	$\lambda_{vk,j}$	vapor thermal conductivity of the k-th bubble shell during the j-th time interval (W × m^{-1} × K^{-1})
α_v	vapor thermal diffusivity (m^2 × s^{-1})	$\alpha_{vk,j}$	vapor thermal diffusivity of the k-th bubble shell during the j-th time interval (m^2 × s^{-1})
Δr_k	radial thickness of the k-th bubble shell (m)	$\Delta r_{k,sat}$	radial thickness of the k-th bubble shell in saturated conditions (m)
V_k	volume of the k-th bubble shell (m^3)		
Ω_j	characteristic temperature of the process during the j-th time interval (K)	$\Psi_{kn,j}$	eigenfunction of the thermal problem associated to the n-th thermal eigenvalue during the j-th time interval (dimensionless)
$\beta_{n,j}$	n-th eigenvalue of the thermal problem during the j-th time interval (s$^{-1/2}$)	$c_{n,j}$	thermal solution combination constant associated to the n-th thermal eigenvalue during the j-th time interval (K × m)
$\Psi_{kn,j}$	eigenfunction of the thermal problem associated to the n-th thermal eigenvalue during the j-th time interval (dimensionless)	$B_{kn,j}$	coefficient of the n-th eigenfunction (dimensionless)
$A_{kn,j}$	coefficient of the n-th eigenfunction (dimensionless)	$h_{c,j}$	liquid—vapor convection heat transfer coefficient during the j-th time interval (W × m^2 × K^{-1})
$\dot{m}_{l,j}$	condensed mass flow rate during the j-th time interval (kg × s^{-1})	H_{opt}	optimal vapor depth of injection (m)
$h_{vl,j}$	latent heat of condensation during the j-th time interval (J × kg^{-1})		
p_{des}	intermediate desuperheating pressure (MPa)		
H	vapor depth of injection (m)		

References

1. Stoecker, W.F. *Industrial Refrigeration Handbook*, 1st ed.; McGraw-Hill: New York, NY, USA, 1998; p. 800.
2. Farajisarir, D. Growth and Collapse of Vapour Bubbles in Convective Subcooled Boiling of Water. Ph.D. Thesis, University of British Columbia, Vancouver, BC, Canada, 1993.
3. Van Stralen, S.J.D. The growth rate of vapour bubbles in superheated pure liquids and binary mixtures: Part I: Theory. *Int. J. Heat Mass Transf.* **1968**, *11*, 1467–1489. [CrossRef]
4. Plesset, M.S.; Zwick, S.A. The Growth of Vapor Bubbles in Superheated Liquids. *J. Appl. Phys.* **1954**, *25*, 493–500. [CrossRef]
5. Akiyama, M. Bubble Collapse in Subcooled Boiling. *Bull. JSME* **1973**, *16*, 570–575. [CrossRef]

6. Bergman, T.L.; Incropera, F.P.; DeWitt, D.P.; Lavine, A.S. *Fundamentals of Heat and Mass Transfer*; John Wiley & Sons: New York, NY, USA, 2011; p. 1048.

7. Gumerov, N.A. The heat and mass transfer of a vapor bubble with translatory motion at high Nusselt numbers. *Int. J. Multiph. Flow* **1996**, *22*, 259–272. [CrossRef]

8. Bohdal, T. Bubbly boiling of environment-friendly refrigerating media. *Int. J. Heat Fluid Flow* **2000**, *21*, 449–455. [CrossRef]

9. Sujatha, K.; Mani, A.; Murthy, S.S. Experiments on a bubble absorber. *Int. Commun. Heat Mass Transf.* **1999**, *26*, 975–984. [CrossRef]

10. Wu, X.; Liu, J.; Xu, S.; Wang, W. Effect of vibration parameters on the bubble absorption characteristics of working fluids R134a-DMAC in a vertical tube. *Int. J. Refrig.* **2019**, *99*, 234–242. [CrossRef]

11. Merrill, T.L.; Perez-Blanco, H. Combined heat and mass transfer during bubble absorption in binary solutions. *Int. J. Heat Mass Transf.* **1997**, *40*, 589–603. [CrossRef]

12. Surtaev, A.; Serdyukov, V.; Malakhov, I. Effect of subatmospheric pressures on heat transfer, vapor bubbles and dry spots evolution during water boiling. *Exp. Therm. Fluid Sci.* **2020**, *112*, 109974. [CrossRef]

13. Mimouni, S.; Guingo, M.; Lavieville, J.; Mérigoux, N. Combined evaluation of bubble dynamics, polydispersion model and turbulence modeling for adiabatic two-phase flow. *Nucl. Eng. Des.* **2017**, *321*, 57–68. [CrossRef]

14. Yan, P.; Jin, H.; He, G.; Guo, X.; Ma, L.; Yang, S.; Zhang, R. Numerical simulation of bubble characteristics in bubble columns with different liquid viscosities and surface tensions using a CFD-PBM coupled model. *Chem. Eng. Res. Des.* **2020**, *154*, 47–59. [CrossRef]

15. Lobanov, P.; Pakhomov, M.; Terekhov, V. Experimental and numerical study of the flow and heat transfer in a bubbly turbulent flow in a pipe with sudden expansion. *Energies* **2019**, *12*, 2735. [CrossRef]

16. Saha, P.; Sandilya, P. A dynamic lumped parameter model of injection cooling system for liquid subcooling. *Int. J. Therm. Sci.* **2018**, *132*, 552–557. [CrossRef]

17. Xu, Z.; Zhang, J.; Lin, J.; Xu, T.; Wang, J.; Lin, Z. An overall bubble diameter model for the flow boiling and numerical analysis through global information searching. *Energies* **2018**, *11*, 1297. [CrossRef]

18. Kumar, R.; Trabold, T.A.; Maneri, C.C. Experiments and modeling in bubbly flows at elevated pressures. Journal of Fluids Engineering. *Trans. ASME* **2003**, *125*, 469–478. [CrossRef]

19. Zarate, J.A.; Roy, R.P.; Kang, S.; Laporta, A. Modeling and simulation of subcooled turbulent boiling flow. *Am. Soc. Mech. Eng. Heat Transf. Div.* **2000**, *366*, 263–271.

20. Hahn, D.W.; Özisik, M.N. *Heat Conduction*, 2nd ed.; John Wiley & Sons: New York, NY, USA, 1993; p. 692.

21. Jiji Latif, M. *Heat Conduction*, 3rd ed.; Springer: Berlin/Heidelberg, Germany; New York, NY, USA, 2009; p. 418.

22. Florschuetz, W.; Chao, B. On the Mechanics of Vapor Bubble Collapse. *J. Heat Transf.* **1965**, *87*, 209–220. [CrossRef]

23. Mikhailov, M.D.; Özisik, M.N. *Unified Analysis and Solutions of Heat and Mass Diffusion*, 1st ed.; John Wiley and Sons: New York, NY, USA, 1984; p. 524.

24. *IAPWS R12-08: Release on the IAPWS Formulation 2008 for the Viscosity of Ordinary Water Substance*; IAPWS: Berlin, Germany, 2008.

25. *IAPWS R15-11: Release on the IAPWS Formulation 2011 for the Thermal Conductivity of Ordinary Water Substance*; IAPWS: Plzen, Czech Republic, 2011.

26. *IAPWS R1-76: Revised Release on Surface Tension of Ordinary Water Substance*; IAPWS: Moscow, Russia, 2014.

27. *IAPWS R7-97: Revised Release on the IAPWS Industrial Formulation 1997 for the Thermodynamic Properties of Water and Steam*; IAPWS: Lucerne, Switzerland, 2012.

28. Çengel, Y.A. *Introduction to Thermodynamics and Heat Transfer*; McGraw-Hill: New York, NY, USA, 2009; p. 960. ISBN 978-0071287739.

29. Morrison, F.A. Data Correlation for Drag Coefficient for Sphere. Ph.D. Thesis, Department of Chemical Engineering, Michigan Technological University, Houghton, MI, USA, 2016. Available online: www.chem.mtu.edu/~{}fmorriso/DataCorrelationForSphereDrag2016.pdf (accessed on 3 January 2019).

Article

Finite-Element Simulation for Thermal Modeling of a Cell in an Adiabatic Calorimeter

José Eli Eduardo González-Durán [1,†], Juvenal Rodríguez-Reséndiz [2,*,†], Juan Manuel Olivares Ramirez [3,†], Marco Antonio Zamora-Antuñano [4,†] and Leonel Lira-Cortes [5,†]

[1]   Instituto Tecnológico Superior del Sur de Guanajuato, Guanajuato 38980, Mexico; je.gonzalez@itsur.edu.mx
[2]   Facultad de Ingeniería, Universidad Autónoma de Querétaro, Querétaro 76010, Mexico
[3]   Universidad Tecnológica de San Juan del Río, San Juan del Río 76800, Mexico; jmolivar01@yahoo.com
[4]   Departamento de Ingeniería, Universidad del Valle de Mexico, Querétaro 76230, Mexico; marco.zamora@uvmnet.edu
[5]   Centro Nacional de Metrología, El Marques 76246, Mexico; llira@cenam.mx
*   Correspondence: juvenal@uaq.edu.mx; Tel.: +52-442-192-1200
†   These authors contributed equally to this work.

Received: 3 April 2020; Accepted: 1 May 2020; Published: 6 May 2020

Abstract: This research obtains a mathematical formulation to determine the heat transfer in a transient state, in a calorimeter cell, considering an adiabatic system. The development of the cell was established and the mathematical model was transiently solved, which approximated the physical phenomenon under the cell operation. A numerical method for complex geometries was used to validate performance. The results obtained in the transient heat transfer in a cylinder under boundary and initial conditions were compared using an analytical solution and numerical analysis employing the finite-element method with commercial software. The study from the temperature distribution can afford, selection between a cylindrical and spherical geometry, design criteria that are generated by changing parameters such as dimension, temperature, and working fluids to develop an adiabatic calorimeter to measure the heat capacity in fluids. We show the mathematical solution with its initial and boundary conditions as well as a comparison with a numerical solution for a cylindrical cell with a maximum error from 0.075% in the temperature value, along with a theoretical and numerical analysis for a temperature difference of 1 °C.

Keywords: adiabatic calorimetry; numerical simulation; heat capacity; finite-element method; heat transfer

---

1. Introduction

For engineering design, with heat exchange in, for example, refrigerators, the car engine, the food industry, and the heat treatment of materials, the research or development of new substances as refrigerants or working fluids is necessary. Among these substances is nanofluid, which has been recently introduced and improves heat transfer in geometries such as enclosures, channels, and microchannels [1]. For the engineering applications mentioned above, it is necessary to know the thermophysical properties of the working fluids that are involved in the energy exchange processes. One way to know how efficient the working fluid is for the exchange of energy in the form of heat is through thermal efficiency. However, to calculate the thermal efficiency, it is necessary to know the value of the heat capacity of the working fluid. The amount of heat required to increase the temperature by one degree of a substance is known as heat capacity. Therefore, it is important to measure the heat capacity as accurately as possible for the efficient use and development of the fluids involved in the exchange of heat. In [2], heat capacity is an important variable to calculate the thermal efficiency in any system that involves heat transfer. The most reliable way to obtain the value of heat capacity is by an adiabatic calorimetry technique [3]. Certain researchers have developed their own devices that work under this principle [4–9] with different configurations for the geometry of the cell in their adiabatic calorimeters. The cell is a key element of the calorimeter containing the fluid to measure its heat capacity. It has other elements such as sensors, heaters, inlets, and outlets for the fluid. According to adiabatic calorimetry, the cell is surrounded by a shield called an adiabatic, and its function is to stay at the same temperature of the cell to avoid heat transfer between both [9]. The cell and its elements are located within a cryostat, which is responsible for generating an environment with a constant temperature. The working principle is as follows: A quantity of test fluid is set inside of a cell to evaluate the heat capacity. When the fluid is inside a cell, heat is added and the temperature of the fluid consequently increases. The amount of heat (Q) added, the mass of fluid (m) contained in the cell, and the variation of the temperature (T) of fluid are known [9]. It is possible to calculate heat capacity from the next equation:

$$Q = (mC_p)\Delta T \tag{1}$$

where C_p is the heat capacity of the sample cell, and the heat added is obtained by the electrical energy calculated by

$$Q = IVt \tag{2}$$

where I, V, and t are the current, voltage, and duration of heating, respectively [7], taking into account their respective uncertainties. In the laboratory of thermophysical properties of Centro Nacional de Metrología (CENAM), they are developing an adiabatic calorimeter for measuring the heat capacity of fluids at room conditions. One of the most critical elements of the adiabatic calorimeter is the cell, which contains the fluids to determine the value of the heat capacity. Then to developing the adiabatic calorimeter, it is necessary to start to analyze the temperature distribution in the cell for the thermal design criteria regarding the cell; in the ideal case, the temperature of fluid contained in the cell is constant or equal to any point, but in a real case, temperature gradients are present. Therefore, mathematical models were used to analyze the temperature distribution in the cell from this work, as in other works, e.g., [10], which focused on finding errors in the thermal conductivity measurement process, [11], which studied the impact of the geometry for heat transfer in nanofluids, and [12], which studied a Spray Fluidized Bed Granulator. However, from the literature, no works related to the analysis of the thermal performance to select the geometry of a cell in an adiabatic calorimeter have been found. Attempts to learn the temperature distribution within the cell, the location of the temperature sensor, and the fluid inlets and outlets can be studied by inverse heat conduction problem [13].

Numerical analysis was used because it is a tool to change parameters such as dimensions, geometric shapes, and initial and boundary conditions to analyze the behavior of a mathematical or physical model in less time. However, the mathematical and numerical models provide only approximations to the physical phenomenon, so it is essential to validate the results, and the best way is to use a prototype and to characterize it [12,14–16].

The use of simulation is a fundamental aspect of developing an engineering process, research for scientific development, and the characterization of geometrical parameters, such as in [17]. To verify the accuracy of numerical methodologies, numerical results are compared with analytical solutions [17–19]. In other research, three-dimensional governing equations (continuity, momentum, and energy) along with boundary conditions are solved using the finite volume method and other processes of scientific applications in the study of energies.

Therefore, the following formulations were made: theoretical and numerical analysis by the finite-element method for a cylindrical cell and only a numerical analysis for a spherical cell. The present work performs a comparison among the results of the analytical and numerical solutions for a cylindrical cell, and the results of numerically evaluating a spherical cell are also included. Moreover, the temperature distribution between spherical and cylindrical cells, as well as the effect of varying their dimensions, is compared and evaluated.

The main objective of this work was to select the geometry with the lowest temperature gradients through carrying out an analytical solution of the thermal behavior of the cell from an adiabatic calorimeter to later compare it with a numerical solution, and to determine the error between both methods to provide greater reliability in the simulation of processes, where there is an exchange of energy with geometries of greater complexity, allowing for a smaller number of experimental prototypes that will reduce time and cost in the design of any thermal system. In addition, the calculated information given by the simulation contributes to a criteria design. It permits to choose a spherical geometry instead of a cylindrical, as reported in the state-of-the-art.

This article is organized as follows: Section 1 describes the adiabatic calorimeter, and works reported in the literature on the working principle of this type of device are reported. Section 2 presents the mathematical formulation wherein we establish a partial differential equation in a transient state, which was solved step by step in an analytical way with Bessel functions and their eigenfunctions. The solution was programmed in a Matlab language to compare results with the finite-element method. Section 3 shows the numerical model used to simulate the operation of the cell and the parameters used in the numerical model, and its boundary and initial conditions such as the element type chosen for the solution by ANSYS are also shown. Section 4 depicts a comparison of the results obtained through theoretical analysis and the numerical model used to show the power of using the numerical model to solve this kind of device and process as well as how useful it is to approximate physical phenomena. Section 5 gives an analysis of the numerical tools and its accuracy concerning an analytical model. Finally, Section 6 shows the conclusions based on theoretical and numerical results analysis done with regard to a specific part of an adiabatic calorimeter.

2. Theoretical Analysis

In this section, we show the mathematical model in a transient state established to describe the behavior of a cell from an adiabatic calorimeter, it was simplified by its symmetry along axis z. Several subsections were added, and they include the solution for temporal and spacial coordinates and the application of boundary conditions to find the eigenvalues. The mathematical model was solved step by step in an analytical way, and the final solution was programmed in Matlab to compare the results obtained with the finite-element method in the following section.

Mathematical Formulation

The equation of heat conduction is set in a cylindrical coordinate system for a theoretical analysis because the proposal cell presents this geometry [13].

The differential equation of heat conduction in the cylindrical coordinate system is [20]:

$$\frac{\partial^2 T}{\partial r^2} + \frac{1}{r}\frac{\partial T}{\partial r} + \frac{1}{r^2}\frac{\partial^2 T}{\partial \phi^2} + \frac{\partial^2 T}{\partial z^2} = \frac{1}{\alpha}\frac{\partial T}{\partial t} \tag{3}$$

where

$$\alpha = \frac{k}{\rho c_p}$$

where α is the thermal diffusivity, k is the thermal conductivity, ρ is the density, and c_p is the heat capacity of the material to be analyzed. The cell is a solid of radius b and height c, and its boundaries are a temperature constant T_c [3,5,6,8,21,21,22]. Initially, the entire system is at temperature T_0, where there is no heat generation inside the cell, as Figure 1 shows together with the coordinate system. Under these conditions and because there is azimuthal symmetry, T is independent of ϕ, so the equation for the solution is

$$\frac{\partial^2 T}{\partial r^2} + \frac{1}{r}\frac{\partial T}{\partial r} + \frac{\partial^2 T}{\partial z^2} = \frac{1}{\alpha}\frac{\partial T}{\partial t} \tag{4}$$

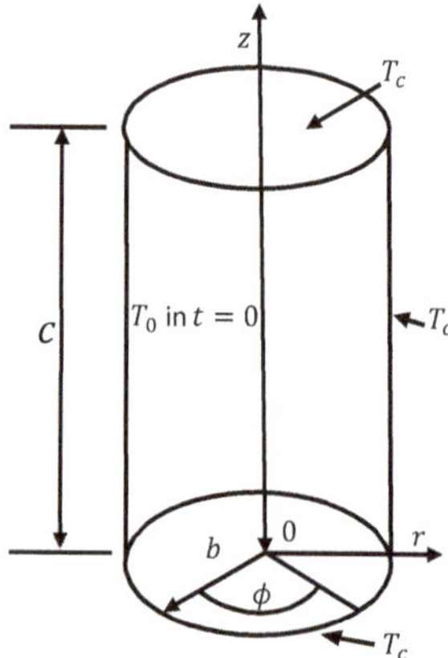

Figure 1. The cylinder with a radius and height specified, under certain initial and boundary conditions in the cylindrical coordinate system.

The initial condition is at $t = 0$ with $T = T_0$ at $0 \leq r \geq b$; $0 \leq z \geq c$ and $t > 0$, a boundary condition $T = T_c$, at $r = 0$, $r = b$, $z = 0$, and $z = c$. This assumes that the variables are separated $T(r,z,t) = R(r)Z(z)\Gamma(t)$. Substituting in Equation (4), simplifying, and dividing by $RZ\Gamma$,

$$\frac{1}{R}\left(\frac{d^2 R}{dr^2} + \frac{1}{r}\frac{dR}{dr}\right) + \frac{1}{Z}\frac{d^2 Z}{dz^2} = \frac{1}{\alpha\Gamma}\frac{d\Gamma}{dt} \tag{5}$$

Thus, in Equation (5), the left side is spatial coordinates, and the right side is time coordinates. The differential equation is equal to a constant

$$\frac{1}{\alpha\Gamma}\frac{d\Gamma}{dt} = -\lambda^2 \tag{6}$$

Therefore, a temporary solution is

$$\Gamma(t) = \Gamma(0)e^{\alpha\lambda^2 t} \tag{7}$$

For the axis z,

$$\frac{1}{Z}\frac{d^2 Z}{dz^2} = -\eta^2 \tag{8}$$

$$\frac{d^2 Z}{dz^2} + Z\eta^2 = 0 \tag{9}$$

and the solution to Equation (9) is

$$Z = A\cos(z\eta) + B\sin(z\eta) \tag{10}$$

where grouping constants $A = (A + B)$ and $B = (iA + iB)$

Solving the spatial part for the axis r, we have

$$\frac{1}{R}\left(\frac{d^2 R}{dr^2} + \frac{1}{r}\frac{dR}{dr}\right) = -\lambda^2 + \eta^2 = -\beta^2 \tag{11}$$

and Equation (11) is a differential equation from Bessel of the order v and is written as

$$\frac{d^2 R_v}{dr^2} + \frac{1}{r}\frac{dR_v}{dr} + \beta^2 R_v = 0 \tag{12}$$

The solutions to Equation (12) are the Bessel functions of order v, which are

$$R_v(\beta r) = \{J_v(\beta r), Y_v(\beta r)\} \tag{13}$$

With the superposition of all solutions, Equation (5) is

$$T(r,z,t) = \sum_{m=1}^{\infty}\sum_{p=1}^{\infty} R_v(\beta_m r)Z(\eta_p,z)\Gamma(t) \tag{14}$$

where $R_v(\beta_m r)$ and $Z(\eta_p, z)$ are the eigenfunctions, the solutions of the equations separated; β_m and η_p are the respective eigenvalues.

To find the value of β, a change of variable is made, and $T^* = T - T_c$ is defined from

$$R_v(\beta r) = A J_v(\beta r) + B Y_v(\beta r) \tag{15}$$

with the boundary condition at $r = 0 \to T = T_c \Rightarrow T^* + T_c = T_c \to T^* = 0$, substituting into Equation (15). Since $r = 0$, $R_v(\beta r)$ is finite:

$$B Y_v(\beta r) = 0 \Rightarrow B = 0$$

From the boundary condition at $r = 0$,

$$A J_v(\beta b) = 0 \tag{16}$$

A cannot be 0, therefore the values of β are the positive roots of

$$J_v(\beta_m b) = 0 \tag{17}$$

Since the problem includes the origin, $Y_v(\beta r)$ implies that $v = 0$, which is divergent.

To find η, the equation

$$Z = A\cos(z\eta) + B\sin(z\eta) \tag{18}$$

and the boundary condition at $Z(z = 0) = T^* = 0$ implies that $A = 0$. The boundary condition at $Z(z = c) = T^* = 0$ entails

$$B\sin(c\eta) = 0 \tag{19}$$

As a result of Equation (19), B cannot be 0. The values of η are

$$\eta_p = \frac{p\pi}{c} \tag{20}$$

The solution of the equation is

$$T^*(r,z,t) = \sum_{m=1}^{\infty}\sum_{p=1}^{\infty} C_m J_0(\beta_m r)\sin(\eta_p z)e^{-\alpha\beta^2 t} \tag{21}$$

Applying the initial condition to Equation (21), we have

$$T^*(r, z, t = 0) = T_0 - T_c = \sum_{m=1}^{\infty} \sum_{p=1}^{\infty} C_m J_0(\beta_m r) \sin(\eta_p z) \tag{22}$$

To find the constant C_m,

$$C_m = \frac{1}{N(\beta_m)N(\eta_p)} = \int_0^b \int_0^c r J_0(\beta_m r') \sin(\eta_p z') dr' dz' \tag{23}$$

To find the norms, for $N(\beta_m)$,

$$N(\beta_m) = \int_0^b r J_0^2(\beta_m r) \tag{24}$$

From Özisik [20], we have

$$N(\beta_m) = \frac{b^2}{2} \left[J_v'^2(\beta_m b) + \left(1 - \frac{v}{\beta_m^2 b^2} \right) J_v^2(\beta_m b) \right] \tag{25}$$

With $v = 0$,

$$N(\beta_m) = b^2 [J_0'^2(\beta_m b)] \tag{26}$$

To find $N(\eta_p)$,

$$N(\eta_p) = \int_0^c \sin(\eta_p z) \sin(\eta_p z) dz \tag{27}$$

Solving the previous equation, we have

$$N(\eta_p) = \frac{c}{2} \tag{28}$$

Substituting in Equation (14), we have

$$T^*(r, z, t) = \sum_{m=1}^{\infty} \sum_{p=1}^{\infty} \frac{1}{N(\beta_m)N(\eta_p)} J_0(\beta_m r) \sin(\eta_p z) e^{-\alpha \beta^2 t} \int_0^b \int_0^c r J_0(\beta_m r') \sin(\eta_p z') dr' dz' \tag{29}$$

Finally, the solution according to the initial and boundary conditions posed is

$$T^*(r, z, t) = \sum_{m=1}^{\infty} \sum_{p=1}^{\infty} \frac{J_0(\beta_m r) \sin(\frac{p\pi}{c} z) e^{-\alpha[\beta_m^2 + (\frac{p\pi}{c})^2]t}}{\frac{1}{2} c b^2 [J_0'^2(\beta_m b)]}$$
$$(T_0 - T_c) \left[\frac{b}{\beta_m} J_1(\beta_m b) \right] \left[-\frac{c}{p\pi}(-1)^p + \frac{c}{p\pi} \right] + T_c \tag{30}$$

Equation (30) was programmed in Matlab, and the input data required are: the initial temperature T_0, the temperature at the border T_c, the thermal diffusivity value α is according to the fluid to be analyzed, the radius r, the height c, and the number of divisions required in the discretization, the initial and final time and the time steps. All that data are used to obtain the temperature in coordinates r, z and any time t. In the solution obtained convergence is achieved with 65 values for p and for m were set to 200. The finite-element method was used with commercial software to compare the results obtained from the analytical solution [1,23].

3. Numerical Solution

This section describes the numerical model used for comparison with the analytical solution, and all parameters such as boundary and initial conditions, and the dimensions of the cylinder were used the same as for analytical solution. The numerical solution was solved via the finite-element method by commercial software.

Model to Compare the Analytical Solution

We used the ANSYS Mechanical Parametric Design Language (APDL) user interface 18.0, which uses the finite-element method to approximate the solution of transient heat transfer conduction by Equation (4) shown above [10,23] for the numerical solution, running in a workstation with a Xeon processor with 16 cores and 16 GB of RAM.

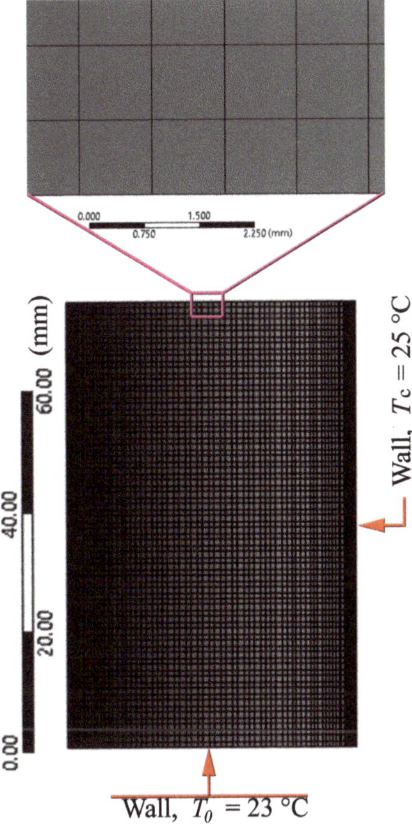

Figure 2. Mesh for the numerical model.

The model used in commercial software is two-dimensional and axisymmetric, which allows for the condition of azimuthal symmetry established for the analytical problem. The type of element that was used is a plane 77, which is an 8-node element for thermal analysis that accepts axisymmetry conditions and transient or steady-state analysis. Taking into account the numerical solutions, the grid number was investigated as a parameter that affects the accuracy of the obtained results. Based on the variations of these parameters in a grid number of 5610 nodes, we can be assured that the obtained results are accurate, because several simulations were computed varying mesh size and time step for convergence until get the minimum percentage which was 0.075% error against the theoretical solution. The surface was meshed using a quadrilateral dominant method as displayed in Figure 2. The solution of heat conduction problems is governed by the characteristic time of heat diffusion. The mesh size is less important. To calculate the numerical solution, a cylinder with the following properties was selected: the thermal conductivity essential variable in the model [2], heat capacity, and density of water [24]. The dimension of the cylinder are as follows: radius: 24.5 mm; height:

74.24 mm. The boundary conditions are a constant temperature of $T_c = 25\,^{\circ}\text{C}$. The initial condition T_0 at $t = 0$ was established at $T_0 = 23\,^{\circ}\text{C}$. The study was solved in a transient mode for a physical time of 6000 s with a time step size of 0.1, for a minimum of 0.05, maximum of 1.5, and a maximum numbers of iterations of 100.

4. Analysis Results

This section shows a comparison of the results between the analytical and numerical solutions. Another geometry is included, i.e., the spherical one, and was analyzed only by the numerical method. The results are shown in four different sizes of cylinders with the same sphere of 2.5 cm radius. Where the analytical solution named "CYLINDRICAL" is compared with solutions obtained by the finite-element method, where the cylinder is named "MEF CYLINDRICAL," and the sphere is named "MEF SPHERICAL," representing the point at which the temperature gradients are greatest and are at the centroid of the sphere and the cylinder. To reduce the solution time and the amount of data, a constant of thermal diffusivity value was considered as the unit instead of the water value of $0.14 \times 10^{-6}\ \text{m}^2/\text{s}$.

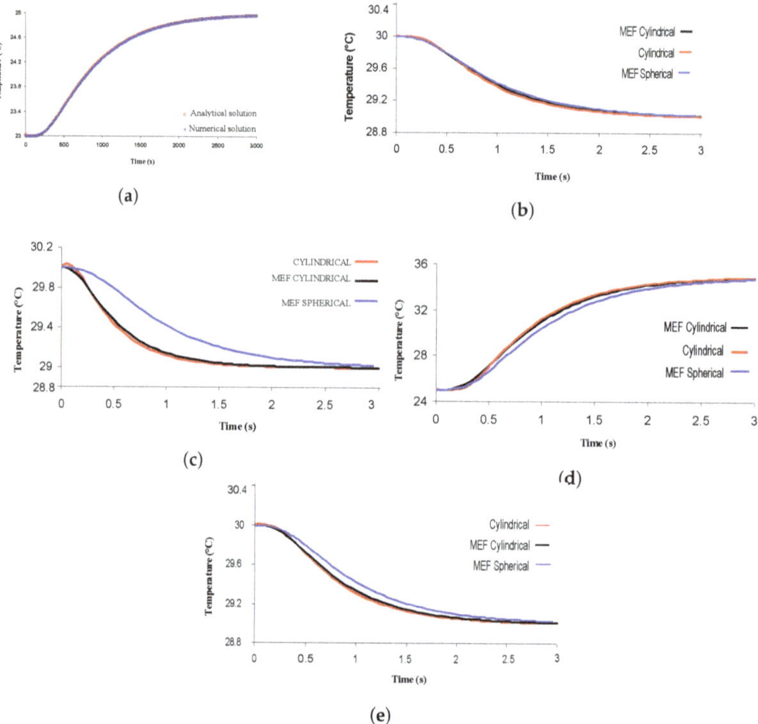

Figure 3. Comparison between analytical and numerical solutions for: (**a**) the centroid of a cylinder, located at a radius of 0.0 cm and a height of 3.71 cm with properties of the water; (**b**) a cylinder with a radius of 2.2 cm and a height of 4.5 cm, compared with the sphere with a 2.5 cm radius; (**c**) a cylinder with a radius of 1.5 cm and a height of 9.9 cm, compared with the sphere with a 2.5 cm radius; (**d**) a cylinder with a radius of 2.2 cm and a height of 4.5 cm, compared with the sphere with a 2.5 cm radius; (**e**) the initial condition of 25 °C and the boundary condition of $T_c = 35\,^{\circ}\text{C}$ for a cylinder with a radius of 2.2 cm and a height of 4.5 cm, compared with the sphere with a 2.5 cm radius.

Figure 3a plots the evolution of temperature according to the boundary conditions mentioned: $T_0 = 23\,^{\circ}\text{C}$ at $t = 0$ and constant temperature $T_c = 25\,^{\circ}\text{C}$ for 6000 s with water, such as the fluid inside the cell. The point analyzed exactly was the centroid of the cylinder, according to the coordinates plotted in Figure 1, the location of centroid is $r = 0$ and $z = 3.712$ cm. The curve depicted in Figure 3a has a maximum error percentage of 0.075% between the analytical and numerical solution with a temperature increase of $2\,^{\circ}$C.

Figure 3b,c,e compares temperatures over time with the initial condition of $T_0 = 30\,^{\circ}$C and with the boundary condition of $T_c = 29\,^{\circ}$C for different cylinder dimensions and a sphere with a radius of 2.5 cm. The dimensions of the cylinder with a smaller temperature gradient have a 1.5 radius and a 9.9 cm height. This reached the steady state at the highest speed with respect to the other cells; at a time of 1.5 s, the cylinder reached a uniform temperature compared with the other configurations, which took almost twice as long [24,25].

For the next case, Figure 3d shows a comparison among analytical and numerical solutions for an initial condition of $T_0 = 25\,^{\circ}$C and a boundary condition of $T_c = 35\,^{\circ}$C. The numerical result of the sphere with a radius of 2.5 cm is included. The error obtained is about 0.86%. Among all analytical and numerical solutions, this was the highest, and it is because there is a big difference in temperature, $10\,^{\circ}$C, in contrast to the $1\,^{\circ}$C difference in the other cases [24,25].

5. Discussion

This work demonstrates the accuracy of analytical and numerical solutions in the development of an adiabatic calorimeter. It is easy to avoid experimental considerations during digital measurements. The adiabatic condition cannot be achieved. It is essential in this apparatus to keep constant temperature of the cryostat and the same temperature between cell and its adiabatic shield. Thus, it is possible to calculate heat leaks in this kind of apparatus to determine the heat capacity more accurately as possible.

An accurate measurement of heat capacity is dependent on the temperature rise measurement and the change-of-volume work adjustment [9]. Thus, this work focuses in the gradients temperature generated by the temperature rise. The main result of this work was shared because [9] chose a spherical geometry unlike [3–8]. It was figured out that spherical geometry has lower gradients that cylindrical one.

Matlab is a high-level tool for the development of applications in engineering and research, as in this work. Based on the operation of the adiabatic calorimeter, was simplified the mathematical model that approximates the operation of a cell in an adiabatic calorimeter and it was programmed in Matlab to analyze the variables that affect the exchange of energy. Different values were considered for it, the temperature, time, and fluid properties. Simulation research provides a reference for engineering applications and excellent tools to evaluate physical phenomena taking account of complex geometries.

Numerical simulations using a finite-element method were performed using the package ANSYS 18.0 (ANSYS, Inc., Canonsburg, PA, USA) to understand the behavior of heat transfer and heat leaks. Through a heating resistor, which was conventionally used during the period of operation, the operating conditions used in numerical simulations can be checked. The performance of the calorimeter can be tested. Measurements can be made to assess improvements in sensitivity. There are sources of heat that cannot be accounted for, e.g., temperature sensors and heat losses through the lid decrease as heat generation and thermal fluid speed increases. Taking into account [10,11], the analytical solution developed in this work is at most a 3% difference for experimental results that could be obtained.

The typical procedure for loading a sample into a calorimetric device is as follows: The pre-packed sample container is filled, and the sample is hermetically sealed on the device. The assembly is reweighed to determine the mass of the sample. The sample vessel is placed in the heater/thermometer assembly. The calorimeter temperature increases stepwise during the heat pulse. This is to minimize its contribution to the total heat load. For most of the equilibrium period, the calorimeter temperature is almost entirely constant because the device deliberately turns on and off every specific time interval. Other studies with calorimeters are the enthalpy-temperature ratio in materials have been conducted by [24,25]. The most recent studies have followed the approach that uses essential thermodynamic relationships [26–28].

The calorimetry allows one to study the thermal behavior in the heat exchange that occurs in a system, managing to understand more clearly the heat transfer mechanism in it, and providing data that helps its characterization in materials and substances. In engineering, its contribution can be extended to different parts of the optimization process, to chemical reactions, and to operating units, i.e., in any process where energy exchange is present. The calorimetry in the food industry, it helps in food preservation by determining the optimal proportions of compounds. It contributes to the study of vegetable crops to analyze germination processes, it aids in the quantification of nanosolids for pharmaceutical use, and it helps in the thermal characterization of nano-structured lipid transporters. This has been made possible in contrast with the works accomplished by [29,30].

6. Conclusions

A strategy was established to improve computational time, as described below. The first analysis shown in Figure 3a, was computed using properties of water to simulate this fluid inside of the cell. For the simulations shown in Figure 3b–e, a thermal diffusivity of value 1 was used to reduce computational time. The time is only 3 s, in contrast with Figure 3a, where time is 3000 s.

The maximum error obtained in this work among the analytical results and the numerical method was around 0.86% for an increase of 10 °C and the minimum was 0.075% for an increase of 1 °C. These results show that there is a good agreement with the finite-element method.

The cylinder of a 1.5 cm radius and a height of 9.9 cm reached in a period time of 1.5 s a uniform temperature inside of the entire cell, which is half the time with respect to other dimensions evaluated. This is because the evaluated cylinders have a smaller radius with respect to the sphere. However, other parameters such as temperature sensor location, the cell material, and the inlets and outlets for fluid to the cell need to be evaluated.

The spherical geometry has better thermal performance than the cylindrical, because the temperature gradients are smaller. The results obtained are that the maximum gradient for a spherical cell is 0.104 °C, and the maximum gradient for a cylindrical cell is 0.208 °C; therefore, geometry affects thermal behavior, as reported by [2]. In real applications, the system in a sphere responds with greater thermal speed than in a system contained within a cylinder. However, the homogeneity in the fluid contained in the cell is the most important variable because accuracy measurements affect the heat capacity value.

When heat is added to a system, increasing the temperature from 23 °C to 25 °C, the difference in temperature between the analytical solution and the numerical solution is 0 as seen in Figure 3a. The maximum temperature variation between the cylinder and the sphere obtained from the simulation occurred at the time of extracting heat from the system, causing a decrease in temperature from 30 °C to 29 °C, generating a temperature variation of 0.08 °C according to Figure 3e.

The error among analytical and numerical results increases with increasing temperature and decreases as the steady state is reached.

Author Contributions: Conceptualization, J.E.E.G.-D. and J.M.O.R.; methodology, J.E.E.G.-D., J.M.O.R., and M.A.Z.-A.; software, M.A.Z.-A., J.M.O.R., and J.E.E.G.-D.; validation, J.E.E.G.-D., J.R.-R., M.A.Z.-A., J.M.O.R., and L.L.-C.; formal analysis, L.L.-C., J.E.E.G.-D., J.R.-R., and M.A.Z.-A.; writing—original draft preparation, M.A.Z.-A., L.L.-C., and J.R.-R.; writing—review and editing, M.A.Z.-A.; supervision, M.A.Z.-A., J.R.-R., and L.L.-C. All authors have read and agreed to the published version of the manuscript.

Funding: This research was funded by the Consejo Nacional de Ciencia y Tecnología (CONACYT) and PRODEP.

Conflicts of Interest: The authors declare that there is no conflict of interest.

Abbreviations

The following abbreviations are used in this manuscript:

Q	Amount of heat
m	Mass
ΔT	Temperature variation
I	Current
V	Voltage
t	Time
$^\circ$C	Celsius grade
α	Thermal diffusivity
k	Thermal conductivity
ρ	Density
C_p	Heat capacity
b	Constant in radial coordinates
c	Constant in z coordinates
T_c	Temperature constant
T^*	Represent a change of variable for T
T_0	Initially temperature
R_v	Bessel functions of order v
$R_v(\beta_m r)$	Eigenfunction
Y_v	Bessel functions of order v
J_v	Solution for Bessel functions of order v
β_b	Eigenvalue evaluated in radius b
$\beta_m b$	Summation of eigenvalue evaluated in radius b
$Z\eta p$	Eigenfunction for z coordinate
η_p	Summation of eigenvalue evaluated in height
C_m	Normal function
J_0	Bessel functions evaluated in 0
CENAM	Centro Nacional de Metrologia

References

1. Hoseinzadeh, S.; Heyns, P.S.; Kariman, H. Numerical investigation of heat transfer of laminar and turbulent pulsating Al_2O_3/water nanofluid flow. *Int. J. Numer. Methods Heat Fluid Flow* **2019**, *30*, 1149–1166. [CrossRef]
2. Sarafraz, M.; Safaei, M.R. Diurnal thermal evaluation of an evacuated tube solar collector (ETSC) charged with graphene nanoplatelets-methanol nano-suspension. *Renew. Energy* **2019**, *142*, 364–372. [CrossRef]
3. Tan, Z.C.; Shi, Q.; Liu, B.P.; Zhang, H.T. A fully automated adiabatic calorimeter for heat capacity measurement between 80 and 400 K. *J. Therm. Anal. Calorim.* **2008**, *92*, 367–374. [CrossRef]
4. Lang, B.E.; Boerio-Goates, J.; Woodfield, B.F. Design and construction of an adiabatic calorimeter for samples of less than 1 cm^3 in the temperature range T = 15 K to T = 350 K. *J. Chem. Thermodyn.* **2006**, *38*, 1655–1663. [CrossRef]
5. Kobashi, K.; Kyômen, T.; Oguni, M. Construction of an adiabatic calorimeter in the temperature range between 13 and 505 K, and thermodynamic study of 1-chloroadamantane. *J. Phys. Chem. Solids* **1998**, *59*, 667–677. [CrossRef]
6. Sorai, M.; Kaji, K.; Kaneko, Y. An automated adiabatic calorimeter for the temperature range 13 K to 530 K The heat capacities of benzoic acid from 15 K to 305 K and of synthetic sapphire from 60 K to 505 K. *J. Chem. Thermodyn.* **1992**, *24*, 167–180. [CrossRef]
7. Tan, Z.C.; Shi, Q.; Liu, X. Construction of High-Precision Adiabatic Calorimeter and Thermodynamic Study on Functional Materials. In *Calorimetry: Design, Theory and Applications in Porous Solids*; IntechOpen: Rijeka, Croatia, 2018; p. 1. [CrossRef]
8. Matsuo, T.; Suga, H. Adiabatic microcalorimeters for heat capacity measurement at low temperature. *Thermochim. Acta* **1985**, *88*, 149–158. [CrossRef]

9. Magee, J.W.; Deal, R.J.; Blanco, J.C. High-temperature adiabatic calorimeter for constant-volume heat capacity measurements of compressed gases and liquids. *J. Res. Natl. Inst. Stand. Technol.* **1998**, *103*, 63. [CrossRef]

10. Sun, M.T.; Chang, C.H. The error analysis of a steady-state thermal conductivity measurement method with single constant temperature region. *J. Heat Transf.* **2007**. [CrossRef]

11. Yari, A.; Hosseinzadeh, S.; Galogahi, M. Two-dimensional numerical simulation of the combined heat transfer in channel flow. *Int. J. Recent Adv. Mech. Eng.* **2014**, *3*, 55–67. [CrossRef]

12. Kaur, G.; Singh, M.; Kumar, J.; De Beer, T.; Nopens, I. Mathematical Modelling and Simulation of a Spray Fluidized Bed Granulator. *Processes* **2018**, *6*, 195. [CrossRef]

13. Al-Najem, N.; Özişik, M. On the solution of three-dimensional inverse heat conduction in finite media. *Int. J. Heat Mass Transf.* **1985**, *28*, 2121–2128. [CrossRef]

14. Renaud, J.; Rossomme, S.; Sarfehnia, A.; Vynckier, S.; Palmans, H.; Kacperek, A.; Seuntjens, J. Development and application of a water calorimeter for the absolute dosimetry of short-range particle beams. *Phys. Med. Biol.* **2016**, *61*, 6602–6619. [CrossRef] [PubMed]

15. Amoabeng, K.O.; Lee, K.H.; Choi, J.M. Modeling and Simulation Performance Evaluation of a Proposed Calorimeter for Testing a Heat Pump System. *Energies* **2019**, *12*, 4589. [CrossRef]

16. Karvinen, H.; Hasani Aleni, A.; Salminen, P.; Minav, T.; Vilaça, P. Thermal Efficiency and Material Properties of Friction Stir Channelling Applied to Aluminium Alloy AA5083. *Energies* **2019**, *12*, 1549. [CrossRef]

17. Mohammed, H.A.; Narrein, K. Thermal and hydraulic characteristics of nanofluid flow in a helically coiled tube heat exchanger. *Int. Commun. Heat Mass Transf.* **2012**. [CrossRef]

18. Pitarch, J.L.; Sala, A.; de Prada, C. A Systematic Grey-Box Modeling Methodology via Data Reconciliation and SOS Constrained Regression. *Processes* **2019**, *7*, 170. [CrossRef]

19. Moreno-Piraján, J.C.; Giraldo, L. Isoperibolic Titration Calorimetry as a Tool for the Prediction of Thermodynamic Properties of Cyclodextrins. *Energies* **2008**, *1*, 93–104. [CrossRef]

20. Özisik, N.; Bayazitoglu, Y.A. *Elements of Heat Transfer*; McGraw-Hill: New York, NY, USA, 1988.

21. Zhi-Cheng, T.; Jinchun, Y.; Yi, S.; Shuxia, C.; Lixing, Z. An adiabatic calorimeter for heat capacity measurements in the temperature range 300–600 K and pressure range 0.1–15 MPa. *Thermochim. Acta* **1991**, *183*, 29–38. [CrossRef]

22. Zhong, Q.; Dong, X.; Zhao, Y.; Wang, J.; Zhang, H.; Li, H.; Guo, H.; Shen, J.; Gong, M. Adiabatic calorimeter for isochoric specific heat capacity measurements and experimental data of compressed liquid R1234yf. *J. Chem. Thermodyn.* **2018**, *125*, 86–92. [CrossRef]

23. Gadalla, M.; Ghommem, M.; Bourantas, G.; Miller, K. Modeling and Thermal Analysis of a Moving Spacecraft Subject to Solar Radiation Effect. *Processes* **2019**, *7*, 807. [CrossRef]

24. Zhu, H.; Sun, B.; Jiang, J.; Xu, W. Measurement of hazardous reactions under extreme conditions with a house-built high-performance adiabatic calorimeter. *J. Therm. Anal. Calorim.* **2020**. [CrossRef]

25. Kossoy, A. An in-depth analysis of some methodical aspects of applying pseudo-adiabatic calorimetry. *Thermochim. Acta* **2020**, *683*. [CrossRef]

26. Choi, Y.; Jeon, K.; Park, Y.; Hyun, S. Numerical simulation of heat-loss compensated calorimeter. *Int. J. Comput. Methods Exp. Meas.* **2019**, *7*, 285–296. [CrossRef]

27. Ding, J.; Yu, L.; Wang, J.; Xu, Q.; Yang, S.; Ye, S. A symmetric dual-channel accelerating rate calorimeter with the varying thermal inertia consideration. *Thermochim. Acta* **2019**, *678*. [CrossRef]

28. Ivsic, B.; Dadic, M.; Malaric, R.; Martinovic, Z. Thermal considerations on adiabatic coaxial line for microcalorimeter measurements. In Proceedings of the 2019 2nd International Colloquium on Smart Grid Metrology (SMAGRIMET), Split, Croatia, 9–12 April 2019. [CrossRef]

29. Sarafraz, M.; Arya, A.; Nikkhah, V.; Hormozi, F. Thermal performance and viscosity of biologically produced silver/coconut oil nanofluids. *Chem. Biochem. Eng. Q.* **2016**, *30*, 489–500. [CrossRef]

30. Sarafraz, M.M.; Tlili, I.; Tian, Z.; Bakouri, M.; Safaei, M.R.; Goodarzi, M. Thermal Evaluation of Graphene Nanoplatelets Nanofluid in a Fast-Responding HP with the Potential Use in Solar Systems in Smart Cities. *Appl. Sci.* **2019**, *9*, 2101. [CrossRef]

 *energies*

Article

Triple Solutions of Carreau Thin Film Flow with Thermocapillarity and Injection on an Unsteady Stretching Sheet

Kohilavani Naganthran, Ishak Hashim and Roslinda Nazar *

Department of Mathematical Sciences, Faculty of Science and Technology, Universiti Kebangsaan Malaysia, Bangi 43600 UKM, Selangor, Malaysia; kohi@ukm.edu.my (K.N.); ishak_h@ukm.edu.my (I.H.)
* Correspondence: rmn@ukm.edu.my

Received: 21 May 2020; Accepted: 16 June 2020; Published: 19 June 2020

Abstract: Thin films and coatings which have a high demand in a variety of industries—such as manufacturing, optics, and photonics—need regular improvement to sustain industrial productivity. Thus, the present work examined the problem of the Carreau thin film flow and heat transfer with the influence of thermocapillarity over an unsteady stretching sheet, numerically. The sheet is permeable, and there is an injection effect at the surface of the stretching sheet. The similarity transformation reduced the partial differential equations into a system of ordinary differential equations which is then solved numerically by the MATLAB boundary value problem solver bvp4c. The more substantial effect of injection was found to be the reduction of the film thickness at the free surface and development of a better rate of convective heat transfer. However, the increment in the thermocapillarity number thickens the film, reduces the drag force, and weakens the rate of heat transfer past the stretching sheet. The triple solutions are identified when the governing parameters vary, but two of the solutions gave negative film thickness. Detecting solutions with the most negative film thickness is essential because it implies the interruption in the laminar flow over the stretching sheet, which then affects the thin film growing process.

Keywords: thin film; boundary layer; thermocapillarity; triple solutions; Carreau fluid

---

1. Introduction

The discovery of the boundary layer by Prandtl [1] remarked the highest achievement in the development of fluid mechanics and created a proper basis to understand the dynamics of the real fluid. The Prandtl boundary layer is found in a wide range of aerodynamics and engineering applications. The boundary layer idea then evolved into the theoretical works on the boundary layer flow over a stretching surface which was contributed by the following literature: Sakiadis [2,3], Crane [4], and Carragher and Crane [5]. These works were highly appreciated due to its significance in the extrusion process. As time progressed, the boundary layer flow has been probed under various circumstances to improve the quality of the extrusion process, and hence many works have been reported accordingly (see [6–19]). Cast film extrusion is an emerging industrial process that produces cast films which are widely used for food and textiles packaging and coating substrates in the extrusion coating process [20]. The demand for this process is still high, and research activities are conducted so that it can be improved from time to time. The theoretical work in thin film flow was started by Wang [21] by investigating the behavior of thin liquid film flow past an impermeable stretching surface and it was found that the similarity solutions were absent when the value of the unsteadiness parameter (S) falls within the range of $S > 2$. Usha and Sridharan [22] reexamined the work of [21] by considering the asymmetric flow and justified that the similarity solutions become unavailable when the value of the unsteadiness parameter fell within the range of $S > 4$. Shortly thereafter, the subject of non-Newtonian fluid captured the attention of the researchers in the thin film flow since non-Newtonian fluid such as paint also used in the extrusion coating process. Thus, the problem of thin film flow towards an unsteady stretching sheet in a generalized non-Newtonian fluid was solved by Andersson et al. [23] and concluded that at a higher value of the unsteadiness parameter, the impact of the power-law index is more significant. The researchers then realized that the heat transfer aspect is also equally important as the fluid flow characteristics in a thin film flow so that the complications in the design of various heat exchangers and chemical processing equipment can be surmounted. Andersson et al. [24] came forward to investigate the heat transfer features in a thin film flow past an unsteady stretching surface and discovered that, at low Prandtl numbers ($Pr < 1$), the effect of the unsteadiness parameter on the rate of heat transfer is highly significant. Chen [25] also was interested in studying the heat transfer characteristics in the thin film flow and hence revisited the work of [23] to examine the heat transfer feature in the power-law model considered in [23]. On the other hand, Wang [26] attempted to solve the thin film problem in [24] by using a different form of similarity transformation and generated the analytic solutions via the homotopy analysis method (HAM). The work of [26] proved that the analytic method also could be employed in the theoretical investigation of the thin film flow mechanism past a stretching sheet. After that, the thin film flow and heat transfer past an unsteady stretching sheet were investigated under various settings, for instance, see [27–32].

Thermocapillarity is a physical effect which formed through the thermally induced surface-tension gradients over a fluid–fluid interface [33]. Researchers then developed this effect in the thin film flow as the surface-tension gradients able to produce interfacial flow which can be found in the continuous casting process. The thermocapillarity effect seemed to be significant in crystal growth melts [34] and nucleate boiling [35]. In the theoretical work of the thin film flow, Dandapat and Ray [36] explored the effect of thermocapillarity on thin film flow past a rotating disc and exposed that thermocapillarity force at the free surface reduces the film thickness when the disc is cooled. Then, Dandapat et al. [37] explored the effect of thermocapillarity in the problem solved by [24] and concluded that the impact of thermocapillarity thickens the film and enhances the rate of heat transfer along the stretching sheet. Meanwhile, Chen [38] tested the thermocapillarity effect in the power-law thin film flow past a stretching sheet and found that thermocapillarity causes thin film to thicken. Later, researchers investigated the impact of thermocapillarity on thin film flow and heat transfer along with other settings such as magnetic field [39], nanofluid [40], thermal radiation [41], and suction/injection effects at the surface of the stretching sheet [42]. Recently, Rehman et al. [43] solved the thin film flow, heat, and mass transfer problem with several physical effects such as thermocapillarity, heat generation/absorption, mixed convection, chemical reaction, and magnetohydrodynamics (MHD) past an unsteady stretching sheet. Another interesting work by Rehman et al. [44] that incorporated the effect of thermocapillarity in the MHD Casson thin film flow past an unsteady stretching sheet under the slip and variable fluid properties influences. Meanwhile, the theoretical work of the thin film flow and heat transfer in a Carreau fluid was initiated by Myers [45], and in fact [45] had proposed the application of several non-Newtonian models, such as power-law model, Ellis model and Carreau model to thin film flow. The idea in [45] then encouraged Tshehla [46] to explore the Carreau thin film flow and heat transfer over an inclined surface. Ashwinkumar and Sulochana [47] examined the Carreau thin nano-liquid film flow and heat transfer with magnetohydrodynamics (MHD) dissipative over an unsteady stretching sheet. Khan et al. [48] extended the work of [47] by probing the impact of an inclined magnetic field in Carreau nano-liquid thin film flow and its heat transfer characteristics with graphene nanoparticles. Recently, Iqbal et al. [49] solved the problem of Carreau magneto-nanofluid thin film flow past an unsteady stretching sheet. So far, in the previous theoretical investigations of the Carreau thin film flow, no one had considered the effects of thermocapillarity and injection in their models and presented multiple solutions.

Therefore, the present study is devoted to analyzing the influences of the thermocapillarity and injection in the thin film flow over an unsteady permeable stretching sheet in a Carreau fluid, theoretically. The present study also employs similarity transformations, which had been proposed by Wang [26], and the reduced model is solved numerically by a collocation method, namely the bvp4c function in MATLAB. To date, no one has reported on the triple solutions in the thin film flow problem, and this study has successfully identified the triple solutions. The presences of these triple solutions are found to be important in detecting the flaw in the flow system by unveiling the most negative film thickness.

2. Problem Formulation

We are focused with an incompressible two–dimensional unsteady Carreau fluid flow confined by a thin liquid film of uniform thickness, $h(t)$ and a horizontal elastic sheet which is stretching from a narrow slit at the origin of the Cartesian coordinate system, as illustrated by Figure 1. The setup of the Cartesian coordinates is in a way where the $y-$ coordinate is normal to the $x-$ coordinate. The state of stretching sheet induce the Carreau fluid to flow within the thin film as the sheet stretched at $U_w(x,t) = \frac{bx}{(1-\alpha t)}$, where both b and α are positive constants with dimension time^{-1}, $\alpha t \neq 1$, and $b > 0$ indicates the rate of stretching. The setup of $\alpha > 0$ yields the constructive stretching rate, $b/(1 - \alpha t)$ to upsurge with time. The surface of the sheet is permeable, and hence there is the mass velocity which is denoted by V_w, wherein $V_w > 0$ signifies suction while $V_w < 0$ injection. The wall temperature is denoted by T_w and $h(t)$ labels the film width. It is assumed that the end effects and gravity are negligible and the thickness of the film $h(t)$ is stable and uniform. It is worth mentioning that the boundary layer approximation is valid if, and only if, the thickness of the liquid film maintains its position without overlapping with the boundary layer thickness [40].

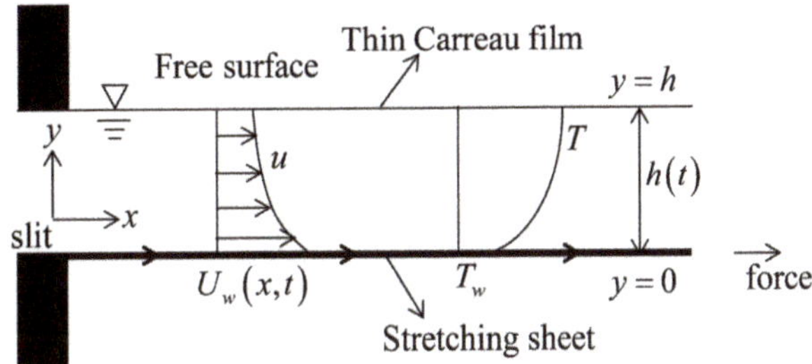

Figure 1. Schematic diagram of the thin film flow along a stretching sheet.

The Cauchy stress tensor for the Carreau fluid can be expressed as [45]

$$\tau = -p\mathbf{I} + \eta\mathbf{A}_1, \tag{1}$$

where

$$\eta = \eta_\infty + (\eta_0 - \eta_\infty)\left[1 + \left(\lambda\dot{\gamma}\right)^2\right]^{\frac{n-1}{2}}, \tag{2}$$

wherein τ is the Cauchy stress tensor, p is the pressure, $\mathbf{I}$ signifies the identity tensor, η_0 implies the zero-shear-rate viscosity, η_∞ is the infinite-shear-rate viscosity, λ denotes the material time constant, and n represents the power–law index. The shear rate which is symbolized by $\dot{\gamma}$ can be conveyed as

$$\dot{\gamma} = \sqrt{\frac{1}{2}\sum_j\sum_j \dot{\gamma}_{ij}\dot{\gamma}_{ij}} = \sqrt{\frac{1}{2}\Pi} = \sqrt{\frac{1}{2}\mathrm{tr}(\mathbf{A}_1)}. \tag{3}$$

The second invariant strain rate tensor denoted by Π and $\mathbf{A}_1$ which implies the Rivlin–Ericksen tensor can be defined as

$$\mathbf{A}_1 = (\mathrm{grad}\mathbf{V}) + (\mathrm{grad}\mathbf{V})^{\mathrm{T}}. \tag{4}$$

As been recommended by Boger [50], we consider the most practical cases where $\eta_0 \gg \eta_\infty$. Normally, the value of η_∞ is determined by the extrapolation procedure or chosen to be zero (suggested theoretical value) [50]. We take the value of η_∞ to be zero and thus Equation (1) simply becomes

$$\tau = -p\mathbf{I} + \eta_0 \left[1 + \left(\lambda \dot{\gamma} \right)^2 \right]^{\frac{n-1}{2}} \mathbf{A}_1.$$ (5)

The Carreau model exhibits the shear thinning or pseudoplastic characteristics when the value of the power–law index lies within $0 < n < 1$. The Carreau model reveals the shear thickening or dilatant feature when $n > 1$. Under these assumptions, the governing liquid film flow of the Carreau fluid can be written as

$$\frac{\partial u}{\partial x} + \frac{\partial v}{\partial y} = 0,$$ (6)

$$\frac{\partial u}{\partial t} + u\frac{\partial u}{\partial x} + v\frac{\partial u}{\partial y} = v\frac{\partial^2 u}{\partial y^2}\left[1 + \lambda^2\left(\frac{\partial u}{\partial y}\right)^2 \right]^{\frac{n-1}{2}} + v(n-1)\lambda^2\frac{\partial^2 u}{\partial y^2}\left(\frac{\partial u}{\partial y}\right)^2\left[1 + \lambda^2\left(\frac{\partial u}{\partial y}\right)^2 \right]^{\frac{n-3}{2}},$$ (7)

$$\frac{\partial T}{\partial t} + u\frac{\partial T}{\partial x} + v\frac{\partial T}{\partial y} = \kappa\frac{\partial^2 T}{\partial y^2},$$ (8)

where u and v are the velocity components along the $x-$ and $y-$ directions, respectively, v is the kinematic viscosity, λ is a material time constant, n signifies the power-law index, T is the temperature, $\kappa = \frac{k}{\rho C_p}$, with k is the thermal conductivity, ρ is the density, and C_p is the specific heat. The Equations (6)–(8) are accompanied by the boundary conditions as

$$u = U_w(x,t), \ v = V_w, \ T = T_w \ \text{at} \quad y = 0,$$
$$\mu\frac{\partial u}{\partial y} = \frac{\partial \chi}{\partial x}, \ \frac{\partial T}{\partial y} = 0, \ v = \frac{dh}{dt} \ \text{at} \quad y = h,$$ (9)

where χ is the surface tension which varies linearly with temperature [37] given as

$$\chi = \chi_0[1 - \gamma(T - T_0)],$$ (10)

where χ_0 is the surface tension at temperature T_0 and γ is the positive fluid property. The wall temperature (T_w) is defined as [37]

$$T_w = T_0 - T_{ref}\left(\frac{bx^2}{2v}\right)(1 - \alpha t)^{-\frac{3}{2}},$$ (11)

where T_0 signifies the temperature at slit, and T_{ref} embodies the reference temperature which can be taken as a constant reference temperature or as a constant temperature difference. The definition of T_w imitates the circumstance where the sheet temperature decreases from T_0 at the slit in proportion to x^2 and the amount of temperature reduction along the sheet increases with time [37]. The expression for $U_w(x,t)$ and T_w only valid for time $t < \alpha^{-1}$. The boundary condition when $y = h$ enforces a kinematic constraint of the fluid motion. Next, we introduce the similarity transformations as follows [26]:

$$\psi = \sqrt{\frac{vb}{1-\alpha t}}\frac{x}{\beta}f(\zeta), \quad u = \frac{\partial \psi}{\partial y} = \frac{bx}{1-\alpha t}f'(\zeta),$$
$$v = -\frac{\partial \psi}{\partial x} = -\sqrt{\frac{vb}{1-\alpha t}}\beta f(\zeta), \quad T = T_0 - T_{ref}\left(\frac{bx^2}{2v}\right)\frac{1}{\left(\sqrt{1-\alpha t}\right)^3}\theta(\zeta),$$ (12)

$$\zeta = \sqrt{\frac{b}{v}}\frac{y}{\beta\sqrt{1-\alpha t}},$$ (13)

where prime designates the derivative with respect to ζ, $\psi(x, y, t)$ is the stream function, and β, the nonzero constant, is an unknown parameter representing the dimensionless film thickness. At the free surface, set $\zeta = 1$ and (13) will take the form

$$\beta = \sqrt{\frac{b}{v(1 - \alpha t)}} h(t), \tag{14}$$

which eventually gives

$$\frac{dh}{dt} = -\frac{\alpha \beta}{2} \sqrt{\frac{v}{b(1 - \alpha t)}}. \tag{15}$$

Employing the similarity conversion as in (12) and (13) into the governing model (6)–(9) satisfies the continuity equation and the remaining equations are transformed as

$$\left[1 + \frac{nWe^2(f'')^2}{J}\right]\left[1 + \frac{We^2(f'')^2}{J}\right]^{\frac{n-3}{2}} f''' + J\left(ff'' - \frac{\sigma\zeta}{2}f'' - f'^2 - \sigma f'\right) = 0, \tag{16}$$

$$\theta'' + \Pr J\left(f\theta' - 2f'\theta - \frac{\sigma\zeta\theta'}{2} - \frac{3\sigma\theta}{2}\right) = 0, \tag{17}$$

along with the boundary conditions

$$\begin{aligned} f(0) &= S, \quad f'(0) = 1, \quad \theta(0) = 1, \\ f(1) &= \sigma/2, \quad f''(1) = M\theta(1), \quad \theta'(1) = 0, \end{aligned} \tag{18}$$

while letting the constant mass transfer parameter, $S = -\frac{V_w}{\beta}\sqrt{\frac{1-\alpha t}{vb}}$, with the setting $S > 0$ connotes suction and $S < 0$ indicates the injection condition, $We^2 = \frac{\lambda^2 b^3 x^2}{v(1-\alpha t)^3}$, is the local Weissenberg number, $J = \beta^2$ is an unknown constant to be calculated as a part of the problem, $M = \frac{\gamma \chi_0 T_{ref}\beta}{\mu\sqrt{bv}}$ is the thermocapillarity number, $\Pr = \frac{v}{\kappa}$ is the Prandtl number, and $\sigma = \frac{a}{b}$ is the dimensionless measure of unsteadiness to the stretching rate. It has to be noted that when $n = 1$ and $We = 0$, the Carreau fluid model will reveal the Newtonian characteristics. Besides that, setting $S = M = We = 0$ and $n = 3$ in (16)–(18) reduces the present model to the thin film flow problem considered by Wang [26]. Also, by fixing $S = We = 0$ and $n = 3$ in (16)–(18), the thermocapillarity effect in a thin film flow problem solved by Noor and Hashim [39] is recoverable if the value of the Hartmann number in Equation (14) of [39] set to be zero. The physical quantities of interest in the present work are the local skin friction coefficient $\left(C_{fx}\right)$ and the local Nusselt number $\left(Nu_x\right)$ which can be defined as

$$C_{fx} = \frac{\tau_w}{\rho(U_w)^2/2}, \quad Nu_x = \frac{q_w x}{kT_{ref}}, \tag{19}$$

where the wall shear stress (τ_w) and the heat flux from the surface of the sheet (q_w) are given by [51]

$$\tau_w = \left\{\mu_0 \frac{\partial u}{\partial y}\left[1 + \lambda^2\left(\frac{\partial u}{\partial y}\right)^2\right]^{\frac{n-1}{2}}\right\}_{y=0}, \quad q_w = -k\left(\frac{\partial T}{\partial y}\right)_{y=0}, \tag{20}$$

By employing (12)–(13) and inducing (20) into (19) provides the following expression

$$C_{fx}\mathrm{Re}_x^{1/2} = 2f''(0)\left\{1 + \frac{We^2}{J}[f''(0)]^2\right\}^{\frac{n-1}{2}}, \quad 2Nu_x\mathrm{Re}_x^{-3/2}\beta(1 - \alpha t)^{1/2} = \theta'(0). \tag{21}$$

The local Reynolds number defined as $\mathrm{Re}_x = \frac{xU_w(x,t)}{\nu}$.

3. Computational Scheme

The developed boundary value problem of (16)–(18) in the previous section was solved numerically via a built-in method in the MATLAB software, that is the bvp4c function. The MATLAB solver bvp4c solver, which was originated by Shampine et al. [52], incorporates the finite-difference program that prompts the three stages of Lobatto IIIa rule. The robust Lobatto IIIa rule belongs to the implicit Runge-Kutta methods and exercised the collocation method associated with the implicit trapezoidal method. Thus, the MATLAB solver bvp4c denotes the collocation method, which presents the C1-continuous solution with a fourth-order precision uniformly in the interval where the function is integrated. In the present work, the bvp4c routine can commence by defining the following new variables of

$$y(1) = f, \ y(2) = f', \ y(3) = f'',$$
$$y(4) = \theta, \text{ and } y(5) = \theta'. \tag{22}$$

By employing the variables in (22), the system of ordinary differential equations (16)–(18) can be written in the terms

$$
\begin{aligned}
f' &= y(2), \\
f'' &= y(3), \\
f''' &= \frac{J \cdot \sigma \cdot \zeta \cdot 0.5 \cdot y(3) + J \cdot (y(2))^2 + J \cdot \sigma \cdot y(2) - J \cdot y(1) \cdot y(3)}{\left[1 + \frac{n \cdot We^2 \cdot (y(3))^2}{J}\right]\left[1 + \frac{We^2 \cdot (y(3))^2}{J}\right]^{\frac{n-3}{2}}}, \\
\theta' &= y(5), \\
\theta'' &= 2 \cdot \mathrm{Pr} \cdot J \cdot y(2) \cdot y(4) + 0.5 \cdot \mathrm{Pr} \cdot J \cdot \sigma \cdot \zeta \cdot y(5) + 1.5 \cdot \mathrm{Pr} \cdot J \cdot \sigma \cdot y(4) - \mathrm{Pr} \cdot J \cdot y(1) \cdot y(5),
\end{aligned}
\tag{23}
$$

Accompanied with the boundary conditions (18) which is written as

$$
\begin{aligned}
y_a(1) - S &= 0, \ y_a(2) - 1 = 0, \ y_a(4) - 1 = 0, \\
y_b(1) - \sigma/2 &= 0, \ y_b(3) - M \cdot y_b(4) = 0, \ y_b(5) = 0.
\end{aligned}
\tag{24}
$$

The subscripts '*a*' and '*b*' describe the position at the surface of the stretching sheet and the free surface, respectively. The relative tolerance has been fixed to 1×10^{-10} throughout the computation process. The bvp4c function eases the computation process of the boundary value problems involving unknown parameters, and efficient in solving the boundary value problems even with the poor guesses [52]. However, a good initial guess is requisite to obtain multiple solutions. By using this clue, the present work has generated three different numerical solutions by using three sets of different initial guesses. The existence of the non-uniqueness solutions is common in a boundary value problem because the nonlinearity of the mathematical model in (16)–(18) and the variation of the respective governing parameters may lead to bifurcations in solutions which encourages the presences of the multiple solutions [53]. In this work, non-uniqueness solutions have been identified and are classified as the first, second, and third solutions. The first solution is a numerical solution which converged with a thin boundary layer whereas the second and third solutions converged with the thicker boundary layer. These solutions satisfy the boundary condition (18). Before the present numerical results discussed, it is important to validate the mathematical model as in (16)–(18) and measure the efficiency of the collocation method. Tables 1 and 2 show the comparison of the numerical results with the previous analytic solutions, and there is a good agreement. This validates the present model and proves the accuracy of the collocation method in solving a boundary layer problem as the present results able to withstand the homotopy analysis method (HAM) which has been employed by Wang [26] and Noor and Hashim [39].

Table 1. Comparison values of β and $f''(0)$ when $\mathrm{Pr} = n = 1$ and $We = S = M = 0$.

S	β		$f'(0)$	
	Wang [26]	**Present Study**	**Wang [26]**	**Present Study**
0.6	3.13125	3.131710	−3.74233	−3.742786
0.7	2.57701	2.576995	−3.14965	−3.149614
0.8	2.15199	2.151994	−2.68094	−2.680966
0.9	1.81599	1.815987	−2.29683	−2.296825
1.0	1.54362	1.543616	−1.97238	−1.972385

Table 2. Comparison values of β, $f''(0)$ and $-\theta'(0)$ when $n = 3$, $\mathrm{Pr} = 1$, $We = S = 0$ and $M = 1$.

S	β		$f'(0)$		$-\theta'(0)$	
	Noor and Hashim [39]	**Present Study**	**Noor and Hashim [39]**	**Present Study**	**Noor and Hashim [39]**	**Present Study**
0.6	3.077525	3.180970	−3.648093	−3.798725	4.946056	5.077439
0.8	2.238349	2.230600	−2.777115	−2.770430	3.729232	3.733874
1.0	1.654829	1.653841	−2.097875	−2.095761	2.884973	2.886293
1.2	1.271942	1.271892	−1.598689	−1.598526	2.297023	2.297249
1.4	1.002731	1.002729	−1.199091	−1.199085	1.856868	1.856914
1.6	0.803335	0.803335	−0.856981	−0.856982	1.507663	1.507691

4. Results and Discussion

This section presents the numerical results in the form of the velocity and temperature profiles with a variation of the unsteadiness parameter (σ), the thermocapillarity number (M), and the constant mass transfer parameter (S) within the range of $(0.8 \leq \sigma \leq 1.6)$, $(0.01 \leq M \leq 2.0)$, and $(-0.3 \leq S \leq -1.0)$, respectively. In the present study, the Prandtl number (Pr) has a fixed value of 30 as the interest is about the molten polymer. Meanwhile, the values of the local Weissenberg number (We) and the power-law index (n) are remained as 0.04 and 0.6, respectively. This section also organized in a way where the discussion of the first and second solutions is given initially, while the explanations of the third solution end up the section.

Figure 2 demonstrates the velocity profiles of the Carreau fluid when σ varies past a stretching surface under the influence of injection at the rate of −0.3. The first and second solutions express an increment in the fluid velocity when the value of σ increases from 0.8 to 1.6. Vajravelu et al. [54] have reported the unique solution with the similar result trend. Based on Table 3, the increment in σ gradually reduces the film thickness of the first solution. The decrement in the film thickness at the free surface increases the fluid velocity past the permeable stretching sheet, which then enhances the wall shear stress along with the stretching sheet. Consequently, the value of the reduced skin friction coefficient $\left(C_{fx}\mathrm{Re}_x^{1/2}\right)$ in Table 4, increases and higher values of $C_{fx}\mathrm{Re}_x^{1/2}$ implies the growth in the frictional drag exerted on the stretching surface, which is not favorable in sustaining the laminar boundary layer flow. Intriguingly, the second solution in Table 3 which revealed the increment in the film thickness also records the increment in the values of $C_{fx}\mathrm{Re}_x^{1/2}$ like the first solution. Based on the values of $C_{fx}\mathrm{Re}_x^{1/2}$ for the first and second solutions in Table 4, a decrement in σ elucidates that the stretching sheet found to exert the drag force on the Carreau fluid by expressing the negative values of $C_{fx}\mathrm{Re}_x^{1/2}$. Next, Figure 3 displays the temperature profiles, and based on the first and second solutions, when the value of σ increases, the fluid temperature increases and the heat flux from the surface of the sheet rises. Eventually, the thermal conductivity past the stretching sheet depreciates and increases the value of the local Nusselt number.

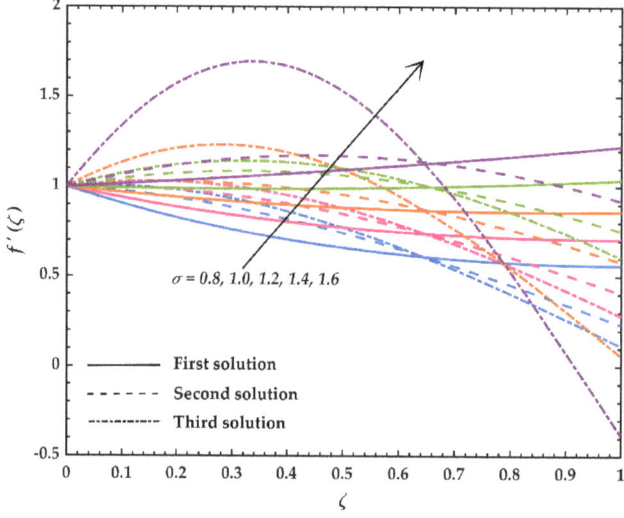

Figure 2. Velocity profiles, $f'(\zeta)$ when $\mathrm{Pr} = 30, We = 0.2, n = 0.6, S = -0.3$, and $M = 2$ as σ varies.

Table 3. Value of J as σ varies at $\mathrm{Pr} = 30, We = 0.2, n = 0.6, S = -0.3$ and $M = 2$.

σ	J		
	First Solution	**Second Solution**	**Third Solution**
0.8	1.0059106	−0.9334169	−1.5042712
1.0	0.4682307	−0.7821466	−1.2449714
1.2	0.2058215	−0.6771271	−1.9809900
1.4	0.0946128	−0.5994718	−0.9362696
1.6	0.0502582	−0.5395827	−2.3099532

Table 4. Value of $C_{fx}\mathrm{Re}_x^{1/2}$ and $\theta'(0)$ as σ varies at $\mathrm{Pr} = 30, We = 0.2, n = 0.6, S = -0.3$ and $M = 2$.

σ	$C_{fx}\mathbf{Re}_x^{1/2}$			$\theta'(0)$		
	FS [1]	**SS** [2]	**TS** [3]	**FS** [1]	**SS** [2]	**TS** [3]
0.8	−1.980577	−0.296318	0.369629	−6.407399	−15.643947	25.767820
1.0	−1.277002	0.141808	0.802004	−5.299440	−14.000526	27.249769
1.2	−0.665062	0.592954	3.288306	−4.064348	−12.656062	−101.360411
1.4	−0.149305	1.053797	1.746104	−3.054334	−11.503536	29.040625
1.6	0.304406	1.524980	8.210864	−2.385140	−10.487733	33.146586

[1] FS = First solution. [2] SS = Second solution. [3] TS = Third solution.

Table 5 overviews the effect of the thermocapillarity number (M) over $C_{fx}\mathrm{Re}_x^{1/2}$ and $\theta'(0)$. The first solution in Table 5 exhibits the gradual increment in the film thickness (J) when the values of M increases from 0.01 to 2.0. The increment in the film thickness then affects the fluid velocity to decrease, and this is depicted in Figure 4. The decreasing fluid velocity then reduces the wall shear stress past the stretching sheet and results in the decrement of $C_{fx}\mathrm{Re}_x^{1/2}$, as shown in Table 6. Noor and Hashim [39], and Rehman et al. [44] also have reported the decrement of the skin friction coefficient in their work. On the other hand, the second solution indicates the decrement in the value of J past the unsteady stretching sheet. No one has reported such a trend before, and this suggests that an increment in the surface tension gradient have the potential to enhance the film thickness. The gradual increment in the film thickness triggers the speed of the fluid flow to increase, which then increases the wall shear stress and eventually, the value of $C_{fx}\mathrm{Re}_x^{1/2}$ increases. Besides that, the first solution in Figure 5 illustrates the decreasing trend of the fluid temperature when the thermocapillarity effect dominates near the free surface. When the fluid temperature decreases, the heat flow rate intensity from the surface of the stretching sheet declines. Thus, the thermal conductivity over the stretching surface increases and inducing the rate of heat transfer to decrease when M increases (see Table 6). These trends are in accordance with the findings given by [44]. The negative values of $\theta'(0)$ elucidate that the heat energy is transferred from the fluid to the stretching sheet. Conversely, the second solution shows the decrement in J, and the fluid temperature at the free surface found be increasing when M increases after encountering some mild fluctuations across the boundary layer. As a result, the rate of convection heat transfer becomes inconsistent with the trend of increasing, decreasing, and increasing.

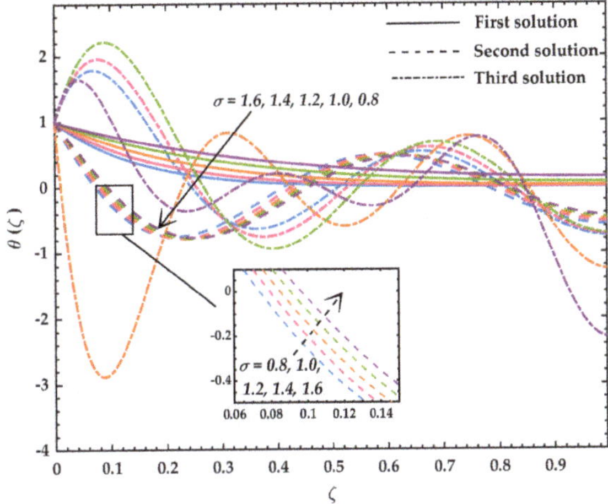

Figure 3. Temperature profiles, $\theta(\zeta)$ when $Pr = 30$, $We = 0.2$, $n = 0.6$, $S = -0.3$ and $M = 2$ as σ varies.

Table 5. Value of J as M varies at $Pr = 30$, $We = 0.2$, $n = 0.6$, $S = -0.3$, and $\sigma = 1.2$.

M	J		
	First Solution	**Second Solution**	**Third Solution**
0.01	0.1654261	−0.0237977	−0.1987100
0.1	0.1681247	−0.1855381	−1.1277911
0.5	0.1785531	−0.5951102	−3.2025872
1.0	0.1891674	−0.6234975	−1.1007880
2.0	0.2058215	−0.6771271	−1.0672621

Table 6. Value of $C_{fx}Re_x^{1/2}$ and $\theta'(0)$ as M varies at $Pr = 30$, $We = 0.2$, $n = 0.6$, $S = -0.3$ and $\sigma = 1.2$.

M	$C_{fx}Re_x^{1/2}$			$\theta'(0)$		
	FS [1]	**SS** [2]	**TS** [3]	**FS** [1]	**SS** [2]	**TS** [3]
0.01	−0.613689	−0.372742	−0.127939	−3.714041	−39.195759	294.841558
0.1	−0.617140	−0.146453	1.377446	−3.739231	25.692494	782.037891
0.5	−0.630449	0.462043	8.923835	−3.833974	−40.763726	−40.729483
1.0	−0.643954	0.506981	1.326987	−3.926454	−22.672611	68.706145
2.0	−0.665062	0.592954	1.265005	−4.064348	−12.656062	28.245961

[1] FS = First solution. [2] SS = Second solution. [3] TS = Third solution.

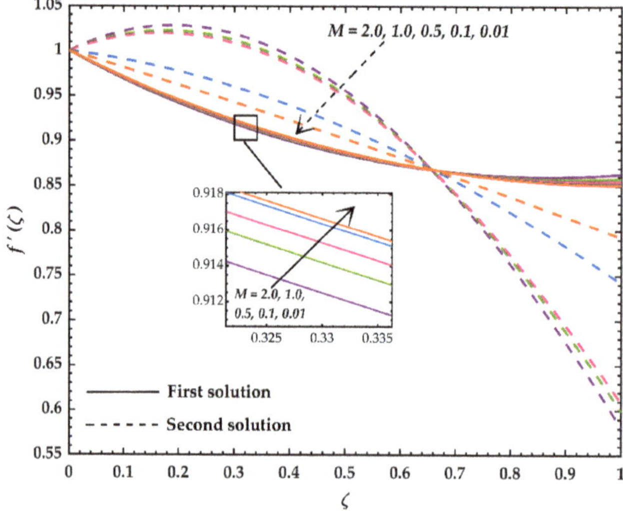

Figure 4. Velocity profiles, $f'(\zeta)$ when $\mathrm{Pr} = 30, We = 0.2, n = 0.6, S = -0.3$ and $\sigma = 1.2$ as M varies.

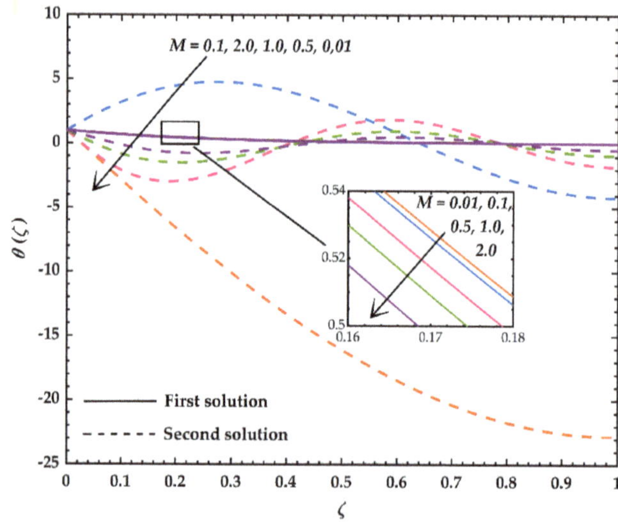

Figure 5. Temperature profiles, $\theta(\zeta)$ when $\mathrm{Pr} = 30, We = 0.2, n = 0.6, S = -0.3$ and $\sigma = 1.2$ as M varies.

Table 7 displays the values of J when the values S decreases from -0.3 to -1.0. The decrement in S explicates the increment in the injection intensity at the surface of the stretching sheet. The first solution shows that an increment in the injection strength reduces the film thickness at the free surface and affects the fluid velocity to increase steadily (see Figure 6). Rehman et al. [44] also has obtained the similar trend when S decreases. The improvement in the fluid velocity, as shown in Figure 6, increases the skin friction at the surface of the stretching sheet, which then enhances the value of $C_{fx}\mathrm{Re}_x^{1/2}$ when S decreases from -0.3 to -1.0 (see Table 8). Interestingly, the second solution in Table 8 found to increase the film thickness at the free surface, and no one has reported this observation before. Even though the second solution displays the increment in J, it does increases the value of $C_{fx}\mathrm{Re}_x^{1/2}$ when the impact of injection increases. From the aspect of the heat transfer, the first solution shows that the decrement in S increases the fluid temperature and is disclosed in Figure 7. The effect of injection found to increases the amount of heat flux emitted from the stretching sheet, and this substantially augments the rate of convective heat transfer. Therefore, the values of $\theta'(0)$ increase when the value of S decreases. The second solution postulates the increment of the fluid temperature at the free surface after confronted the high oscillation across the boundary layer. Hence, the value of $\theta'(0)$ decreases when S varies from -0.3 to -0.7, and then upsurges when S varies from -0.7 to -1.0. Stronger injection effect at the surface of the stretching sheet increases the strength of the reverse or overshoot flow, which is shown in Figures 7 and 8c [55]. The observed inflection points in Figures 7 and 8c signify the flow instability, and this may yield the irregular rate of heat transfer past the stretching sheet.

Table 7. Value of J as S varies at $\mathrm{Pr} = 30, We = 0.2, n = 0.6, M = 2$ and $\sigma = 1.2$.

	J		
S	**First Solution**	**Second Solution**	**Third Solution**
-0.3	0.2058215	-0.6771271	-1.9809900
-0.5	0.0603109	-0.5696448	-1.1092403
-0.7	0.0246653	-0.5309704	-1.2876096
-1.0	0.0072303	-0.2978425	-0.1952172

Table 8. Value of $C_{fx}\mathrm{Re}_x^{1/2}$ and $\theta'(0)$ as S varies at $\mathrm{Pr} = 30, We = 0.2, n = 0.6, M = 2$ and $\sigma = 1.2$.

	$C_{fx}\mathrm{Re}_x^{1/2}$			$\theta'(0)$		
S	FS [1]	SS [2]	TS [3]	FS [1]	SS [2]	TS [3]
-0.3	-0.665062	0.592954	3.288306	-4.064348	-12.656062	-101.360411
-0.5	0.301583	1.430791	2.632399	-2.271308	-27.057939	287.943093
-0.7	1.052951	2.453767	4.824057	-1.471507	-50.930799	3901.196761
-1.0	1.535432	3.604139	3.337833	-0.733154	-4.841747	-8.748436

[1] FS = First solution. [2] SS = Second solution. [3] TS = Third solution.

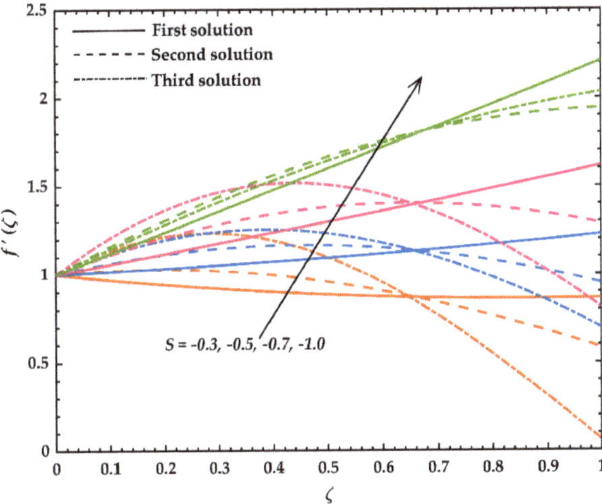

Figure 6. Velocity profiles, $f'(\zeta)$ when $\mathrm{Pr} = 30, We = 0.2, n = 0.6, M = 2$ and $\sigma = 1.2$ as S varies.

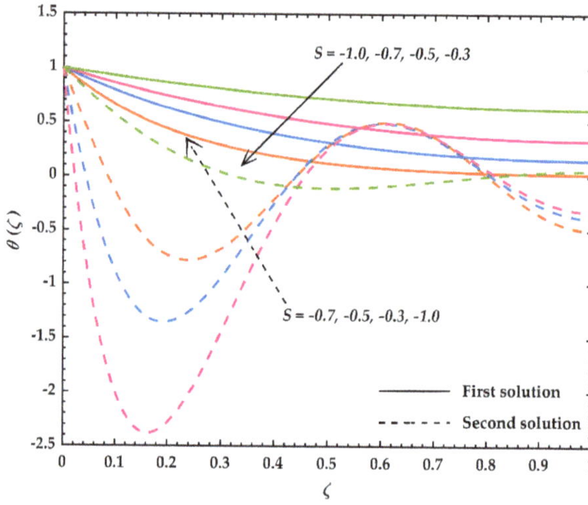

Figure 7. Temperature profiles, $\theta(\zeta)$ when $\mathrm{Pr} = 30, We = 0.2, n = 0.6, M = 2$ and $\sigma = 1.2$ as S varies.

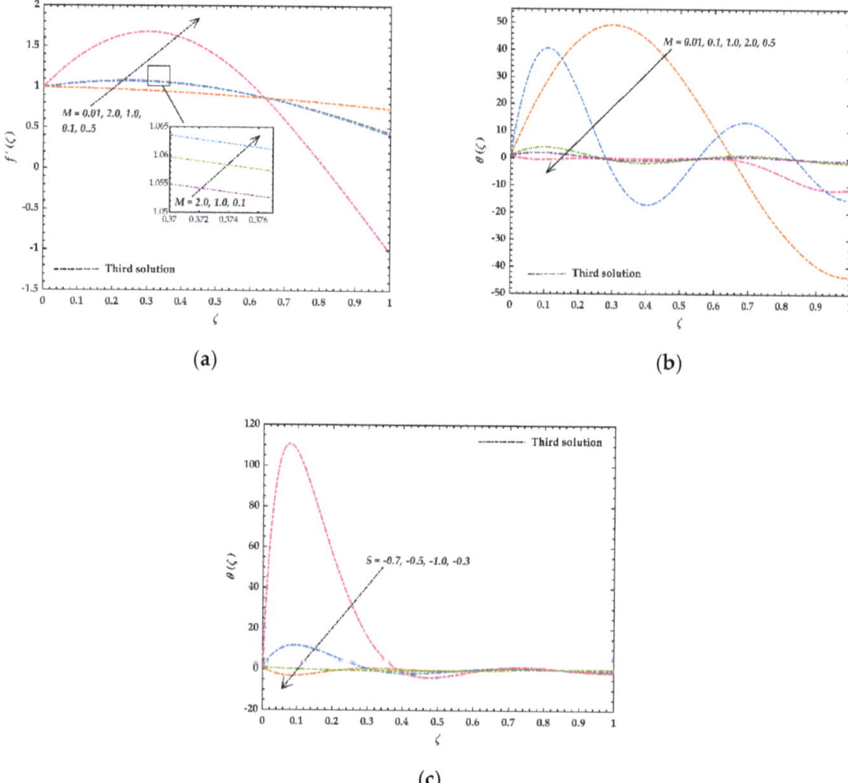

Figure 8. (**a**) Velocity profiles, $f'(\zeta)$ of the third solution when $\mathrm{Pr} = 30$, $We = 0.2$, $n = 0.6$, $S = -0.3$ and $\sigma = 1.2$ as M varies; (**b**) Temperature profiles, $\theta(\zeta)$ of the third solution when $\mathrm{Pr} = 30$, $We = 0.2$, $n = 0.6$, $S = -0.3$ and $\sigma = 1.2$ as M varies; (**c**) Temperature profiles, $\theta(\zeta)$ of the third solution when $\mathrm{Pr} = 30, We = 0.2, n = 0.6, M = 2$ and $\sigma = 1.2$ as S varies.

On the other hand, the remaining solution (third solution) yields the most negative values of J when $\sigma, M,$ and S varies, and the most negative values can be set to zero [56]. When J is set to zero, the momentum equation for the hydrodynamic boundary layer, as shown in Equation (16) is undefined and deleted from the problem matrix at this nodal point. Meanwhile, in the coating activities, the growth of the film is strictly dependent on the substrate flow over the stretching sheet (which is the Carreau fluid flow in the present work), and any changes occurring in the substrate vicinity will be reflected in the thin film thickness. Defects in the thin film flow such as a strong flow of divergence over the stretching sheet, or saddle points of attachment due to effect of surface tension, unsteadiness, or injection may yield the most negative film thickness. Therefore, when the most negative film thickness appeared, one can predict a rupture in the process and contact occurs between the film and stretching sheet surfaces [56]. Based on the results of the physical quantities and profiles (see Figures 2, 3, 6 and 8) of the third solution in the present work, it is clear that there is an abrupt increase in the velocity profiles and irregular rising and falling in the fluid temperature which results in the uneven trend (increase, decrease and increase) of $C_{fx}\mathrm{Re}_x^{1/2}$ and $\theta'(0)$, respectively. Even the reported values of J for the third solution showed a changing trend; for instance, the thickness increase, decrease, increase. These essentially suggest the defects (as been explained above) within the boundary layer region, which have the potential to interrupt the film growing process.

5. Conclusions

The present investigation was conducted to observe the effects of thermocapillarity and injection in the Carreau thin film flow and heat transfer past an unsteady stretching sheet. The appropriate similarity variables transformed the governing boundary layer model which was in the form of the partial differential equations into a system of ordinary differential equations to ease the computational process. The reduced form of the mathematical model was solved numerically by a collocation method, namely bvp4c function in the MATLAB software. Interestingly, this study reported triple solutions in thin-film flow problems for the first-time. However, only one solution (first solution) can be physically reliable since the other two solutions (second and third solutions) are not reliable with negative film thickness. An increase in the unsteadiness parameter and higher intensity of injection reduces the film thickness, which then increases the value of the skin friction coefficient and improves the rate of convective heat transfer past a permeable stretching sheet. Besides that, an increase in the thermocapillarity number increases the film thickness, which then depreciates the value of the skin friction coefficient and deteriorates the rate of convective heat transfer past a permeable stretching sheet. This remarkable contribution could be useful in improving the manufacturing and materials processing.

Author Contributions: Conceptualization, K.N., I.H., and R.N.; Methodology, K.N.; Software, K.N.; Validation, K.N., I.H., and R.N.; Formal analysis, K.N.; Investigation, K.N., I.H., and R.N.; Writing—original draft preparation, K.N.; Writing—review and editing, I.H. and R.N.; Funding acquisition, R.N. All authors have read and agreed to the published version of the manuscript.

Funding: This research was funded by Universiti Kebangsaan Malaysia, grant number GUP–2019–034 and the APC was funded by the research university grant (GUP–2019–034) from Universiti Kebangsaan Malaysia.

Acknowledgments: The authors are thankful to the honorable reviewers for their constructive suggestions to improve the quality of the paper.

Conflicts of Interest: The authors declare no conflict of interest. The funders had no role in the design of the study; in the collection, analyses, or interpretation of data; in the writing of the manuscript, or in the decision to publish the results.

Nomenclature

$\mathbf{A}_1$	Rivlin–Ericksen tensor (Pa)
b	Stretching rate $\left(s^{-1}\right)$
C_{fx}	Local skin friction coefficient $(-)$
C_p	Specific heat $\left(Jkg^{-1}\,K^{-1}\right)$
$h(t)$	Liquid thin film thickness (m)
$\mathbf{I}$	Identity tensor
J	Unknown constant $(-)$
k	Thermal conductivity $\left(Wm^{-1}K^{-1}\right)$
M	Thermocapillary number $(-)$
n	Power–law index $(-)$
Nu_x	Local Nusselt number $(-)$
p	Pressure (Pa)
Pr	Prandtl number $(-)$
q_w	Heat flux at the wall $\left(J\,s^{-1}m^{-2}\right)$
Re_x	Local Reynolds number $(-)$
S	Constant mass transfer parameter $(-)$
t	Time (s)
T	Temperature (K)
T_0	Temperature at the slit (K)
T_w	Surface temperature (K)
T_{ref}	Reference temperature (K)
u,v	$x-$ and $y-$ components of velocity $\left(m\,s^{-1}\right)$
$U_w(x,t)$	Stretching surface velocity $\left(m\,s^{-1}\right)$
V_w	Uniform surface mass flux $\left(m\,s^{-1}\right)$
We	Local Weissenberg number $(-)$
x,y	Cartesian coordinates (m)

Greek Symbols

α	Positive constant $\left(s^{-1}\right)$
β	Dimensionless film thickness $(-)$
$\dot{\gamma}$	Shear rate (s)
γ	Positive fluid property $\left(K^{-1}\right)$
ζ	Similarity variable $(-)$
η_0	Zero-shear-rate viscosity $\left(kg\,m^{-1}\,s^{-1}\right)$
η_∞	Infinite-shear-rate viscosity $\left(kg\,m^{-1}\,s^{-1}\right)$
κ	Thermal diffusivity $\left(m^2\,s^{-1}\right)$
σ	Unsteadiness parameter $(-)$
λ	Material time constant (s)
μ	Dynamic viscosity $\left(kg\,m^{-1}\,s^{-1}\right)$
ν	Kinematic viscosity $\left(m^2\,s^{-1}\right)$
ρ	Density $\left(kg\,m^{-3}\right)$
τ	Cauchy stress tensor (Pa)
τ_w	Wall shear stress $\left(kg\,m^{-1}\,s^{-2}\right)$
χ	Surface tension $\left(kg\,s^{-2}\right)$
χ_0	Surface tension at temperature T_0 $\left(kg\,s^{-2}\right)$
$\psi(x,y,t)$	Stream function

References

1. Prandtl, L. Über Flussigkeitsbewegungen bei sehr kleiner Reibung. *Verhandlg. III Intern. Math.* **1904**, 484–491. Available online: https://ci.nii.ac.jp/naid/20000989592/ (accessed on 12 June 2020).
2. Sakiadis, B.S. Boundary-layer behavior on continuous solid surfaces: I. Boundary-layer equations for two-dimensional and axisymmetric flow. *AICH J.* **1961**, *7*, 26–28. [CrossRef]
3. Sakiadis, B.S. Boundary-layer behavior on continuous solid surfaces: II. The boundary layer on a continuous flat surface. *AICH J.* **1961**, *7*, 221–225. [CrossRef]
4. Crane, L.J. Flow past a stretching plate. *Z. Angew. Math. Phys.* **1970**, *21*, 645–647. [CrossRef]
5. Carragher, P.; Crane, L.J. Heat transfer on a continuous stretching sheet. *Z. Angew. Math. Phys.* **1982**, *62*, 564–565. [CrossRef]
6. Krishna, M.V.; Chamkha, A.J. Hall and ion slip effects on MHD rotating boundary layer flow of nanofluid past an infinite vertical plate embedded in a porous medium. *Results Phys.* **2019**, *15*, 102652. [CrossRef]
7. Basha, H.T.; Sivaraj, R.; Reddy, A.S.; Chamkha, A.J. SWCNH/diamond-ethylene glycol nanofluid flow over a wedge, plate and stagnation point with induced magnetic field and nonlinear radiation–solar energy application. *Eur. Phys. J. Spec. Top.* **2019**, *228*, 2531–2551. [CrossRef]
8. Rasool, G.; Zhang, T.; Chamkha, A.J.; Shafiq, A.; Tlili, I.; Shahzadi, G. Entropy generation and consequences of binary chemical reaction on MHD Darcy–Forchheimer Williamson nanofluid flow over non-linearly stretching surface. *Entropy* **2020**, *22*, 18. [CrossRef]
9. Gorla, R.S.R.; Chamkha, A. Natural convective boundary layer flow over a nonisothermal vertical plate embedded in a porous medium saturated with a nanofluid. *Nanosc. Microsc. Therm.* **2011**, *15*, 81–94. [CrossRef]
10. Magyari, E.; Chamkha, A.J. Combined effect of heat generation or absorption and first-order chemical reaction on micropolar fluid flows over a uniformly stretched permeable surface: The full analytical solution. *Int. J. Therm. Sci.* **2010**, *49*, 1821–1828. [CrossRef]
11. Chamkha, A.J. Solar radiation assisted natural convection in uniform porous medium supported by a vertical flat plate. *J. Heat Transf.* **1997**, *119*, 89–96. [CrossRef]
12. Chamkha, A.J.; Al-Mudhaf, A. Unsteady heat and mass transfer from a rotating vertical cone with a magnetic field and heat generation or absorption effects. *Int. J. Therm. Sci.* **2005**, *44*, 267–276. [CrossRef]
13. Takhar, H.S.; Chamkha, A.J.; Nath, G. Combined heat and mass transfer along a vertical moving cylinder with a free stream. *Heat Mass Transf.* **2000**, *36*, 237–246. [CrossRef]
14. Chamkha, A.J.; Takhar, H.S.; Soundalgekar, V.M. Radiation effects on free convection flow past a semi-infinite vertical plate with mass transfer. *Chem. Eng. J.* **2001**, *84*, 335–342. [CrossRef]
15. Takhar, H.S.; Chamkha, A.J.; Nath, G. Unsteady three-dimensional MHD-boundary-layer flow due to the impulsive motion of a stretching surface. *Acta Mech.* **2001**, *146*, 59–71. [CrossRef]
16. Chamkha, A.J.; Al-Mudhaf, A.F.; Pop, I. Effect of heat generation or absorption on thermophoretic free convection boundary layer from a vertical flat plate embedded in a porous medium. *Int. Commun. Heat Mass* **2006**, *33*, 1096–1102. [CrossRef]
17. Ghalambaz, M.; Behseresht, A.; Behseresht, J.; Chamkha, A. Effects of nanoparticles diameter and concentration on natural convection of the Al_2O_3-water nanofluids considering variable thermal conductivity around a vertical cone in porous media. *Adv. Powder Technol.* **2015**, *26*, 224–235. [CrossRef]
18. Chamkha, A.J. Hydromagnetic natural convection from an isothermal inclined surface adjacent to a thermally stratified porous medium. *Int. J. Eng. Sci.* **1997**, *35*, 975–986. [CrossRef]
19. Reddy, M.G.; Rani, M.S.; Kumar, K.G.; Prasannakumar, B.C.; Chamkha, A.J. Cattaneo-Christov heat flux model on Blasius-Rayleigh-Stokes flow through a transitive magnetic field and Joule heating. *Phys. A* **2020**, *548*, 123991. [CrossRef]
20. Cotto, D.; Duffo, P.; Haudin, J.M. Cast film extrusion of polypropylene films. *Int. Polym. Proc.* **1989**, *4*, 103–113. [CrossRef]
21. Wang, C.Y. Liquid film on an unsteady stretching surface. *Q. Appl. Math.* **1990**, *48*, 601–610. [CrossRef]
22. Usha, R.; Sridharan, R. The axisymmetric motion of a liquid film on an unsteady stretching surface. *J. Fluids Eng.* **1995**, *117*, 81–85. [CrossRef]
23. Andersson, H.I.; Aarseth, J.B.; Braud, N.; Dandapat, B.S. Flow of a power-law fluid film on an unsteady stretching surface. *J. Non-Newton. Fluid Mech.* **1996**, *62*, 1–8. [CrossRef]

24. Andersson, H.I.; Aarseth, J.B.; Dandapat, B.S. Heat transfer in a liquid film on an unsteady stretching surface. *Int. J. Heat Mass Transf.* **2000**, *43*, 69–74. [CrossRef]

25. Chen, C.H. Heat transfer in a power-law fluid film over a unsteady stretching sheet. *Heat Mass Transf.* **2003**, *39*, 791–796. [CrossRef]

26. Wang, C. Analytic solutions for a liquid film on an unsteady stretching surface. *Heat Mass Transf.* **2006**, *42*, 759–766. [CrossRef]

27. Noor, N.F.M.; Abdulaziz, O.; Hashim, I. MHD flow and heat transfer in a thin liquid film on an unsteady stretching sheet by the homotopy method. *Int. J. Numer. Meth. Fluids* **2010**, *63*, 357–373. [CrossRef]

28. Aziz, R.C.; Hashim, I. Liquid film on unsteady stretching sheet with general surface temperature and viscous dissipation. *Chin. Phys. Lett.* **2010**, *27*, 110202. [CrossRef]

29. Aziz, R.C.; Hashim, I.; Abbasbandy, S. Flow and heat transfer in a nanofluid thin film over an unsteady stretching sheet. *Sains Malys.* **2018**, *47*, 1599–1605. [CrossRef]

30. Aziz, R.C.; Hashim, I.; Alomari, A.K. Thin film flow and heat transfer on an unsteady stretching sheet with internal heating. *Meccanica* **2011**, *46*, 349–357. [CrossRef]

31. Kumar, K.A.; Sandeep, N.; Sugunamma, V.; Animasaun, I.L. Effect of irregular heat source/sink on the radiative thin film flow of MHD hybrid ferrofluid. *J. Therm. Anal. Calorim.* **2020**, *139*, 2145–2153. [CrossRef]

32. Tassaddiq, A.; Amin, I.; Shutaywi, M.; Shah, Z.; Ali, F.; Islam, S.; Ullah, A. Thin film flow of couple stress magneto-hydrodynamics nanofluid with convective heat over an inclined exponentially rotating stretched surface. *Coatings* **2020**, *10*, 337. [CrossRef]

33. Tan, M.J.; Bankoff, S.G.; Davis, S.H. Steady thermocapillary flows of thin liquid layers. I. Theory. *Phys. Fluids A* **1990**, *2*, 313–321. [CrossRef]

34. Arafune, K.; Hirata, J. Thermal and solutal Marangoni convection in ln-Ga-Sb system. *J. Crystal Growth* **1999**, *197*, 811–817. [CrossRef]

35. Christopher, D.M.; Wang, B.X. Similarity simulation for Marangoni convection around a vapor bubble during nucleation and growth. *Int. J. Heat Mass Transf.* **2001**, *44*, 799–810. [CrossRef]

36. Dandapat, B.S.; Ray, P.C. The effect of thermocapillarity on the flow of thin liquid film on a rotating disc. *J. Phys. D Appl. Phys.* **1994**, *27*, 2041–2045. [CrossRef]

37. Dandapat, B.S.; Santra, B.; Andersson, H.I. Thermocapillarity in a liquid film on an unsteady stretching surface. *Int. J. Heat Mass Transf.* **2003**, *46*, 3009–3015. [CrossRef]

38. Chen, C.H. Marangoni effects on forced convection of power-law liquids in a thin film over a stretching surface. *Phys. Lett. A* **2007**, *370*, 51–57. [CrossRef]

39. Noor, N.F.M.; Hashim, I. Thermocapillarity and magnetic field effects in a thin liquid film on an unsteady stretching surface. *Int. J. Heat Mass Transf.* **2010**, *53*, 2044–2051. [CrossRef]

40. Maity, S.; Ghatani, Y.; Dandapat, B.S. Thermocapillary flow of a thin nanoliquid film over an unsteady stretching sheet. *J. Heat Trans-T ASME* **2016**, *138*, 042401. [CrossRef]

41. Sarojamma, G.; Vajravelu, K.; Sreelakshmi, K. A study on entropy generation on thin film flow over an unsteady stretching sheet under the influence of magnetic field, thermocapillarity, thermal radiation and internal heat generation/absorption. *Commun. Numer. Anal.* **2017**, *2*, 141–156. [CrossRef]

42. Maity, S. Thermocapillary flow of thin liquid film over a porous stretching sheet in presence of suction/injection. *Int. J. Heat Mass Transf.* **2014**, *70*, 819–826. [CrossRef]

43. Rehman, S.; Rehman, S.U.; Khan, A.; Khan, Z. The effect of flow distribution on heat and mass transfer of MHD thin liquid film flow over an unsteady stretching sheet in the presence of variational physical properties with mixed convection. *Phys. A* **2020**, *551*, 124120. [CrossRef]

44. Rehman, S.; Idrees, M.; Shah, R.A.; Khan, Z. Suction/injection effects on an unsteady MHD Casson thin film flow with slip and uniform thickness over a stretching sheet along variable flow properties. *Bound. Value Probl.* **2019**, *1*, 26. [CrossRef]

45. Myers, T.G. Application of non-Newtonian models to thin film flow. *Phys. Rev. E* **2005**, *72*, 066302. [CrossRef]

46. Tshehla, M.S. The flow of a Carreau fluid down an incline with a free surface. *Int. J. Phys. Sci.* **2011**, *6*, 3896–3910.

47. Ashwinkumar, G.P.; Sulochana, C. Numerical simulation of heat transfer characteristics in thin film flow of MHD dissipative Carreau nanofluid past a stretching sheet with $CoFe_2O_4$ nanoparticles. *Int. J. Res. Eng. Technol.* **2016**, *5*, 18–25.

48. Khan, N.S.; Gul, T.; Kumam, P.; Shah, Z.; Islam, S.; Khan, W.; Zuhra, S.; Sohail., A. Influence of inclined magnetic field on Carreau nanoliquid thin film flow and heat transfer with graphene nanoparticles. *Energies* **2019**, *12*, 1459. [CrossRef]

49. Iqbal, K.; Ahmed, J.; Khan, M.; Ahmad, L.; Alghamdi, M. Magnetohydrodynamic thin film deposition of Carreau nanofluid over an unsteady stretching surface. *Appl. Phys. A-Mater.* **2020**, *126*, 105. [CrossRef]

50. Boger, D.V. Demonstration of upper and lower Newtonian fluid behaviour in a pseudoplastic fluid. *Nature* **1977**, *265*, 126–128. [CrossRef]

51. Khan, M.; Azam, M. Unsteady boundary layer flow of Carreau fluid over a permeable stretching surface. *Results Phys.* **2016**, *6*, 1168–1174. [CrossRef]

52. Shampine, L.F.; Gladwell, I.; Thompson, S. *Solving ODEs with MATLAB*; Cambridge University Press: New York, NY, USA, 2003; p. 166.

53. Schlicthing, H.; Gersten, K. *Boundary-Layer Theory*, 9th ed.; Springer Nature: Berlin, Germany, 2017; p. 99.

54. Vajravelu, K.; Prasad, K.V.; Ng, C.O. Unsteady flow and heat transfer in a thin film of Ostwald-de Waele liquid over a stretching surface. *Commun. Nonlinear Sci. Numer. Simulat.* **2012**, *17*, 4163–4173. [CrossRef]

55. Ali, M.E. The effect of suction or injection on the laminar boundary layer development over a stretched surface. *J. King Saud Univ.* **1996**, *8*, 43–58. [CrossRef]

56. Dowson, D.; Priest, M.; Dalmaz, G.; Lubrecht, A.A. *Tribological Research and Design for Engineering Systems*; Elsevier: Amsterdam, The Netherlands, 2003; p. 581.

Article

Numerical Simulation of the Flow and Heat Transfer in an Electric Steel Tempering Furnace

Iván D. Palacio-Caro, Pedro N. Alvarado-Torres and Luis F. Cardona-Sepúlveda *

Grupo de Materiales Avanzados y energía (MATyER), Facultad de Ingeniería,
Instituto Tecnológico Metropolitano, Campus Fraternidad, Calle 54a No 30-1, Medellín 050013, Colombia;
ivan.palacio.caro@gmail.com (I.D.P.-C.); pedroalvarado@itm.edu.co (P.N.A.-T.)
* Correspondence: luiscardona@itm.edu.co; Tel.: +57-4-4600727

Received: 6 March 2020; Accepted: 9 June 2020; Published: 15 July 2020

Abstract: Heat treatments, such as steel tempering, are temperature-controlled processes. It allows ferrous steel to stabilize its structure after the heat treatment and quenching stages. The tempering temperature also determines the hardness of the steel, preferably to its optimum working strength. In a tempering furnace, a heat-resistant fan is commonly employed to generate moderate gas circulation to obtain adequate temperature homogeneity and heat transfer. Nevertheless, there is a tradeoff because the overall thermal efficiency is expected to reduce because of the high rotating speed of the fan. Therefore, this study numerically investigates the thermal efficiency changes of an electric tempering furnace due to changes in the rotating speed of the fan and the effects on temperature homogeneity and the heat transfer rate to the load. Heat losses through the walls were calculated from the external temperature measurement of the furnace. Four different speeds were simulated: 720, 990, 1350, and 1800 rpm. Thermal homogeneity was improved at higher rotating speeds; this is because the recirculation zone caused by the fan improved the flow mixing and the heat transfer. However, it was found that the thermal efficiency of the tempering furnace decreased as the rotating speed values increased. Therefore, these characteristics should be modulated to obtain a profit when controlling the rotating speed. For example, although thermal efficiency decreases by 20% when the rotating speed is doubled, the heat transfer rate to load is increased by up to 50%, which can be beneficial in decreasing the process of tempering times.

Keywords: tempering; heat treatment; electric furnace; CFD simulation; thermal efficiency

---

1. Introduction

Industrial furnaces have become vitally important equipment since they are involved in the production of many consumer products, such as food, beverages, containers, machining tools, infrastructure materials, amongst other applications [1]. In such furnaces, heat generation technology is mainly based on gas heating as well as electricity. Operation costs related to energy consumption are significant in most high-temperature furnaces; therefore, there is potential for energy savings [2,3].

Furnaces should uniformly perform their heat transfer process. Non-uniform temperature distribution inside the heating chamber may lead to low-quality products, thereby increasing production costs. In some applications, such as the heat treatment of metals, it is of utmost importance that the temperature is uniform inside the furnace [4,5].

For example, in a typical hardening process, steel parts are heated to a preset temperature between from 1000 K to 1470 K. Then, steel is quenched in special dielectric oil or by other means, such as molten salts or polymers, to rapidly remove heat. Quench-hardened steel is usually too brittle and must be tempered. The main goal of tempering is to decrease hardness and to increase the toughness of the ferrous material by relieving the tension that accumulates in the lattice of the metal [4,5]. This is usually achieved by heating the steel right after the heat treatment, following a controlled temperature rise to a preset level in the 670–970 K range [6]. If the temperature is not uniform during the heating processes, some parts of the heat-treated steel might have an undesirable structure and mechanical properties. This might be unacceptable for highly demanding applications—in the aeronautics industry for example.

Temperature uniformity inside heat treatment furnaces is commonly rated according to the DIN 17052-1 or the AMS2750 standards [7,8]. Both standards define "furnace classes" based on the maximum temperature deviation within the workspace of the furnace concerning the temperature setpoint. The most stringent class only allows for a ±3 K temperature difference in such a uniformity range. This range is determined by following a time-consuming and expensive experimental survey that only measures temperature by a few points inside the furnace.

Convective furnaces use fans to recirculate a hot fluid and to achieve a homogeneous temperature. Previous research has shown that such furnaces require a preheating stage where energy consumption is the highest and it is proportional to the preheating time [1]. The results of that study also showed that using baffles or other obstacles inside the furnace help reduce the preheating time.

Many studies concerning the optimization of furnaces have been oriented to maximize the quality of the product while minimizing the amount of energy required. In the food industry, for example, the quality of cooking can be controlled by keeping the uniformity of the temperature in the oven [9]. In this case, it was shown that the heating coil placement on a side position guaranteed the homogeneity of the temperature inside the furnace. Reference [10] reported that the uniformity of temperature in a laboratory drying oven was improved experimentally through changes in the ventilation system. By using CFD simulations, the influence of the position of the heaters and deflectors to redirect the air-flow more effectively was analyzed. The heated air was distributed directly on the device's storage space and the authors analyzed the effect of fan speed on the temperature distribution. Four different values were tested, i.e., 1500 rpm, 2000 rpm (existing fan), 2500 rpm, and 3000 rpm. The results led to the conclusion that the higher the rotating speed of the fan, the lower the obtained temperature difference. However, the achieved improvement was only about 0.1 K, which can be considered to be a very low value. These studies recommended the use of CFD to create geometric prototypes that guarantee the uniformity of temperature in convection ovens.

Simulations of a furnace chamber during heat treatment using natural gas combustion have been reported in the literature, where a boundary condition model for the estimation of heat flux through the walls was proposed and validated [11]. However, a finite-element method, rather than a finite-volume method, was used, and only the energy transport equation was solved. In another study [12], an electric nitriding furnace was simulated using a commercial CFD package and velocity uniformity was evaluated using a spatial criterion based on the mean and actual values of the velocity magnitude. There is also another simulation study regarding a nitriding furnace, where a conjugated heat transfer approach was used [13].

In this research, the temperature distribution, fluid dynamics, and heat transfer behavior in an electric tempering furnace are studied. Thus far we know that there is limited research concerning these characteristics in this specific kind of furnace. Numerical simulations have conducted using a relevant design factor as the rotating speeds of the fan. In addition, thermal efficiency has been calculated when the rotating speed is changed. Taking into account the fact that internal velocities and temperatures fields inside the furnace are difficult to directly measure, the numerical approach allows for the identification of the main phenomena governing the thermal and fluid behavior and could contribute to the furnace design for process improvements.

2. Numerical Simulations

2.1. System Description

The study investigates a top loading furnace commonly referred to as a pit-type furnace (Figure 1). Pit furnaces have been widely used for steel tempering (394 K to 1033 K) due to the easy loading and unloading of parts, which is done by lifting the lid away and hooking the parts out of the furnace.

The electric furnace is composed of two chambers: the heating chamber, where 22 metal coils are electrically heated, and the process chamber, where the steel parts are stored and tempered. The heating chamber is internally covered with insulating refractory bricks, classified as group 23 according to the ASTM C155 standard [14]. The process chamber is a cylinder with a 0.3 m radius and 1.2 m height. This space is covered with refractory concrete and AISI 304 stainless steel plates with a thickness of about 7.94 mm (5/16 inches) for mechanical and thermal protection. The gross load capacity is rated at 400 kg. Such a load can be put into metallic baskets and inserted from the top of the process chamber.

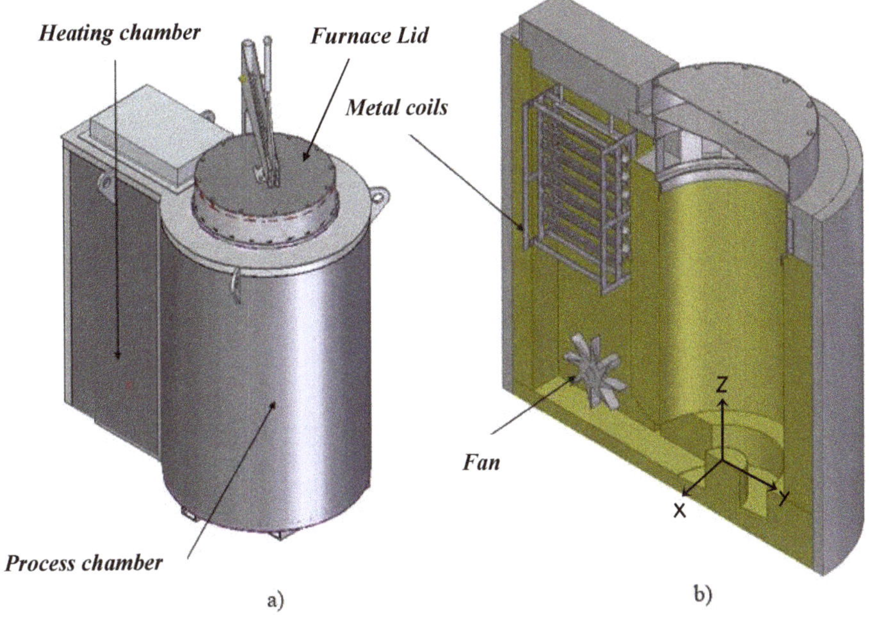

Figure 1. Tempering furnace studied in this work. (**a**) Isometric external view of the furnace and (**b**) cut view of the furnace with the coordinate system.

The furnace is powered by a 220 V three-phase system at a 60 Hz frequency, with a total nominal power of up to 50 kW and a maximum operating temperature of 1120 K, while the temperature in the heating coil can exceed 1370 K. The furnace has a fan at the bottom of the heating chamber that operates at a fixed rotating speed of 990 rpm, powered by a 2.24 kW (3 hp) electric motor. The fan was manufactured using AISI 310 stainless steel. The fan moves hot gases from the hot coils downward inside the heating chamber and then forces them upwards into the workspace and through the load. Then, gases are recirculated through the coils bank. The recirculated gas inside the equipment is a fixed mass of air, with no inlets or outlets. This furnace has two thermocouples to control the temperature: one located at the entrance of the gases to the heating chamber, near the top of the furnace (controlling the process temperature). The second thermocouple is located at the heating chamber itself, and its main function is to avoid coil overheating. Both inputs work together in controlling the heating power delivered to the metallic load.

Figure 2 presents the temporal evolution of the temperature setpoint inside the furnace during the tempering process studied in this work. This heat treatment follows a three-stage process: preheating, cooling, and sustaining. In the preheating stage, the furnace lid is opened, and a loaded charging basket is introduced using an overhead crane. Afterward, the furnace is closed, and a temperature of 973 K is fixed as the setpoint for the heating chamber. One of the aims of this stage is to remove any residual humidity on the steel parts. After two hours of preheating, the cooling stage starts. The setpoint is fixed at 773 K, which is a temperature where the tempering stress relief starts [5]. After one hour, the setpoint is gradually increased to 20 K for half an hour. This process is repeated two times to obtain adequate steel hardness.

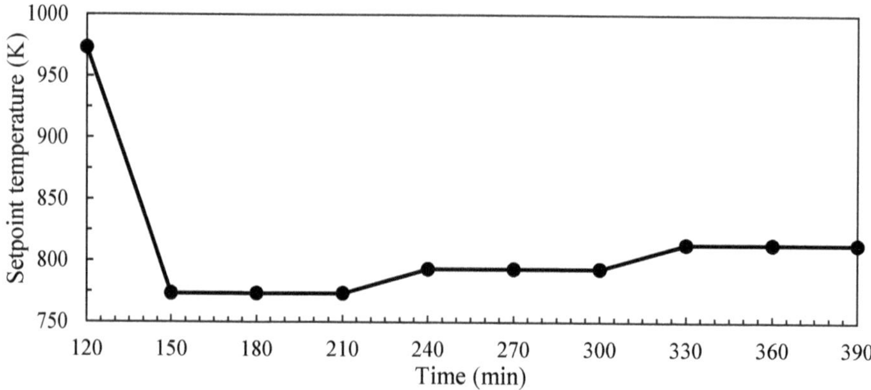

Figure 2. Setpoint temperature at the cooling (120–150 min) and sustain stage (150–390 min) of the tempering process.

2.2. Numerical Models

Steady-state operation of the furnace at the end of the tempering process is considered. It corresponds to a turbulent flow occurring in a low Mach number regime. The transport equations governing fluid flow for this kind of problem are the Favre-averaged equations transport equations [15]. First, the continuity equation is

$$\frac{\partial \overline{\rho}}{\partial t} + \nabla \cdot (\overline{\rho} \widetilde{v}) = 0,$$ (1)

where $\overline{\rho}$ and $\widetilde{v}$ are the average density and velocity. The momentum equation is also solved:

$$\frac{\partial}{\partial t}(\overline{\rho} \widetilde{v}) + \nabla \cdot (\overline{\rho} \widetilde{v} \widetilde{v}) = \nabla \cdot \left\{ \mu_l \left[(\nabla \widetilde{v} + \nabla \widetilde{v}^T) - \frac{2}{3} \nabla \cdot \widetilde{v} I \right] \right\} - \nabla \cdot (\overline{\rho v' v'}) - \nabla \overline{p} + \overline{\rho} g,$$ (2)

where $\mu_l, \overline{p}$ and g are laminar viscosity, averaged pressure, and gravity, respectively. In this equation, I is the identity matrix. The Reynolds stresses $-\nabla \cdot (\overline{\rho v' v'})$ is a term that must be modeled. In this work, the Boussinesq hypothesis was followed:

$$-\overline{\rho v' v'} = \mu_t (\nabla \widetilde{v} + \nabla \widetilde{v}^T) - \frac{2}{3} (\overline{\rho} \widetilde{k} + \mu_t \nabla \cdot \widetilde{v} I),$$ (3)

where $\widetilde{k}$ is the averaged turbulent kinetic energy:

$$\widetilde{k} = \frac{1}{2} \left[\widetilde{(u)^2} + \widetilde{(v)^2} + \widetilde{(w)^2} \right] = \frac{1}{2} \sum_{i=1}^{3} \widetilde{(v'_i)^2}.$$ (4)

The turbulent viscosity μ_t is obtained from the following equation:

$$\mu_t = \overline{\rho} C_\mu \frac{\widetilde{k}^2}{\widetilde{\varepsilon}},$$ (5)

where $\widetilde{\varepsilon}$ is the averaged dissipation of turbulent kinetic energy and C_μ is a modeling constant equal to 0.09, taken from the standard k-epsilon turbulence model. This turbulence model has been used successfully in the past for simulations of thermal treatment furnaces [16]. In this model, two additional transport equations for $\tilde{k}$ and $\tilde{\varepsilon}$ must be solved:

$$\frac{\partial}{\partial t}\left(\bar{\rho}\tilde{k}\right) + \nabla\cdot\left(\overline{\rho\tilde{v}}\tilde{k}\right) = \nabla\cdot\left[\left(\mu_l + \frac{\mu_t}{\sigma_k}\right)\nabla\tilde{k}\right] - \overline{\rho v'v'} : \nabla\tilde{v} - \bar{\rho}\tilde{\varepsilon},\tag{6}$$

$$\frac{\partial}{\partial t}\left(\bar{\rho}\tilde{\varepsilon}\right) + \nabla\cdot\left(\overline{\rho\tilde{v}}\,\tilde{\varepsilon}\right) = \nabla\cdot\left[\left(\mu_l + \frac{\mu_t}{\sigma_\varepsilon}\right)\nabla\tilde{\varepsilon}\right] - C_{\varepsilon1}\bar{\rho}\frac{\tilde{\varepsilon}}{k}\overline{v'v'} : \nabla\tilde{v} - C_{\varepsilon2}\bar{\rho}\frac{\tilde{\varepsilon}^2}{k},\tag{7}$$

where the model constants $C_{\varepsilon1}, C_{\varepsilon2}, \sigma_k, \sigma_\varepsilon$ are 1.44, 1.92, 1, and 1.3, respectively. Additionally, standard wall functions for modeling the velocity profile near the walls were employed in this work [17]. The energy transport equation is also solved in this work:

$$\frac{\partial}{\partial t}\left(\bar{\rho}\overline{E}\right) + \nabla\cdot\left(\tilde{v}\left(\bar{\rho}\overline{E} + \bar{p}\right)\right) = \nabla\cdot\left(k_t\nabla\overline{T}\right) + S_{rad},\tag{8}$$

where k_t is the thermal conductivity of the air, $\overline{T}$ is the average temperature and S_{rad} is the source term due to radiation. The total non-chemical energy E is defined as

$$E = h - \frac{p}{\rho} + \frac{1}{2}v^2,\tag{9}$$

where h in this case stands for enthalpy. In this work, radiation was modeled using a discrete ordinate (DO) model. The DO model solves the radiative heat transfer equation (RTE) for a finite number of discrete angles, each one associated with a vector direction $\vec{s}$ in a polar system (θ, φ, z):

$$\frac{\partial}{\partial s}I\left(\vec{s}\right) + (a + \sigma_s)I\left(\vec{s}\right) = an^2\sigma\frac{T^4}{\pi} + \frac{\sigma_s}{4\pi}\int_0^{4\pi} I\left(\vec{s}\right)\phi\left(\vec{s}, \vec{s}'\right)d\Omega,\tag{10}$$

where $I\left(\vec{s}\right)$ is the spectral radiation intensity, a is the absorption coefficient, σ_s is the dispersion coefficient, Ω is the solid angle, ϕ is the scattering phase function and σ is the Stefan–Boltzmann constant (5.67×10^{-8} Wm^{-2} K^{-4}). This radiation model requires θ and φ angles to be discretized in a finite number of subdivisions; the more subdivisions, the better the results, but the computational time increases substantially. In this work, 2×2 subdivisions were used. Additionally, because isotropic scattering was assumed in this work, the scattering phase function $\phi\left(\vec{s}, \vec{s}'\right)$ in the RTE is unity. Finally, the source term in the energy equation is [18]:

$$S_{rad} = aG - 4an^2\sigma T^4,\tag{11}$$

where n is the refractive index and G is the incident radiation, which is evaluated as

$$G = \int_{4\pi} Id\Omega.\tag{12}$$

The baseline operation simulation considers the rotating speed of the fan to be 990 rpm. Simulations at 720, 1350, and 1800 rpm were also performed. The fan is configured as a rotating wall using a moving-reference frame (MRF) method with a frozen rotor approach—as presented in Figure 3. MRF is an acceptable modeling approach that has been used successfully in axial [19], centrifugal [20], and other kinds of fans [21]. Additionally, a previous study using a nitrocarburation furnace with a fan [22] was modeled using the MRF method.

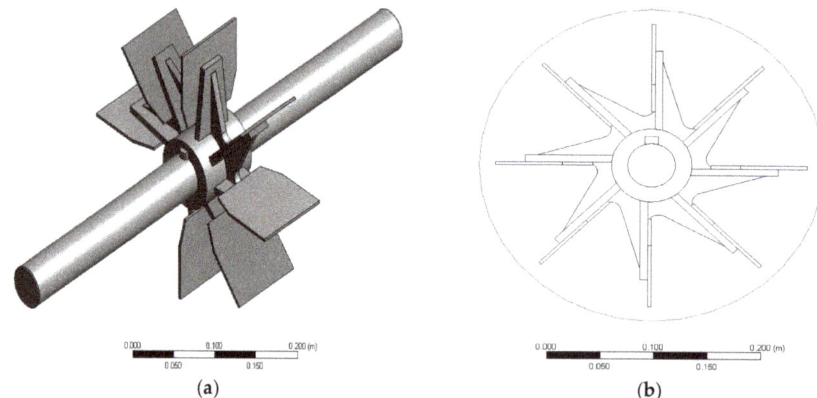

Figure 3. Fan modeled in this work. (**a**) Isometric view; (**b**) cylindrical boundary and domain configured with rotating speed using a moving-reference frame (MRF).

2.3. Boundary Conditions

To properly configure boundary conditions in the simulations, the heat flux lost through each wall was estimated with the experimentally measured external wall temperature. Temperatures were measured using a type K thermocouple of 1 mm point diameter with a UT320 UNI-T digital indicator. Measurements were made at 24 points on the outside of the furnace, as shown in Figure 4. This figure specifies the location of each temperature measuring point using a dot. The temperature was measured three times at each point and then they were averaged for the whole surface. The maximum standard deviation of the measurement for any point was 12.4 K, while the average standard deviation was 4.5 K.

Boundary conditions are presented in Table 1. The domain simulated in this work corresponds to a closed system with no inlets or outlets. Due to its symmetry, only half of the fluid domain was simulated. The heat flux presented in Table 1 was calculated from temperature measurements and was then configured as the heat transfer boundary condition for each outer wall of the furnace.

Table 1. Wall boundary conditions for baseline operation.

Boundary Name	Energy	Momentum
Heating_elements	$T = 879.5$ K, $\varepsilon = 0.85$	
Side_heating_chamber	$\dot{q} = -418$ W/m^2, $\varepsilon = 0.65$	
Back_heating_chamber	$\dot{q} = -228$ W/m^2, $\varepsilon = 0.65$	Stationary wall, No-slip condition,
Workspace_chamber	$\dot{q} = -269$ W/m^2, $\varepsilon = 0.65$	(zero gradient)
Top_heating_chamber	$\dot{q} = -500$ W/m^2, $\varepsilon = 0.65$	
All other walls	$\dot{q} = 0$ W/m^2, $\varepsilon = 0.65$	
Fan_walls	$\dot{q} = 0$ W/m^2, $\varepsilon = 0.85$	Moving wall, No-slip condition, Rotating speed = 990 rpm

In this work, the heat flux $\dot{q}$ was assumed to be due to a combination of convection and radiation:

$$\dot{q} = \dot{q}_{conv} + \dot{q}_{rad}. \tag{13}$$

The tempering furnace studied in this work is located inside a room with no significant airflow coming in or out. Therefore, a 5 Wm^{-2} K^{-1} natural convective coefficient h_{conv} was assumed [23], and the convective heat flux $\dot{q}_{conv}$ was calculated with Newton's law of cooling:

$$\dot{q}_{conv} = h_{conv}(T_s - T_{surr}), \tag{14}$$

where the surrounding temperature T_{surr} is 300 K on average and T_s is the external wall surface temperature. Ts were calculated as the temperature average of every section of the oven (Figure 4): side heating chamber (4 data), back heating chamber (2 data), workspace chamber (6 data) and, top heating chamber (12 data). To obtain a measurement of the radiation losses from the walls, the heat flux was calculated with the following expression [24]:

$$\dot{q}_{rad} = \varepsilon\sigma\left(T_s^4 - T_{surr}^4\right), \tag{15}$$

where ε is the surface emissivity. All surfaces were considered opaque with a diffusive fraction equal to 1. Refractory emissivity was assumed to be 0.65, and steel emissivity was assumed to be 0.85. Air was modeled as an ideal gas with variable properties depending on temperatures, such as density or viscosity. Pressure-momentum coupling was achieved using a coupled algorithm for faster convergence. Convective terms in all transport equations had second-order accuracy and convergence criteria were established at 1×10^{-6} for all residuals. Simulations were carried out in ANSYS FLUENT® (Canonsburg, PA, USA); v19 in a computer equipped with an Intel Core i7 8700k® processor with 16 GB of RAM.

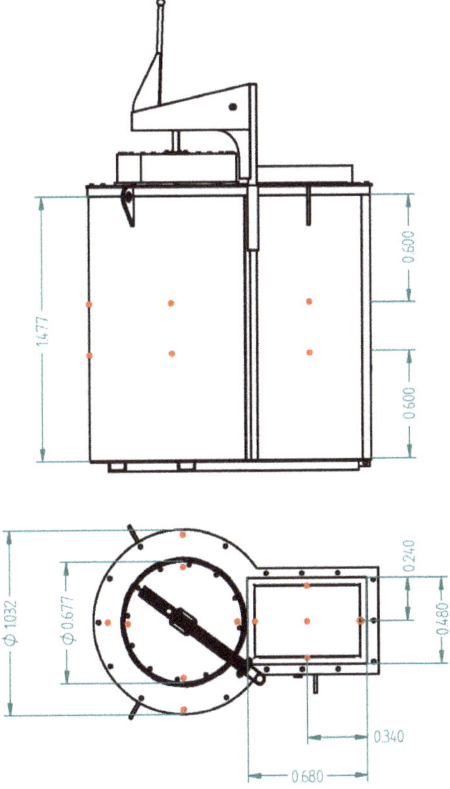

Figure 4. Points selected for temperature measurement on the exterior of the furnace walls (measurements in m).

2.4. Mesh Independence Study

The geometry was constructed in a the-dimensional computational domain, which comprised components such as a heating chamber, fan, metal coils, and process chamber. The geometry was taken from a real pit type furnace. The metal coils were constructed as cylindrical metal bars. Additionally, the present study constructed non-structured full tetrahedral meshes using ANSYS Meshing® (Canonsburg, PA, USA), as shown in Figure 5. After a grid-independent test, the coarse mesh with 1.5 million cells was rejected, due to the simulations not capturing the thermal behavior of the furnaces well, mainly near the wall zones.

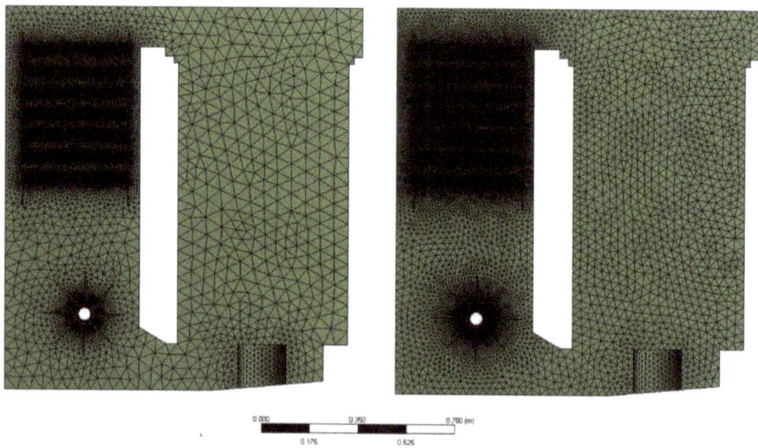

Figure 5. Grid systems (**a**) coarse mesh, 1.5 million cells, and (**b**) fine mesh, 3 million cells.

Two other meshes were considered in this work: one fine-mesh approximately double the size of the coarse mesh (3 million cells) and a very fine mesh composed of about 6 million cells. Figure 6 shows the results of the temperature as a function of the radial distance of the workspace or process chamber of the furnace (at Z = 0.5 m). This horizontal line is located at the middle of the vertical symmetry plane passing through the center of the process chamber. In this case, the baseline operation for the three meshes shows that there is less than a 2% difference in the results between the fine and the very fine meshes. Therefore, it can be argued that the fine mesh is accurate enough for this study and the 3 million-cell mesh is chosen as the mesh for all the results that are reported in the next section. This mesh represents a good compromise between detail and computational time. Simulations are initialized at 700 K and convergence is reached at close to 10 thousand iterations. The computational time for each run was close to 100 h.

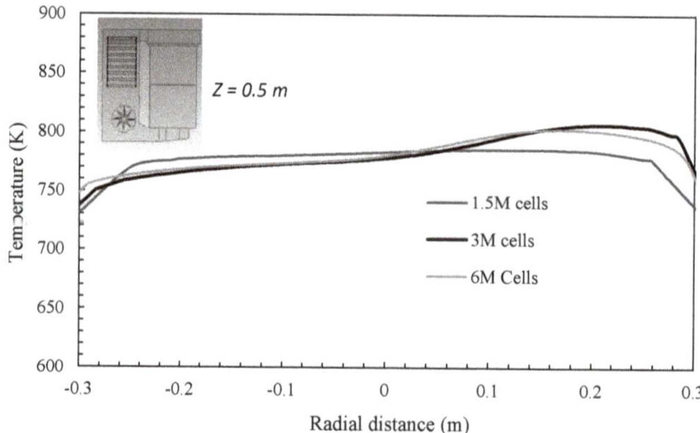

Figure 6. Temperature profile at a horizontal line located in the middle of the workspace for the three different meshes studied at 990 rpm.

2.5. Convective Coefficient to the Load and Energy Balance

Simulations were performed with and without the metallic charging basket needed for hanging or supporting the steel parts to be treated inside the process chamber of the furnace. An empty and fully loaded basket are presented in Figure 7. Such a basket could affect the volumetric flow supplied by the fan, and therefore, its influence on air-flow was analyzed as well. The basket is made of AISI 304 stainless steel, with thermal conductivity of 16.6 W/(m K), a density of 7900 kg/m^3, and specific heat of 477 J/(kg K) [23]. Calculated from density and volume, the mass of this basket is 54.7 kg.

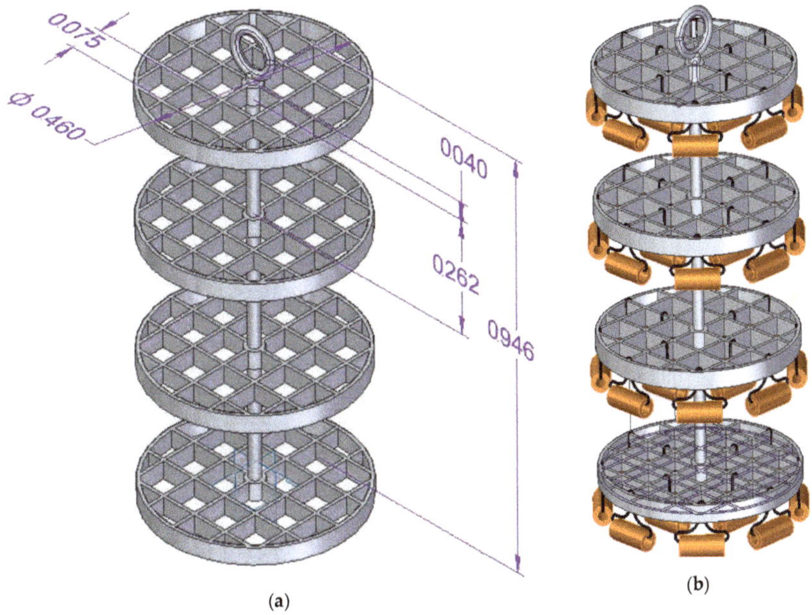

Figure 7. Stainless steel charging basket. (**a**) Empty basket; (**b**) loaded with parts to be tempered.

In this work, the heating of a cylindrical steel pin, which is a typical part of this kind of furnace and process, was considered. Each pin is 300 mm long with a 50.8 mm external and 25.4 mm internal diameter, made of carbon steel, with thermal conductivity of 56.7 W/(m K), a density of 7854 kg/m^3, and specific heat of 487 J/(kg K) [23]. Overall, the heating of 180 kg of these pins was considered in this work.

The results predicted by simulations allowed for the calculation of heat transfer to the load by convection using three well-known empirical correlations for the average non-dimensional Nusselt number Nu. The maximum value of axial speed predicted by the CFD software, as well as the process temperature, were the main inputs for these correlations. Then, the average of the correlations is reported in the results. The first correlation considered in this work is the one developed by Hilpert [23,25]:

$$Nu = C\,Re^m Pr^{1/3}\,, \tag{16}$$

where parameters C and m are obtained from Reference [23] and they depend on the Reynolds number Re, with the pin diameter as the characteristic length. Prandtl number Pr is obtained with the surface-fluid average temperature, using air as a fluid.

The second correlation used was developed by Žukauskas [26]:

$$Nu = C\,Re^m Pr^n\!\left(\frac{Pr}{Pr_s}\right)^{1/4}\,, \tag{17}$$

where parameters C and m are obtained from Reference [23] and the Pr_s is the Prandtl number based on the surface temperature, considered in this work to be 773.15 K.

In addition to these correlations, the expression of Churchill and Bernstein [27], where all properties are evaluated at the temperature of the film, was also considered:

$$Nu = 0.3 + \frac{0.62 Re^{1/2} Pr^{1/3}}{[1 + (0.4/Pr)^{2/3}]^{1/4}} \left[1 + \left(\frac{Re}{282,000}\right)^{5/8}\right]^{4/5}. \tag{18}$$

The energy balance for a closed furnace during the heating stage has recently been studied in Reference [28]. In this case, energy input is not due to the combustion of natural gas, but rather electricity, and thermal efficiency η can therefore be calculated as

$$\eta = \frac{Heat\ transfer\ rate\ to\ the\ load}{Electric\ power\ input}, \tag{19}$$

where the heat transfer to the load corresponds to the sensible heat demanded by the basket and the cylindrical steel pins during a processing time of 2 h, and the electric power input is calculated with the energy required by the fan and the metal coils. In theory, the fan studied in this work should follow the similarity law [24]:

$$\left(\frac{\dot{Q}}{\omega D^3}\right)_1 = \left(\frac{\dot{Q}}{\omega D^3}\right)_2 = \Phi, \tag{20}$$

where $\dot{Q}$ is the volumetric flow rate, ω is the angular velocity, and D is the fan's outer diameter. This equation states that, for any two different operational conditions 1 and 2, the dimensionless flow coefficient Φ is assumed to be equal. If the fans are also assumed to have the same dimensionless power coefficient C_P, then the following similarity law also should apply:

$$\left(\frac{\dot{W}_{shaft}}{\rho \omega^3 D^5}\right)_1 = \left(\frac{\dot{W}_{shaft}}{\rho \omega^3 D^5}\right)_2 = C_P, \tag{21}$$

where $\dot{W}_{shaft}$ is the shaft power. These laws are derived from dimensional analysis following Buckingham's theorem π. An analysis of the validity of these similarity laws of the fan modeled in this work is presented in the results.

3. Results and Discussion

The numerical results include a baseline operation and the effect of the fan rotating speed in the tempering furnace. We focus on the temperature and fluid dynamic fields inside the furnace, as well as the heat transfer behavior and efficiencies when the speed is changed.

3.1. Baseline Operation

The tempering system is characterized by its capacity to achieve a homogeneous temperature field throughout the process chamber, with the help of an internal fan. Figure 8 shows the temperature distribution at the vertical symmetry plane passing through the center of the heating and the process chambers of the tempering furnace. This corresponds to a baseline operation that has a fan rotating speed of 990 rpm. From the figure, it can be seen that the global thermal behavior of both chambers is well captured by the numerical calculations: the highest temperatures are obtained at the metal coils, the heat transfer is forced by the fan from the higher temperature zone to the lower temperature zone, and the thermal homogeneity occurs in the process chamber. Additionally, the experimental temperature data inside the furnace is in agreement with the numerical data. For example, the temperature of the process chamber was 813.1 K (last stage of Figure 2) and the simulated temperatures correspond to 806.4 K.

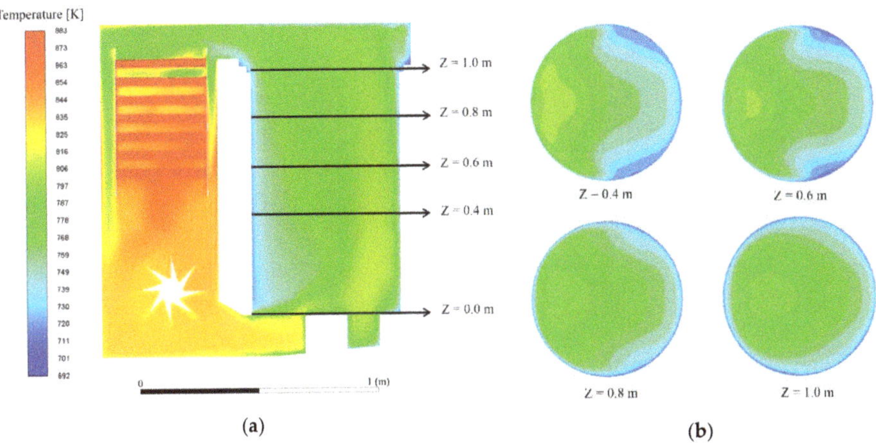

Figure 8. Temperature distribution of the baseline operation without the charging basket. (**a**) Cut view of the symmetry plane; (**b**) cut views at different heights.

The temperatures obtained inside the workspace chamber are between 750 K and 800 K. Therefore, the temperature uniformity range of the furnace is about 50 K, according to the DIN 17052-1 standard. Nevertheless, the four cut views presented at different heights in Figure 8 show that, as fluid flows from the bottom to the top of the chamber, the outer temperature next to the walls becomes closer to the center level and can be regarded as having a better temperature distribution, albeit equal in uniformity according to the norm. The results also show that the rotor is exposed to about 800 K during operation. The shaft of this rotor must be actively cooled to avoid heat damage to the engine; however, this heat loss was considered to be negligible, compared to the heat losses through the walls, and therefore was not taken into account (the adiabatic wall was configured as presented in Table 1).

Figure 9 presents streamlines for the baseline operation without a charging basket. Flow velocity near the rotor reaches up to 16.8 m/s. In contrast, velocity inside the process chamber is considerably lower, less than 2 m/s in most of the chamber. Despite the furnace being of convective type, this result indicates that the actual velocity, in the zone where the steel parts are exposed to hot air, is relatively low. As was presented in the methodology, convection heat transfer is proportional to the convective coefficient. This coefficient increases with the non-dimensional Reynolds number, which in turn is proportional to the fluid velocity. A low velocity within the workspace means a low heat transfer rate to the load and higher processing times.

Streamlines in Figure 9 also show an important recirculation zone inside the process chamber of the furnace. Recirculation helps increase the residence time of the hot gas in the workspace. This inner swirl has a similar function as the inner baffles reported by other authors [1] because it helps to augment the time that the hot gases remain in the process chamber of the furnace. In addition, some eddies are shown at the upper corners of the heating and process chambers, which could hinder the fluid transport inside the furnace. Unlike the main recirculation zone, these eddies occur in regions where they are not relevant to the process.

Besides, from Figure 9 it can be seen that a significant amount of fluid coming from the process chamber is directed to the left side wall of the metal coils, instead of the metal coils. This accords with the original design of the furnace, which does not allow for optimally guiding the fluid to the coils. A geometrical modification at the upper heating chamber zone could improve the heat transfer from the metal coils to the recirculating air.

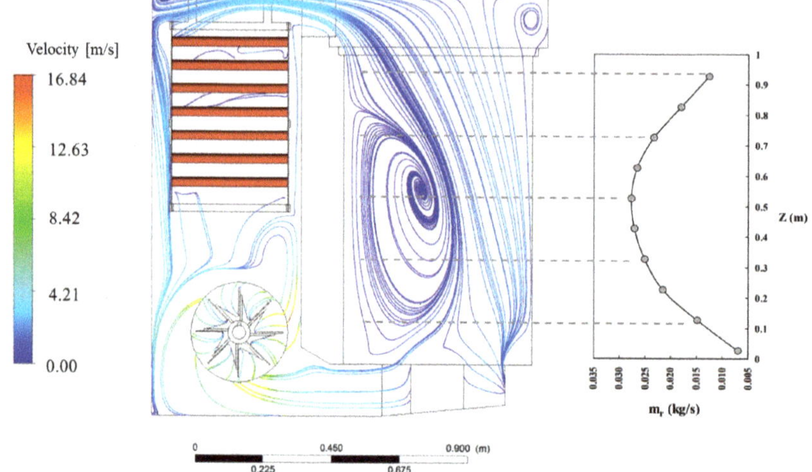

Figure 9. Streamlines inside the furnace for baseline operation and air mass recirculation behavior. Velocity legend applies to the recirculating gases.

The occurrence of a wide recirculation zone is evident in Figure 9. The figure also shows the mass recirculation at several heights of the process chamber (Z-direction). This was calculated taking into account the principle of mass conservation. That is, the mass flux in the Z-direction of the process chamber remains constant. This fact allows the use of negative air velocities to calculate the air mass recirculation in different transversal planes and establish a profile of recycled gases that increases progressively until reaching a maximum near the center of the vortex and then diminishes (right side of Figure 9). This result is similar to the one found by a previous study [29] using a non-reactive coflow jet. This type of calculation has been used with reactive gases [30,31], providing valuable fluid dynamics information of combustion chambers. In addition, it can be observed that even at high Z-direction of the process chamber, there is an air mass recirculation. This can be explained due to the airflow hits the top wall of the camera, and then come back.

Prediction of the behavior of the fluid near the walls is relevant, as shown above. Figure 10 displays the distribution of the Y+ on the inner walls of the furnace. Y+ is a non-dimensional measure of the first element size on the wall of the computational mesh. The maximum value of Y+ is about 65 on the walls next to the fan, which is lower than 100 and can be considered an acceptable value for the standard wall function of the k-epsilon turbulence model [17]. Therefore, this result confirms that the mesh resolution is adequate near the walls.

Figure 11 shows the temperature distribution in the symmetry boundary of the furnace for the baseline operation (990 rpm) with the metallic charging basket inside. In this case, the temperature uniformity range in the process chamber (the right side of the figure) is higher, at about 70 K; meanwhile, in the case without the charging basket, it is about 50 K (Figure 8). This homogeneity is quantified by applying the concept of temperature uniformity, which represents the difference in temperature between different points in the system. In this case, an acceptable calculation of temperature uniformity is obtained as the maximum value of the temperature difference between the central point of the furnace and the corner points [10]. As can be seen in the Figure, the gases near the left lateral wall of the process chamber are cooler than the rest of the gases in the chamber. Besides, the upper part of the basket is cooler than the lower part. In a real process, treatments with noticeable differences in temperature within the chamber could impact the optimum steel working strength. Therefore, strategies to improve the thermal homogeneity of the furnace (e.g., fan rotating speed variation, as proposed in this study) should be developed.

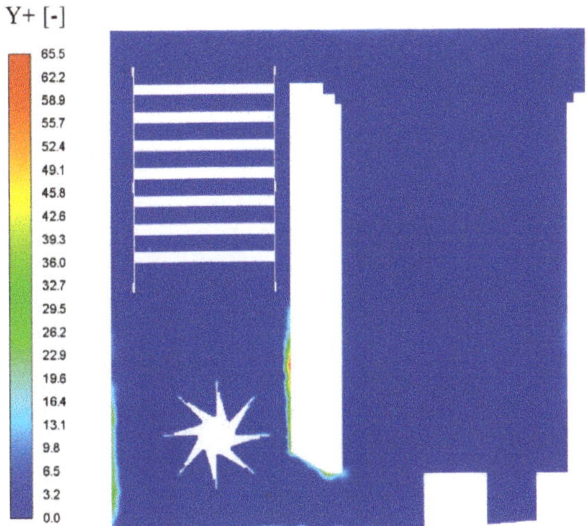

Figure 10. Distribution of Y+ on the walls for the baseline operation without the charging basket. The Y+ legend applies to the gases.

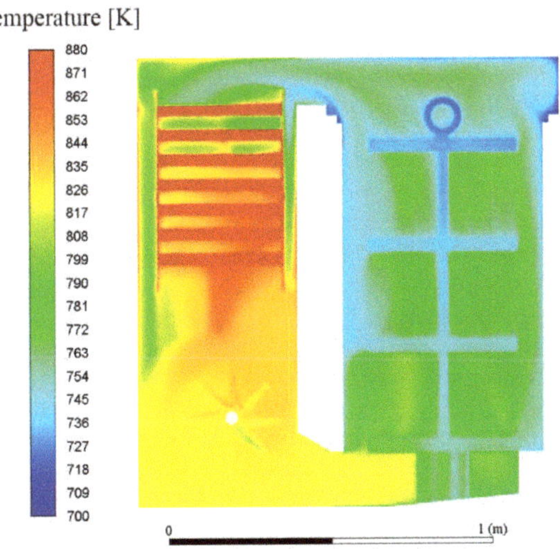

Figure 11. Temperature distribution of the baseline operation with a charging basket.

Figure 12 presents streamlines for the baseline operation with a charging basket. The maximum velocity is about 17 m/s in the rotor zone, while velocities inside and around the basket are closer to 2 m/s in most of the chamber. It should be noted that the inner recirculation zone identified in the case without a basket is still present. Additionally, a significant amount of fluid coming from the process chamber, directed to the left side wall of the metal coils, is obtained. Therefore, velocity values, as well as distributions, are very similar in cases with and without a metallic charging basket. This is due to the basket geometry configuration allowing for the flow of gases to pass through it and it hardly offering axial resistance. An analogy can be drawn with the temperature case profile, where some differences in thermal homogeneity were found. In that case, the basket offers thermal resistance during the preheating stage, and there are energy losses through the walls during the steady-state stage.

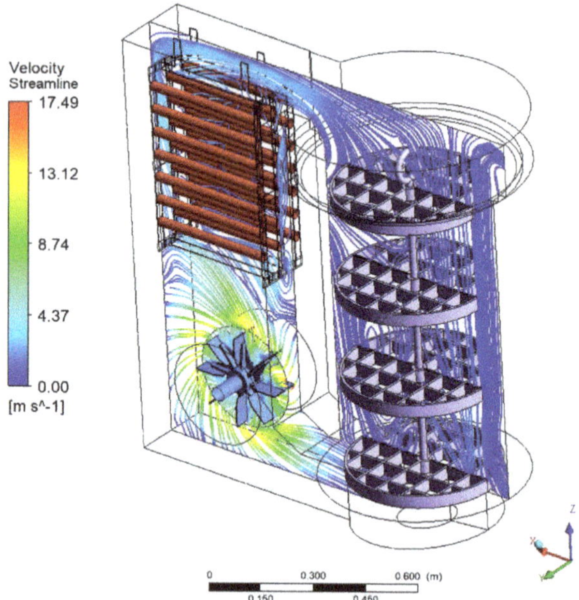

Figure 12. Streamlines inside the furnace for baseline operation with a charging basket. The velocity streamline legend applies to the recirculating gases.

Figure 13 presents energy inflows and outflows for the furnace for the baseline operation of 990 rpm, using two Sankey diagrams—one during the preheating stage and the other at the steady-state. It should be noted that input energy demand is at its maximum during the preheating stage. When the load reaches the steady-state condition, electricity input diminishes to the minimum level required to maintain the temperature inside the process chamber, because heat is being lost through the walls. In this furnace, the energy loss through the walls is roughly half of the energy required for heating the load, suggesting that furnace insulation should be improved. It should also be noted that heat recirculation of hot air between chambers accounts for up to 83.7% of all energy involved in the system. This flow is generated by the fan at a low energy expense of 1.34 kW.

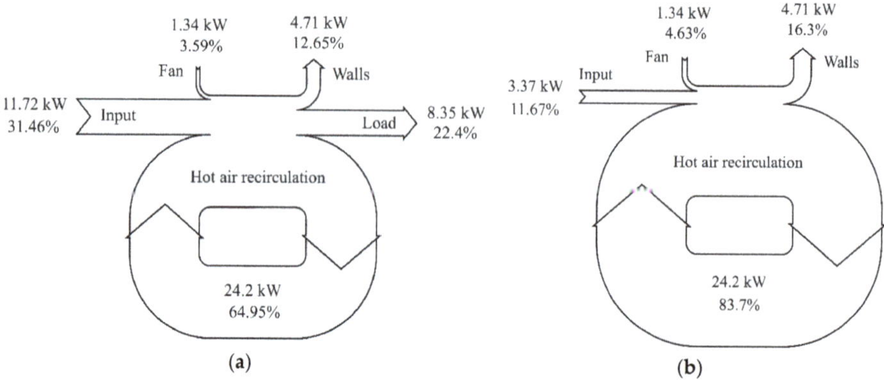

Figure 13. Sankey diagram for energy flows during the heating stage of the tempering process for the baseline operation of 990 rpm. (**a**) During the preheating stage; (**b**) at steady state.

If the rotating speed is increased from 990 rpm, fan energy demand would also increase, and efficiency would, therefore, be decreased. However, the heat transfer rate to the load could be improved, and shorter tempering times could be achieved. As pointed out in Reference [5], steel quality is not negatively affected by longer tempering times, but shorter processing times are desirable in a furnace that operates continuously. Therefore, the heat transfer rate could benefit from increased rotating speed. This behavior is studied and presented in the following section.

3.2. Effect of the Fan Rotating Speed

To estimate the temperature uniformity induced by the flow mixing in the furnace, the temperature distributions were analyzed. The temperature distribution for different fan rotating speeds at a horizontal line located in the middle of the process chamber (at Z = 0.5 m) is presented in Figure 14 for the case of an empty furnace. The higher the rotating speed, the higher the average temperature inside the chamber. This trend could be explained because the tempering furnace was analyzed as a closed system with no inlets or outlets. As the rotating speed increases, the required fan power also increases; therefore, as the losses through the walls were similar to each other, the internal energy inside the chamber, and the temperature increased.

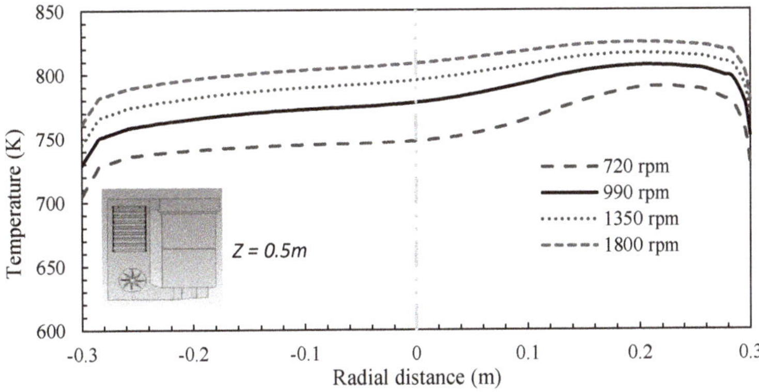

Figure 14. Temperature profile at a horizontal line located in the middle of the workspace for the five different angular rotation rates studied.

From Figure 14, the maximum temperature difference of the baseline operation of 990 rpm, between radial distances of −0.25 and 0.25 m, is 47.5 K. This difference is reduced to 41.2 and 34.6 K at 1350 and 1800 rpm, respectively. Therefore, a higher rotating speed of the fan promotes temperature homogeneity. Additionally, it should be noted that the temperature profile is not symmetric around the process chamber centerline, and in all cases, the temperature decreases when the radial distance is very close to the walls (0.3 or −0.3 m).

Figure 15 shows the axial velocity profile at a horizontal line located in the middle of the workspace for the four different rotating speeds studied (same line presented in Figure 14). Flow is not symmetric around the chamber centerline and both positive and negative velocities appear. A negative velocity axial means that the flow is reversing from top to bottom, which is a result that, combined with the positive velocity values, indicates the presence of a swirl. Therefore, there is an inner recirculation zone within the process chamber due to a swirl that intensifies with higher rpms, and as axial velocities differentials become larger. As shown before, for the 1800 rpm case, temperature homogeneity is enhanced, and the velocity profiles suggest that this is due to a stronger swirl than in other cases. The fan thus plays an important role in enhancing the flow mixing and convective heat transfer in the furnace. The following analyses quantify the fan rotating speed effect in the convective heat transfer.

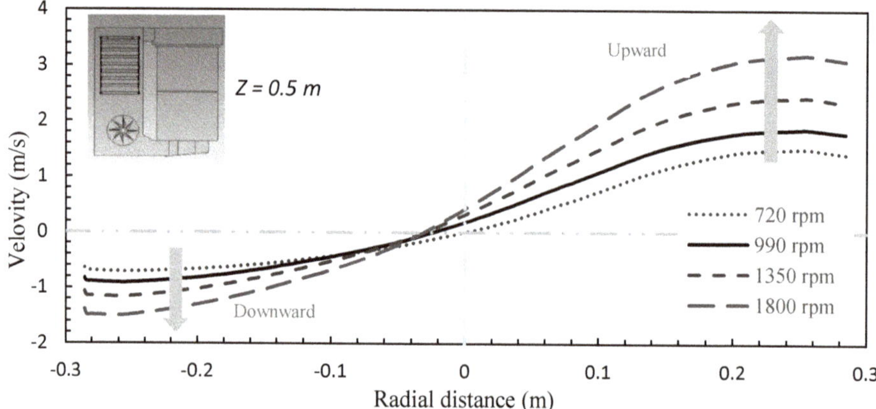

Figure 15. Axial velocity profile at a horizontal line located in the middle of the workspace for the five different rotating speeds studied.

The use of negative air velocities allowed the calculation of the mass recirculation (mr) within the process chamber. The results of this calculation can be seen in Figure 16a. A mass recirculation profile is displayed for each rotating speed which shows that as the gases proceed upwards (i.e., z increases) more gases are entrained into the upcoming gases until they reach a maximum located near the center of the vortex (see Figure 9). Two aspects are interesting to point out: as the rotating speed increases the mass recirculation also increases and the maximum moves to lower heights. In other words, more mass is recirculated, and the center of the vortex changes its Z-direction with the rotating speed increase. The maximum mr values are obtained in the range of 0.4 to 0.7 m. At the particular case of 720 rpm rotating speed, the maximum mr is located near a Z–direction of 0.7 m; and the mr values can be even higher than others rpms (i.e., 990 and 1350) in the upper zone of the process chamber.

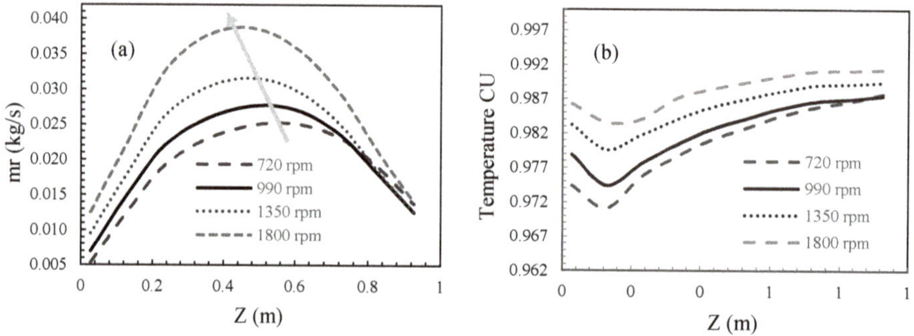

Figure 16. Mass recirculation (**a**) and temperature CU (**b**) at several heights of the process chamber.

The mass recirculation could be related to the thermal homogeneity inside the chamber. To do that we have used a temperature coefficient of uniformity (CU) following the definition presented in a previous study [16]. The coefficient was calculated in different transversal planes taking temperature data of each plane. This definition gives more information than the simple temperature difference presented in Section 3.1.

The CU profiles reported as a function of the Z-direction in Figure 16b evidence an increase in the temperature uniformity inside the process chamber as the rotating speed increases. It should be noted that even a low CU increase (i.e., 0.01) allows improvements in thermal homogeneity. For instance, in Figure 14 the maximum temperature difference at 990 rpm case was 47.5 K and the difference was reduced to 34.6 K in the 1800 rpm case. The temperature uniformity rise to high values indicates a sort of well-mixed condition, which is a valuable characteristic in convection type furnaces. In addition, the 720 rpm case is interesting because the vortex of the recirculation zone is located in the upper zone of the furnace. Therefore, it presents higher values of mr than other rotating speed conditions in that zone, according to Figure 16a. It also allows an improvement of temperature CU in that area, according to Figure 16b. This behavior suggests a correspondence between mass recirculation and thermal homogeneity.

Furthermore, some observations can be made from Figure 16b. At low Z-direction values, the fluid coming from the heating to the process chamber suffers a volume expansion that causes the temperature to drop; as a result, the temperature CU decreases. In Figures 9 and 14 can be seen that temperatures profiles are not symmetric around the process chamber centerline. This probably means that, even though the temperature values at the several locations become close to each other, some hot spots are present in some locations inside the chamber. This asymmetric temperature behavior decreases at higher heights of the chamber. The above is consistent with the rise of the temperature CU showed in Figure 16b.

The relative mass recirculation is presented in Figure 17. It normalizes the maximum mass recirculated (mr_{max}) over the total mass (mt). Maximum mr and mt increase with the rotating speed. However, the total mass rises at a higher rate. As a result, relative mass recirculation tends to decrease. For example, it takes a value of 0.87 at the 720 rpm case and 0.51 at 1800 rpm. These values mean that 87% or 51% of the inlet mass recirculates near to the vortex of the recirculation zone. These relative mass recirculation values are low when compared with others found in the literature [29–31] that use systems with high momentum coflow jets. The above suggests that the fluid dynamics characteristics of the tempering furnace could be improved by means of exploring other ways to increase the air momentum or even the use of secondary air.

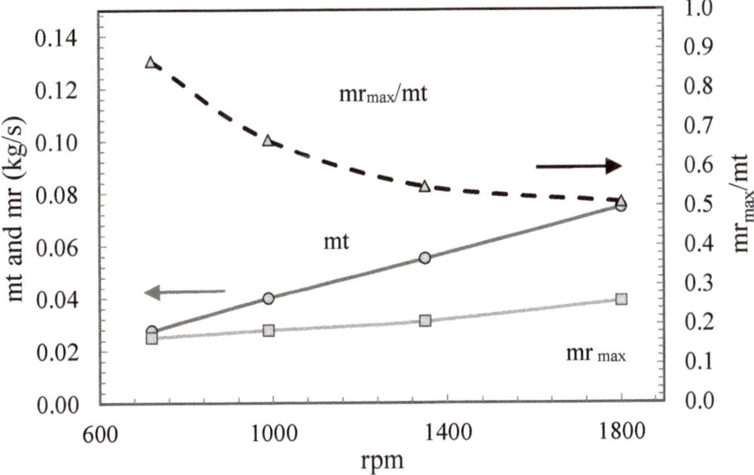

Figure 17. Relative mass recirculation at a different rotating speed of the fan.

3.3. Thermal Efficiency and Heat Transfer to the Load

Figure 18 presents the variation in the hot air flow rate inside the furnace with the fan rotating speed. Simulation results closely follow the first theoretical similarity law for fans, described by Equation (20), which assumes an equal dimensionless flow coefficient Φ for any two different operational conditions. This validates that the fan used in this study follows the classic laws for fans with low uncertainties.

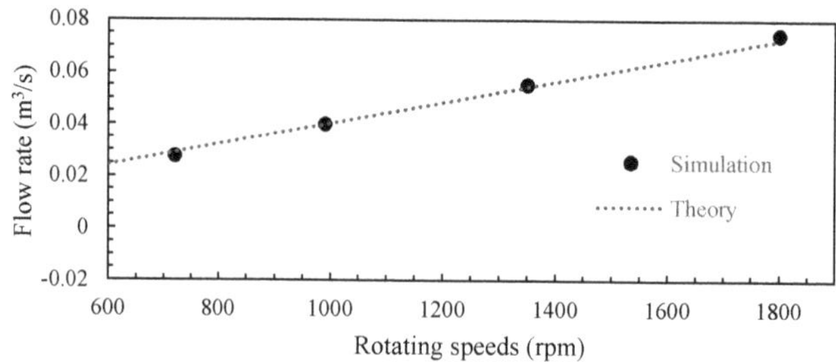

Figure 18. Recirculation flow rate of hot air at different rotating speeds of the fan.

Additionally, because the geometry of the fan remains constant, and the fluid is considered incompressible, the second similarity law, described in Equation (21), applies and assumes that the same power coefficient C_P for any two different operational conditions. The power requirements are estimated in the next equation, which states that the power is proportional to the cube of the rotating speed:

$$\left(\frac{\dot{W}_{shaft}}{\omega^3}\right)_1 = \left(\frac{\dot{W}_{shaft}}{\omega^3}\right)_2 \tag{22}$$

Then, the calculated fan power takes the following values: 0.51, 1.34, 3.38, and 10.72 kW, with rotating speeds of 720, 990, 1350, and 1800 rpms, respectively. From the results, rotating speed variation can negatively affect the thermal efficiency of the furnace; however, it should be contrasted with its positives effect on thermal homogeneity and the convective heat transfer to the load.

Figure 19 presents the heat transfer by convection to the load, represented by the dimensionless Nusselt number, and the thermal efficiency η as a function of the fan rotating speed. Both can be approximated to a quadratic equation as a function of the rpms with a coefficient of determination R^2 of 0.99. At the baseline operation of 990 rpm, the efficiency is 63.96%, while the Nusselt number is 24.89. If the rotating speed is increased to 1800 rpm, the thermal efficiency decreases to 42.23%. However, the Nusselt number goes to 36.74. In other words, a 50% increase in the convection heat transfer rate can be achieved at a 20% drop in efficiency. When a rotating speed of 1350 rpm is used the efficiency is reduced by 8.6% and allows for an increase in the calculated Nusselt number by 15%. The fan thus plays an important role in enhancing the flow mixing and convective heat transfer in the furnace. Additionally, by controlling the fan speed, it is also possible to improve the temperature uniformity inside the chamber.

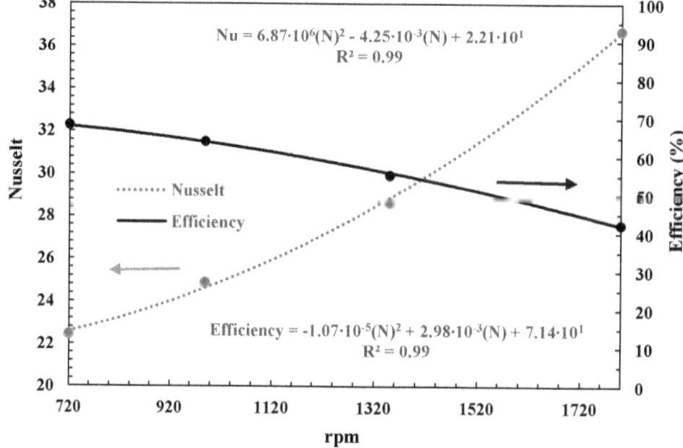

Figure 19. Nusselt number and efficiency variation with the rotating speed of the fan. N = rpm.

4. Conclusions

A numerical investigation has been carried out, presenting the thermal and fluid dynamics behavior of an electric tempering furnace. A set of conclusions have been obtained on both items and are summarized in what follows.

The simulations predict the thermal homogeneity inside the process chamber effectively. Thermal homogeneity was negatively affected by the charging basket use and it was improved by the increase in the rotating speed of the fan. The thermal homogeneity can be associated with the strong recirculation zone formed caused by the rotational force of the fan. The recirculation enhances the flow mixing, increases the residence time of the air in the process chamber, and thus enhances the convective heat transfer.

The calculated air mass flow shows that mass recirculation and the vortex location inside the recirculation zone are affected by the rotating speed variation. When the rotating speed increases more mass is recirculated in the process chamber and the central zone of the recirculation is displaced downward. However, the ratio of mass recirculation to total mass decreases as a result of a higher overall mass flow rate. These values were lower than unity and they were also lower than others found in the literature, suggesting that the fluid dynamics characteristics of the tempering furnace could be further improved.

The thermal efficiency of the tempering furnace could be penalized by an increment in the rotating speed of the fan. This is due to the increase in the fan's mechanical power consumption when the rotating speed increase. Nevertheless, it was shown that thermal efficiency decreases by only 20% when the rpms are doubled, while the heat transfer rate to load is increased by up to 50%. For a furnace operated continuously, this could be translated into lower processing times, and a higher production output, even if that means a lower thermal efficiency operational point. Moreover, it was shown that a higher rotating speed is related to a more homogenous temperature distribution, which is a highly desirable feature in heat treatment equipment.

Author Contributions: Investigation, I.D.P.-C., P.N.A.-T. and L.F.C.-S.; writing—original draft, I.D.P.-C., P.N.A.-T., and L.F.C.-S. All authors have read and agreed to the published version of the manuscript.

Funding: This research received no external funding.

Acknowledgments: Authors want to thank Forjas Bolivar S.A.S for allowing them to carry out the study.

Conflicts of Interest: The authors declare no conflict of interest.

Nomenclature

a	Absorption coefficient
C, m	Constants of Nusselt models
C_u	Constant of k-epsilon turbulence model
$C_{\varepsilon 1}, C_{\varepsilon 2}$	Constants of transport equations
D	Diameter
Y^+	Dimensionless wall distance
h	Enthalpy or heat transfer coefficient
$\tilde{v}$	Favre average velocity
$\tilde{k}$	Favre average turbulent kinetic energy
g	Gravity
$\dot{q}$	Heat transfer
Z	Height of the chamber
I	Identity matrix
G	Incident radiation
mr	Mass recirculation
E	Non-chemical energy
Nu	Nusselt number
$\dot{W}$	Power

Pr	Prandtl number
Pr_s	Prandtl number on the surface temperature
C_p	Power coefficient
n	Refractive index
N	Rotating speed
$\bar{p}$	Reynolds averaged pressure
$\bar{T}$	Reynolds averaged temperature
Re	Reynolds number
S	Source term
$I(\vec{s})$	Spectral radiation intensity
T	Temperature
kt	Thermal conductivity
t	Time
mt	Total mass
$\vec{s}$	Vector direction
$\dot{Q}$	Volumetric flow rate

Symbols

$\widetilde{\rho}$	Average density
$\widetilde{\varepsilon}$	Average dissipation of turbulent kinetic energy
θ, φ, z	Angles of a polar system.
ω	Angular velocity
$\sigma_k, \sigma_\varepsilon,$	Constants of transport equations
Φ	Dimensionless flow coefficient
σ_s	Dispersion coefficient
μ_l	Laminar viscosity
ϕ	Scattering phase function
σ	Stefan Boltzmann constant
Ω	Solid angle
ε	Surface emissivity
η	Thermal efficiency
μ_t	Turbulent viscosity

Subscripts

conv	Convection
rad	Radiation
shaft	Fan shaft
s	surface
surr	Surrounding
max	maximum

Abbreviations

AISI	American Iron and Steel Institute
AMS	Aerospace Material Specification
ASTM	American Society for Testing and Materials
CFD	Computational fluid dynamics
CU	Coefficient of uniformity
DIN	German Institute for Standardization
DO	Discrete ordinate
MRF	Moving reference frame
rpm	Revolutions per minute
RTE	Radiative transfer equation

References

1. Díaz-Ovalle, C.O.; Martínez-Zamora, R.; González-Alatorre, G.; Rosales-Marines, L.; Lesso-Arroyo, R. An approach to reduce the pre-heating time in a convection oven via CFD simulation. *Food Bioprod. Process.* **2017**, *102*, 98–106. [CrossRef]
2. Kluczek, A.; Olszewski, P. Energy audits in industrial processes. *J. Clean. Prod.* **2017**, *142*, 3437–3453. [CrossRef]
3. Chan, D.Y.-L.; Yang, K.H.; Lee, J.D.; Hong, G.B. The case study of furnace use and energy conservation in iron and steel industry. *Energy* **2010**, *35*, 1665–1670.
4. Thelning, K.-E. Fundamental metallographic concepts. In *Steel and its Heat Treatment*; Butterworth-Heinemann: Oxford, UK, 1984; pp. 1–47, ISBN 9780408014243.
5. Bryson, W.E. The Heat-Treating Processes Step 4: Tempering. In *Heat Treatment*; Carl Hanser Verlag GmbH & Co. KG: München, Germany, 2015; pp. 108–111, ISBN 9781569904855.
6. Babu, S. Classification and Mechanisms of Steel Transformation. In *Steel Heat Treatment*; Totten, G.E., Ed.; CRC Press LLC: Boca Raton, FL, USA, 2006; pp. 91–118, ISBN 978-0-8493-8455-4.
7. Deutsches Institut fur Normung E.V. (DIN). *DIN 17052-1 Heat Treatment Furnaces—Part 1: Requirements for Temperature Uniformity*; Deutches Institute fur Normung: Berlin, Germany, 2013; pp. 1–12.
8. *Aerospace Material Specification AMS2750E*; SAE International: Troy MI, USA, 2012.
9. Navaneethakrishnan, P.; Srinivasan, P.S.S.; Dhandapani, S. Numerical and Experimental Investigation of Temperature Distribution Inside a Heating Oven. *J. Food Process. Preserv.* **2010**, *34*, 275–288. [CrossRef]
10. Smolka, J.; Bulinski, Z.; Nowak, A.J. The experimental validation of a CFD model for a heating oven with natural air circulation. *Appl. Therm. Eng.* **2013**, *54*, 387–398. [CrossRef]
11. Hadała, B.; Malinowski, Z.; Rywotycki, M. Energy losses from the furnace chamber walls during heating and heat treatment of heavy forgings. *Energy* **2017**, *139*, 298–314. [CrossRef]
12. Esteves, M.; Planchon, O.; Lapetite, J.M.; Silvera, N.; Cadet, P. The "EMIRE" large rainfall simulator: Design and field testing. *Earth Surface Process. Landf.* **2000**, *25*, 681–690. [CrossRef]
13. Gu, J.; Wang, J.; Pan, J. Numerical Simulation of Heat Treatment Based on the Model With Expanded Solution Domain. *Mater. Perform. Charact.* **2012**, *1*, 1–16.
14. ASTM International. *ASTM C155-97: Standard Classification of Insulating Firebrick*; ASTM International: West Conshohocken, PA, USA, 2013.
15. Poinsot, T.; Veynante, D. Theoretical and Numerical Combustion. *Combust. Flame* **2005**, *32*, 534.
16. Wang, J.; Chen, N.; Shan, X. Numerical Simulation and Measurement of Velocity Distribution in a Gas Nitriding Furnace. *Solid State Phenom* **2006**, *118*, 331–336. [CrossRef]
17. ANSYS Inc. *ANSYS FLUENT Theory Guide*; ANSYS Inc.: Canonsburg, PA, USA, 2019.
18. Modest, M.F. Fundamentals of Thermal Radiation. In *Radiative Heat Transfer*, 3rd ed.; Elsevier: Amsterdam, The Netherlands, 2013; ISBN 978-0-12-386944-9.
19. Peng, W.; Li, G.; Geng, J.; Yan, W. A strategy for the partition of MRF zones in axial fan simulation. *Int. J. Vent.* **2018**, *3315*, 1–15. [CrossRef]
20. Patil, S.R.; Chavan, S.T.; Jadhav, N.S.; Vadgeri, S.S. Effect of Volute Tongue Clearance Variation on Performance of Centrifugal Blower by Numerical and Experimental Analysis. *Mater. Today* **2018**, *5*, 3883–3894. [CrossRef]
21. Gebrehiwot, M.G.; De Baerdemaeker, J.; Baelmans, M. Numerical and experimental study of a cross-flow fan for combine cleaning shoes. *Biosyst. Eng.* **2010**, *106*, 448–457. [CrossRef]
22. Lee, H.J.; Park, J.; Lee, S.H. Numerical Investigation on Influence of Fan Speed and Swirling Gas Injection on Thermal-Flow Characteristics in Nitrocarburizing Furnace. *Mater. Trans.* **2017**, *58*, 1322–1328. [CrossRef]
23. Incropera, F.P.; DeWitt, D.P.; Bergman, T.L.; Lavine, A.S. *Fundamentals of Heat and Mass Transfer*, 6th ed.; John Wiley & Sons: New York, NY, USA, 2006; ISBN 978-0471457282.
24. Gerhart, P.M.; Gerhart, A.L.; Hochstein, J.I. *Munson, Young and Okiishi's Fundamentals of Fluid Mechanics*, 8th ed.; Wiley: Hoboken, NJ, USA, 2016; ISBN 978-1119080701.
25. Hilpert, R. Wärmeabgabe von geheizten Drähten und Rohren im Luftstrom. *Forsch. auf dem Geb. des Ing.* **1933**. [CrossRef]
26. Žukauskas, A. Heat Transfer from Tubes in Crossflow. *Adv. Heat Transf.* **1987**, *8*, 93–160.
27. Churchill, S.W.; Bernstein, M.A. Correlating Equation for Forced Convection From Gases and Liquids to a Circular Cylinder in Crossflow. *J. Heat Transf.* **1977**, *99*, 300–306. [CrossRef]

28. Gomez, R.S.; Porto, T.R.N.; Magalhães, H.L.F.; Moreira, G.; André, A.M.M.C.N.; Melo, R.B.F.; Lima, A.G.B. Natural gas intermittent kiln for the ceramic industry: A transient thermal analysis. *Energies* **2019**, *12*, 1568–1578. [CrossRef]

29. Lezcano, C.; Amell, A.; Cadavid, F. Numerical calculation of the recirculation factor in flameless furnaces. *Dyna* **2013**, *80*, 144–151.

30. Cavigiolo, A.; Galbiati, M.A.; Effuggi, A.; Gelosa, D.; Rota, R. Mild combustion in a laboratory-scale apparatus, Combust. *Sci. Technol.* **2003**, *175*, 1347–1367.

31. Mi, J.; Li, P.; Zheng, C. Numerical Simulation of Flameless Premixed Combustion with an Annular Nozzle in a Recuperative Furnace. *Chin. J. Chem. Eng.* **2010**, *18*, 10–17. [CrossRef]